Advances in Measurement and Data Analysis of Surfaces with Functionalized Coatings

Advances in Measurement and Data Analysis of Surfaces with Functionalized Coatings

Guest Editor
Przemysław Podulka

Basel • Beijing • Wuhan • Barcelona • Belgrade • Novi Sad • Cluj • Manchester

Guest Editor
Przemysław Podulka
Faculty of Mechanical
Engineering and Aeronautics
Rzeszów University of Technology
Rzeszów
Poland

Editorial Office
MDPI AG
Grosspeteranlage 5
4052 Basel, Switzerland

This is a reprint of the Special Issue, published open access by the journal *Coatings* (ISSN 2079-6412), freely accessible at: www.mdpi.com/journal/coatings/special_issues/functionalized_coatings.

For citation purposes, cite each article independently as indicated on the article page online and using the guide below:

Lastname, A.A.; Lastname, B.B. Article Title. *Journal Name* **Year**, *Volume Number*, Page Range.

ISBN 978-3-7258-3430-3 (Hbk)
ISBN 978-3-7258-3429-7 (PDF)
https://doi.org/10.3390/books978-3-7258-3429-7

© 2025 by the authors. Articles in this book are Open Access and distributed under the Creative Commons Attribution (CC BY) license. The book as a whole is distributed by MDPI under the terms and conditions of the Creative Commons Attribution-NonCommercial-NoDerivs (CC BY-NC-ND) license (https://creativecommons.org/licenses/by-nc-nd/4.0/).

Contents

About the Editor vii

Preface ix

Przemysław Podulka
Advances in Measurement and Data Analysis of Surfaces with Functionalized Coatings
Reprinted from: *Coatings* **2022**, *12*, 1331, https://doi.org/10.3390/coatings12091331 . . . . **1**

Przemysław Podulka
Proposals of Frequency-Based and Direction Methods to Reduce the Influence of Surface Topography Measurement Errors
Reprinted from: *Coatings* **2022**, *12*, 726, https://doi.org/10.3390/coatings12060726 . . . . **6**

Przemysław Podulka
Resolving Selected Problems in Surface Topography Analysis by Application of the Autocorrelation Function
Reprinted from: *Coatings* **2022**, *13*, 74, https://doi.org/10.3390/coatings13010074 . . . . **27**

Zhenghui Ge, Qifan Hu, Rui Wang, Haolin Fei, Yongwei Zhu and Ziwei Wang
Machine Learning-Driven Optimization of Micro-Textured Surfaces for Enhanced Tribological Performance: A Comparative Analysis of Predictive Models
Reprinted from: *Coatings* **2024**, *14*, 1539, https://doi.org/10.3390/coatings14121539 . . . . **46**

Tomasz Trzepieciński, Krzysztof Szwajka and Marek Szewczyk
Analysis of Surface Topography Changes during Friction Testing in Cold Metal Forming of DC03 Steel Samples
Reprinted from: *Coatings* **2023**, *13*, 1738, https://doi.org/10.3390/coatings13101738 . . . . **65**

Dongdong Ye, Zhou Xu, Houli Liu, Zhijun Zhang, Peiyong Wang and Yiwen Wu et al.
Terahertz Non-Destructive Testing of Porosity in Multi-Layer Thermal Barrier Coatings Based on Small-Sample Data
Reprinted from: *Coatings* **2024**, *14*, 1357, https://doi.org/10.3390/coatings14111357 . . . . **86**

Zhenghui Ge, Qifan Hu, Haitao Zhu and Yongwei Zhu
Surface Tribological Properties Enhancement Using Multivariate Linear Regression Optimization of Surface Micro-Texture
Reprinted from: *Coatings* **2024**, *14*, 1258, https://doi.org/10.3390/coatings14101258 . . . . **102**

Vladimir Syasko, Alexey Musikhin and Igor Gnivush
Improvement of Methods and Devices for Multi-Parameter High-Voltage Testing of Dielectric Coatings
Reprinted from: *Coatings* **2024**, *14*, 427, https://doi.org/10.3390/coatings14040427 . . . . **117**

Di Yun, Cheng Tang, Ulf Sandberg, Maoping Ran, Xinglin Zhou and Jie Gao et al.
A New Approach for Determining Rubber Enveloping on Pavement and Its Implications for Friction Estimation
Reprinted from: *Coatings* **2024**, *14*, 301, https://doi.org/10.3390/coatings14030301 . . . . **132**

Aniket Kumar, Bapun Barik, Piotr G. Jablonski, Sanjiv Sonkaria and Varsha Khare
Functionalized and Biomimicked Carbon-Based Materials and Their Impact for Improving Surface Coatings for Protection and Functionality: Insights and Technological Trends
Reprinted from: *Coatings* **2022**, *12*, 1674, https://doi.org/10.3390/coatings12111674 . . . . **154**

Zhuoyue Li, Cheng Wang, Haijuan Ju, Xiangrong Li, Yi Qu and Jiabo Yu
Prediction Model of Aluminized Coating Thicknesses Based on Monte Carlo Simulation by X-ray Fluorescence
Reprinted from: *Coatings* **2022**, *12*, 764, https://doi.org/10.3390/coatings12060764 . . . . **189**

About the Editor

Przemysław Podulka

Przemysław Podulka studied his Eng. and M.Sc. in Technical and Informatic Sciences at the University of Rzeszów, Rzeszów, Poland. Before his graduation in 2009 and 2013, respectively, he received three scholarships from the Polish Minister of Higher Education in the 2006/2007, 2007/2008, and 2008/2009 academic years. In 2017, he received his Ph.D. in Mechanical Engineering from the Faculty of Mechanical Engineering and Aeronautics, Rzeszów University of Technology, Rzeszów. From 2017 till now, he has been affiliated with the Department of Manufacturing Processes and Production Engineering, Rzeszów University of Technology in Rzeszów, where he was a Research Associate Professor. He received three awards, being classified as 2% of authors from an updated science-wide author database of standardized citation indicators (Scopus) in 2022, 2023, and 2024.

His research activities have encompassed such fields as surface engineering, surface metrology, surface finishing, machining, quantitative characterization of surface topography, analysis of surface topography measurement noise of various rough materials, characterization of coating materials, and the relationship between structure and properties of various engineering materials.

Preface

Studies of various material surfaces must be detailed when evaluating their topography. Analysis can be classified as comprehensive when measuring and analyzing geometrical quantities are performed suitably. From the origin, measurement and evaluation of surface roughness were carried out due to the data collection and study of the received texture parameters. Over the years, when the measuring systems and data processing techniques were evaluated, some requirements for measurement accuracy and user guidance during data processing were requested. Additionally, the industry advances added further requirements to the general principles of surface topography inspection methods. For this reason, the current Special Issue tries to address some of the most current progress regarding surface metrology on the micro-scale objectives. This Special Issue collects recent advances in the measurement and data analysis of various surface topographies and their functional performances.

Przemysław Podulka
Guest Editor

Editorial

Advances in Measurement and Data Analysis of Surfaces with Functionalized Coatings

Przemysław Podulka

Faculty of Mechanical Engineering and Aeronautics, Rzeszow University of Technology, Powstancow Warszawy 8 Street, 35-959 Rzeszow, Poland; p.podulka@prz.edu.pl; Tel.: +48-17-743-2537

Coatings, taking comprehensive studies into account, cannot be considered without their functional performance. When, generally, studying surfaces, a wide analysis of coatings properties is indispensable. Surfaces with functionalized coatings contain characteristic features, affecting the functional properties of many elements. Considering surface, surface roughness or, respectively, surface topography, many aspects can appear in which the whole process of surface studies contains measurement and data analysis, which can be considered as one process or separately.

There are many aspects wherein the effect of surface topography (ST) on functional properties can be large. Generally, the effect of ST on contact, e.g., stiffness [1], conductivity [2], oxidation resistance [3], adhesive [4], improvement with friction [5], lubrication [6], wear [7], corrosion [8], and fatigue [9] properties, is very high. Much relevant information can be received from the analysis of ST.

When studying the ST, even highly precise measurement equipment (devices and whole systems) may not allow receiving reliable results when raw measured data are processed erroneously [10], and properly manufactured parts can be classified as faulty and, unfortunately, are rejected. For that reason, ST can be classified as a basic issue in the process of characterization of the manufactured parts and their properties, which, generally, can support the process of control [11]. Moreover, both the measurement process and data processing have a huge, equal influence on the results obtained.

From the above matter, the motivation for presenting the current advances in measurement and data analysis of surfaces with functional properties must be found. Yet, sophisticated characterization and modeling are required to gain a comprehensive understanding of the mechanical properties of these coatings as well.

This Special Issue (SI) aims to provide a discussion for researchers to share both current and further research findings and help to promote planned research into the studies of surfaces with functionalized coatings, considering manufacturing and measurement (e.g., surface roughness), data analysis, and modeling.

Many factors can affect, unlikely erroneously, the results of ST measurement and data analysis. The main classification can be provided according to the factors that influence the accuracy of the results assessments. In that sense, the ST (measurement and data analysis) errors can be, even roughly, divided into those caused by the measuring method [12], the process of digitization [13], or software data processing [14], and errors caused by the measuring object [15] or other types of errors.

One of the types of errors is facilitated when the measurement process occurs and is often defined as noise [16]. Many types of noises can be considered in this SI, such as instrument noise or instrument white noise [17], random noise [18], phase noise [19], signal-to-ratio noise [20] or, simply, the measurement noise [21]. This was last studied previously in selected domains, e.g., the high-frequency measurement noise [22], which in most cases is caused by vibrations.

Generally, an 'engineering surface' can be analyzed as a surface that is composed of a large number of wavelengths of roughness that are superimposed on each other.

Therefore, analysis of surface topography, based on the frequency methods, when, e.g., a high-frequency noise is separated, is common. Despite many scientific articles on ST analysis being published, valuable information on how to deal with selected types of measurement and data analysis errors is rare and the current state of this area of knowledge is not fully unified.

Another problem, which is often met in the field of surface metrology studies, is the end-effect, or, respectively, edge-effect. This problem is especially crucial when digital filtering is proposed for the analysis of surface roughness. In general, it is extremely difficult to determine the signal received by filtering at both ends of the profile and detail. It is often stated that edge-effect in ST filtration seems to be a considerable problem when applying digital filtering; nevertheless, it was also indicated that this issue affects the analysis of surface roughness with least-square methods, e.g., least square fitted polynomial planes of n-th order [23] and various approaches based on its modifications [24]. Recursive implementation of Gaussian filter seems to be a reasonable alternative [25]. For an areal and profile application, it was assumed in the ISO standards, as well [26]. Moreover, many popular algorithms were proposed for the characterization of ST and, correspondingly, managing the edge-effect problem, such as a two-dimensional discrete spline filter [27] and high-order spline [28]; approaches with a typical example of an extension of the spline filtering [29], many different combinations of boundary conditions of the spline filter [30], wavelets, or its combinations [31] were thoroughly analyzed and compared to provide satisfactory end-data characterization. Some proposals were presented by modification of raw measured data, such as fulfilling the dimples or features generally located near the edge of analyzed detail [32]. It was also found that both surface features' (e.g., dimples) size and distribution can significantly affect the accuracy of the roughness evaluation as well [33]. All of the issues abovementioned can be especially valuable in the reduction of the measurement and data processing errors; nevertheless, all of them require mindful users.

Evaluating the surface coatings' properties is correlated to the characterization of selected features. Very popular in the analysis of ST are, as commonly called, feature-based algorithms [34] or feature-based procedures [35], used for segmentation [36] of the measured surface topography area. Additionally, methods based on multi-scale analysis are also popular in surface metrology [37]. The purpose of using multiscale methods is to more accurately determine the functional correlations between the parameters of machining processes and the created surface topographies [38]. Moreover, many surface topography properties can be received when multi-scale studies are proposed, such as anisotropy that can influence surface function and can be an indication of processing [39].

Surface topography characterization can be supported by many algorithms, schemes, approaches, and procedures, such as those from commercial software. Often used and improved are those methods based on frequency analysis, such as the frequency spectrum (FS [11]) scheme. Further, one of the proposed, commonly available solutions is to analyze the Power Spectral Density (PSD) graph. According to the definition, in its two-dimensional form, PSD has been designated as the preferred means of specifying the surface roughness on the draft international drawing standard for surface texture [40]. The process of dry machining, including cooling using the Minimum Quantity Cooling Lubrication (MQCL), has proven that by using a PSD technique, it is possible to characterize the turning according to the applied cooling methods [41]. The applicability of PSD was also improved with Auto-Correlation Function (ACF) to be especially valuable in the detection of the high-frequency measurement errors of honed cylinder liner surfaces [42].

Reducing errors in measurement and, correspondingly, in the whole process of data analysis, should depend on the type of data studied. Usually, research on the ST is proven directly to the selected types of surfaces. In the proposed SI, in fact, there are no limits on the types of surfaces that can be characterized by their functional performance. Except for tribological properties, such as rolling contact fatigue [43], or, e.g., anti-corrosion properties [44], the surface (topography), and its coatings, are widely studied for implant applications [45]. From the biomedical coatings, many advances indicate increasing in this field of study [46]. Further issues are placed with the environment, where eco-friendly perspectives with biodegradable coatings are searched for by many scholars [47,48]. All of the medical studies seem to be different from those with the tribological point of view that many various requirements must be met, such as antibacterial [49] properties. Another issue, the biocompatibility of implants [50,51], can be mentioned as a major problem that can hinder the clinical application of surfaces. Some actions, such as, e.g., chemical bonding [52], can improve the bioactivity of selected types of surfaces.

Widely studied and presented in the coatings research area are thin film improvements. They are classified as a promising candidate for, e.g., spintronic applications [53], glass substrates [54], considering semiconductors [55] or magnetic and gas sensing [56], and thermoelectric [57] or optical [58] properties. Thin film coatings have a wild range of applications [59] that are suitable for the proposed SI area of study.

Presentation of plenty of algorithms and procedures, currently, requires from surface metrology general guidance on how to deal with different measuring problems without losing the validity of the methods applied. The more approaches appear, the more suggestions on how to use them should be provided. Moreover, increasing the area of a variety of surface topographies considered with functionalized coatings performance makes the SI even more required to be proposed. From that matter, I trust that the issues raised in this editorial, concerning and, respectively, collecting of all of the recent advances in the measurement and data analysis of surfaces with functionalized coatings will be found as useful and provide valuable information for all of the surface metrology areas.

Conflicts of Interest: The author declares no conflict of interest.

References

1. Pruncu, C.I.; Vladescu, A.; Hynes, N.R.J.; Sankaranarayanan, R. Surface Investigation of Physella Acuta Snail Shell Particle Reinforced Aluminium Matrix Composites. *Coatings* **2022**, *12*, 794. [CrossRef]
2. Liu, L.; Chen, X. Effect of surface roughness on thermal conductivity of silicon nanowires. *J. Appl. Phys.* **2010**, *107*, 033501. [CrossRef]
3. Li, Z.; Wang, C.; Ju, H.; Li, X.; Qu, Y.; Yu, J. Prediction Model of Aluminized Coating Thicknesses Based on Monte Carlo Simulation by X-ray Fluorescence. *Coatings* **2022**, *12*, 764. [CrossRef]
4. Zielecki, W.; Pawlus, P.; Perłowski, R.; Dzierwa, A. Surface topography effect on strength of lap adhesive joints after mechanical pre-treatment. *Arch. Civ. Mech. Eng.* **2013**, *13*, 175–185. [CrossRef]
5. Shi, R.; Wang, B.; Yan, Z.; Wang, Z.; Dong, L. Effect of Surface Topography Parameters on Friction and Wear of Random Rough Surface. *Materials* **2019**, *12*, 2762. [CrossRef]
6. Epstein, D.; Keer, L.M.; Wang, Q.J.; Cheng, H.S.; Zhu, D. Effect of Surface Topography on Contact Fatigue in Mixed Lubrication. *Tribol. Trans.* **2003**, *46*, 506–513. [CrossRef]
7. Zheng, M.; Wang, B.; Zhang, W.; Cui, Y.; Zhang, L.; Zhao, S. Analysis and prediction of surface wear resistance of ball-end milling topography. *Surf. Topogr. Metrol. Prop.* **2020**, *8*, 025032. [CrossRef]
8. Szala, M.; Świetlicki, A.; Sofińska-Chmiel, W. Cavitation erosion of electrostatic spray polyester coatings with different surface finish. *Bull. Pol. Acad. Sci. Tech. Sci.* **2021**, *69*, e137519. [CrossRef]
9. Macek, W. Correlation between Fractal Dimension and Areal Surface Parameters for Fracture Analysis after Bending-Torsion Fatigue. *Metals* **2021**, *11*, 1790. [CrossRef]
10. Podulka, P. Proposals of Frequency-Based and Direction Methods to Reduce the Influence of Surface Topography Measurement Errors. *Coatings* **2022**, *12*, 726. [CrossRef]
11. Podulka, P. Selection of Methods of Surface Texture Characterisation for Reduction of the Frequency-Based Errors in the Measurement and Data Analysis Processes. *Sensors* **2022**, *22*, 791. [CrossRef] [PubMed]
12. Pawlus, P.; Wieczorowski, M.; Mathia, T. *The Errors of Stylus Methods in Surface Topography Measurements*; Zapol: Szczecin, Poland, 2014.
13. Pawlus, P. Digitisation of surface topography measurement results. *Measurement* **2007**, *40*, 672–686. [CrossRef]
14. Podulka, P. The effect of valley depth on areal form removal in surface topography measurements. *Bull. Pol. Acad. Sci. Tech. Sci.* **2019**, *67*, 391–400. [CrossRef]
15. Magdziak, M. Selection of the Best Model of Distribution of Measurement Points in Contact Coordinate Measurements of Free-Form Surfaces of Products. *Sensors* **2019**, *19*, 5346. [CrossRef] [PubMed]
16. Muhamedsalih, H.; Jiang, X.; Gao, F. Accelerated Surface Measurement Using Wavelength Scanning Interferometer with Compensation of Environmental Noise. *Procedia CIRP* **2013**, *10*, 70–76. [CrossRef]

17. Jacobs, T.D.B.; Junge, T.; Pastewka, L. Quantitative characterization of surface topography using spectral analysis. *Surf. Topogr. Metrol. Prop.* **2017**, *5*, 013001. [CrossRef]
18. De Groot, P. Principles of interference microscopy for the measurement of surface topography. *Adv. Opt. Photonics* **2015**, *7*, 65. [CrossRef]
19. Servin, M.; Estrada, J.C.; Quiroga, J.A.; Mosino, J.F.; Cywiak, M. Noise in phase shifting interferometry. *Opt. Express* **2009**, *17*, 8789–8794. [CrossRef]
20. Šarbort, M.; Hola, M.; Pavelka, J.; Schovánek, P.; Rerucha, S.; Oulehla, J.; Fořt, T.; Lazar, J. Comparison of three focus sensors for optical topography measurement of rough surfaces. *Opt. Express* **2019**, *27*, 33459–33473. [CrossRef]
21. *ISO WD 25178-600:2014(E)*; Geometrical Product Specifications (GPS)—Surface Texture: Areal—Part 600: Metrological Characteristics for Areal-Topography Measuring Methods (DRAFT). International Organization for Standardization: Geneva, Switzerland, 2014.
22. Podulka, P. Suppression of the High-Frequency Errors in Surface Topography Measurements Based on Comparison of Various Spline Filtering Methods. *Materials* **2021**, *14*, 5096. [CrossRef]
23. Podulka, P. The effect of valley location in two-process surface topography analysis. *Adv. Sci. Technol. Res. J.* **2018**, *12*, 97–102. [CrossRef]
24. Podulka, P. Bisquare robust polynomial fitting method for dimple distortion minimisation in surface quality analysis. *Surf. Interface Anal.* **2020**, *52*, 875–881. [CrossRef]
25. Janecki, D. Edge effect elimination in the recursive implementation of Gaussian filters. *Precis. Eng.* **2012**, *36*, 128–136. [CrossRef]
26. *ISO 16610-28:2016*; Geometrical Product Specifications (GPS)—Surface Texture: Areal—Part 600: Metrological Characteristics for Areal-Topography Measuring Methods (DRAFT). International Organization for Standardization: Geneva, Switzerland, 2016.
27. Goto, T.; Yanagi, K. An optimal discrete operator for the two-dimensional spline filter. *Meas. Sci. Technol.* **2009**, *20*, 125105. [CrossRef]
28. Huang, S.; Tong, M.; Huang, W.; Zhao, X. An isotropic areal filter based on high-order thin-plate spline for surface metrology. *IEEE Access* **2019**, *7*, 116809–116822. [CrossRef]
29. Zhang, H.; Tong, M.; Chu, W. An areal isotropic spline filter for surface metrology. *J. Res. Natl. Inst. Stan.* **2015**, *120*, 64–73. [CrossRef]
30. Tong, M.; Zhang, H.; Ott, D.; Zhao, X.; Song, J. Analysis of the boundary conditions of the spline filter. *Meas. Sci. Technol.* **2015**, *26*, 095001. [CrossRef]
31. Gogolewski, D. Fractional spline wavelets within the surface texture analysis. *Measurement* **2021**, *179*, 109435. [CrossRef]
32. Podulka, P. Edge-area form removal of two-process surfaces with valley excluding method approach. *Matec. Web. Conf.* **2019**, *252*, 05020. [CrossRef]
33. Podulka, P. The Effect of Surface Topography Feature Size Density and Distribution on the Results of a Data Processing and Parameters Calculation with a Comparison of Regular Methods. *Materials* **2021**, *14*, 4077. [CrossRef]
34. Newton, L.; Senin, N.; Smith, B.; Chatzivagiannis, E.; Leach, R. Comparison and validation of surface topography segmentation methods for feature-based characterisation of metal powder bed fusion surfaces. *Surf. Topogr. Metrol. Prop.* **2019**, *7*, 045020. [CrossRef]
35. Podulka, P. Improved Procedures for Feature-Based Suppression of Surface Texture High-Frequency Measurement Errors in the Wear Analysis of Cylinder Liner Topographies. *Metals* **2021**, *11*, 143. [CrossRef]
36. Jiang, X.; Senin, N.; Scott, P.J.; Blateyron, F. Feature-based characterisation of surface topography and its application. *CIRP Ann.-Manuf. Technol.* **2021**, *70*, 681–702. [CrossRef]
37. Sadowski, Ł.; Mathia, T.G. Multi-scale metrology of concrete surface morphology: Fundamentals and specificity. *Constr. Build. Mater.* **2016**, *113*, 613–621. [CrossRef]
38. Guibert, R.; Hanafi, S.; Deltombe, R.; Bigerelle, M.; Brown, C. Comparison of three multiscale methods for topographic analyses. *Surf. Topogr. Metrol. Prop.* **2020**, *8*, 024002. [CrossRef]
39. Bartkowiak, T.; Berglund, J.; Brown, C.A. Multiscale Characterizations of Surface Anisotropies. *Materials* **2020**, *13*, 3028. [CrossRef]
40. Elson, J.M.; Bennett, J.M. Calculation of the power spectral density from surface profile data. *Appl. Opt.* **1995**, *34*, 201–208. [CrossRef]
41. Krolczyk, G.M.; Maruda, R.W.; Nieslony, P.; Wieczorowski, M. Surface morphology analysis of Duplex Stainless Steel (DSS) in Clean Production using the Power Spectral Density. *Measurement* **2016**, *94*, 464–470. [CrossRef]
42. Podulka, P. Fast Fourier Transform detection and reduction of high-frequency errors from the results of surface topography profile measurements of honed textures. *Eksploat. Niezawodn.* **2021**, *23*, 84–89. [CrossRef]
43. Stewart, S.; Ahmed, R. Rolling contact fatigue of surface coatings—A review. *Wear* **2002**, *253*, 1132–1144. [CrossRef]
44. Liu, Y.; Wu, L.; Chen, A.; Xu, C.; Yang, X.; Zhou, Y.; Liao, Z.; Zhang, B.; Hu, Y.; Fang, H. Component Design of Environmentally Friendly High-Temperature Resistance Coating for Oriented Silicon Steel and Effects on Anti-Corrosion Property. *Coatings* **2022**, *12*, 959. [CrossRef]
45. Junker, R.; Dimakis, A.; Thoneick, M.; Jansen, J.A. Effects of implant surface coatings and composition on bone integration: A systematic review. *Clin. Oral Implan. Res.* **2009**, *20*, 185–206. [CrossRef] [PubMed]
46. Yoshida, M.; Langer, R.; Lendlein, A.; Lahann, J. From Advanced Biomedical Coatings to Multi-Functionalized Biomaterials. *J. Macromol. Sci. Pol. R.* **2006**, *46*, 347–375. [CrossRef]

47. Li, L.-Y.; Cui, L.-Y.; Zeng, R.-C.; Li, S.-Q.; Chen, X.-B.; Zheng, Y.; Kannan, M.B. Advances in functionalized polymer coatings on biodegradable magnesium alloys—A review. *Acta Biomater.* **2018**, *79*, 23–36. [CrossRef]
48. Ke, M.; Xie, D.; Tang, Q.; Su, S. Preliminary Investigation on Degradation Behavior and Cytocompatibility of Ca-P-Sr Coated Pure Zinc. *Coatings* **2022**, *12*, 43. [CrossRef]
49. Zhang, D.; Liu, Y.; Liu, Z.; Wang, Q. Advances in Antibacterial Functionalized Coatings on Mg and Its Alloys for Medical Use—A Review. *Coatings* **2020**, *10*, 828. [CrossRef]
50. Yao, S.; Cui, J.; Chen, S.; Zhou, X.; Li, J.; Zhang, K. Extracellular Matrix Coatings on Cardiovascular Materials—A Review. *Coatings* **2022**, *12*, 1039. [CrossRef]
51. Tong, S.; Sun, X.; Wu, A.; Guo, S.; Zhang, H. Improved Biocompatibility of TiO_2 Nanotubes via Co-Precipitation Loading with Hydroxyapatite and Gentamicin. *Coatings* **2021**, *11*, 1191. [CrossRef]
52. Kang, S.; Haider, A.; Gupta, K.C.; Kim, H.; Kang, I. Chemical Bonding of Biomolecules to the Surface of Nano-Hydroxyapatite to Enhance Its Bioactivity. *Coatings* **2022**, *12*, 999. [CrossRef]
53. Siddiqui, S.A.; Hong, D.; Pearson, J.E.; Hoffmann, A. Antiferromagnetic Oxide Thin Films for Spintronic Applications. *Coatings* **2021**, *11*, 786. [CrossRef]
54. Lee, P.-Y.; Widyastuti, E.; Lin, T.-C.; Chiu, C.-T.; Xu, F.-Y.; Tseng, Y.-T.; Lee, Y.-C. The Phase Evolution and Photocatalytic Properties of a Ti-TiO_2 Bilayer Thin Film Prepared Using Thermal Oxidation. *Coatings* **2021**, *11*, 808. [CrossRef]
55. Tsay, C.-Y.; Chiu, W.-Y. Enhanced Electrical Properties and Stability of P-Type Conduction in ZnO Transparent Semiconductor Thin Films by Co-Doping Ga and N. *Coatings* **2020**, *10*, 1069. [CrossRef]
56. Yaragani, V.; Kamatam, H.P.; Deva Arun Kumar, K.; Mele, P.; Christy, A.J.; Gunavathy, K.V.; Alomairy, S.; Al-Buriahi, M.S. Structural, Magnetic and Gas Sensing Activity of Pure and Cr Doped In2O3 Thin Films Grown by Pulsed Laser Deposition. *Coatings* **2021**, *11*, 588. [CrossRef]
57. Latronico, G.; Singh, S.; Mele, P.; Darwish, A.; Sarkisov, S.; Pan, S.W.; Kawamura, Y.; Sekine, C.; Baba, T.; Mori, T.; et al. Synthesis and Characterization of Al- and SnO_2-Doped ZnO Thermoelectric Thin Films. *Materials* **2021**, *14*, 6929. [CrossRef]
58. Potera, P. Special Issue: Optical Properties of Crystals and Thin Films. *Coatings* **2022**, *12*, 920. [CrossRef]
59. Tseluikin, V.; Zhang, L. Carbon and Carbon-Based Composite Thin Films/Coatings: Synthesis, Properties and Applications. *Coatings* **2022**, *12*, 907. [CrossRef]

Article

Proposals of Frequency-Based and Direction Methods to Reduce the Influence of Surface Topography Measurement Errors

Przemysław Podulka

Faculty of Mechanical Engineering and Aeronautics, Rzeszow University of Technology, Powstancow Warszawy 8 Street, 35-959 Rzeszów, Poland; p.podulka@prz.edu.pl; Tel.: +48-17-743-2537

Abstract: Various methods, based on both surface frequency and direction, can be alternatively proposed to reduce the influence of high-frequency measurement and data analysis errors. Various types of details were studied, e.g., cylinder liners after the plateau-honing process, plateau-honed cylinder liners with additionally burnished oil pockets (dimples), turned, ground, milled or laser-textured. They were measured with stylus or non-contact (optic) techniques. It was suggested to support various frequency-based methods, e.g., Frequency Spectrum, Power Spectral Densities or Autocorrelation Function, with direction techniques to provide reduction of errors in both detection and extraction of high-frequency measurement errors. Results can be especially valuable for regular studies when frequency-based measurement errors are difficult to be identified.

Keywords: surface topography; surface texture; measurement; measurement errors; measurement noise; texture direction; frequency spectrum; fast Fourier transform; cylinder liner; oil pockets; grinding; milling; laser-texturing

Citation: Podulka, P. Proposals of Frequency-Based and Direction Methods to Reduce the Influence of Surface Topography Measurement Errors. *Coatings* **2022**, *12*, 726. https://doi.org/10.3390/coatings 12060726

Academic Editor: Matjaz Finsgar

Received: 21 March 2022
Accepted: 23 May 2022
Published: 25 May 2022

Publisher's Note: MDPI stays neutral with regard to jurisdictional claims in published maps and institutional affiliations.

Copyright: © 2022 by the author. Licensee MDPI, Basel, Switzerland. This article is an open access article distributed under the terms and conditions of the Creative Commons Attribution (CC BY) license (https:// creativecommons.org/licenses/by/ 4.0/).

1. Introduction

Assessment of the surface topography, which is received in the last stage of the manufacturing process, can be especially valuable because many surface properties depend on their creating accuracy, like friction [1], sealing [2], lubricant retention [3], wear resistance [4,5], corrosion [6], fatigue [7] or, generally, material contact (tribological) performance [8] and material properties. Processes of analysis of surface topography can be, roughly, divided into several stages. Firstly, the measurement process [9] can be studied, and then, respectively, the accuracy of the data analysis [10] can be carefully considered. Even highly precise measurement equipment may not provide reliable results when data processing errors occur that properly manufactured parts can be classified as a lack and flatly rejected. In practice, both types of errors, that measurement and when raw data are processed, can be placed on an equal footing with each other, however, the impact of each of them individually was not compared previously.

From all errors related to the surface topography measurement and analysis, the main classification can be provided according to the factors that influence the accuracy of the results assessments. In that case, surface topography (measurement and data analysis) errors can be divided into those straightly related to the measuring method(s) [11], caused by the digitisation [12] or data processing [13], software [14], measuring object [15] or other errors [16]. Errors found when the measurement process occurs are defined as noise, measurement noise [17] in particular.

There are many types of measurement noise in surface topography analysis, considering both stylus and non-contact (e.g., optic) methods in the evaluation of surface roughness. They were comprehensively studied and, correspondingly, extensively reviewed in many previous papers [18–21]. Generally, the measurement noise can be defined as the noise added to the output signal when the normal use of the measuring instrument occurs [22]. Moreover, various types of environmental disturbance can introduce noise in different bandwidths [23]. One of the errors caused by the environment of the measuring system is the high-frequency noise [24]. This type of noise can be caused by instability of the mechanics with any influences from the environment or by internal electrical noise, however, in most cases, the high-frequency noise is the result of vibration and, respectively, in real measurement, can greatly affect the stability of slope estimation [25]. Some general strategies to reduce vibration noise are to minimise vibration sources, isolate the vibration sources and/or isolate the instrument [26]. Vibration isolation can be realised by optimising the mechanical structure of the instrument [27]. Another proposal is to compensate for the vibrational effect, like, a piezoelectric transducer has been used for an on-machine wavelength scanning interferometer to compensate for the fluctuations in the optical path length caused by vibration from the surrounding environment [28]. However, extensive studies of environmental noise, such as thermal variation and vibration, in the definition of the accuracy of in-process measurement results, seem to be challenging to develop fast and accurate optical in-process instruments, methods or guidance's [29].

An engineering surface is composed of a large number of wavelengths of roughness that are superimposed on each other [30]. Therefore, analysis of surface topography, based on the frequency methods is common. There are many papers considering the assessment of the surface roughness with an application of a power spectral density (PSD). Introducing, PSD is an alternative method for specifying optical surfaces and quantifying the contribution of each spatial regime to the total surface error [31] and, primarily, the PSD, in its two-dimensional form, has been designated as the preferred quantity for specifying surface roughness on a draft international drawing standard for surface texture [32]. PSD approach, based on Fourier analysis [33] of surface topography data acquired by both stylus profilometry and atomic force microscopy (AFM [34]) was introduced, among others, to distinguish the scale-dependent smoothing effects, resulting in a novel qualitative and quantitative description of surface topography [35].

The utility of the PSD is that it contains statistical information that is unbiased by the particular scan size and pixel resolution chosen by the researcher [36]. Moreover, the application of PSD can be valuable in directly comparing surface roughness measured by various techniques even isolated particles affect the instruments in different ways [37]. Establishing a thorough framework to quantify the similarity of 3D surface topography measurements and determining whether they are from the same surface or not based on the frequency domain representations can be also resolved with PSD and 2D Fourier transformation so, respectively, feature extraction in the frequency domain is received [38]. Some anomalies in the PSD spectra can be related to aspects of the polishing process as well [39].

The power spectral method was used for the fractal characterization of the surface roughness and anisotropy of ground surfaces using the atomic force microscope data obtained from ground specimens [40]. PSD based fractal dimension obtained using AFM data was proposed for the surface characterization of brake plunger manufactured using a rolling process [41]. What is crucial, the surface quality can be compared using PSD for the dry and MQCL (Minimum Quantity Cooling Lubrication) turning process [42]. In some cases, the PSD enabled the derivation of the surface roughness and thus provide useful information on characteristic features which compose the microstructure of the films [43]. PSD can also be used to indicate how the process modifies topography at different scales [44].

Almost all the traditional noise reduction methods have the problem of energy loss [45]. The measurement noise was studied with a fractal dimension and, respectively, it was found that the noise has a significant influence on the amplitude and, correspondingly, the estimated fractal dimension of low-dimensional surfaces, but has relatively less influence on the high-dimensional surfaces. Thus, denoising for low-dimensional surfaces was required prior to fractal analysis [46].

In spectral analysis, circular spectra and/or accumulation spectra are suitable for the analysis of sample surfaces from isotropic machining processes [47]. PSD and autocorrelation function (ACF) are Fourier transform pairs and are related to the frequency or, respectively, spectral analysis. It was found valuable to use frequency-based procedures for analysing turned or ground surfaces with the application of a profile (2D) or areal (3D) performance [48].

From the literature review, it was found that there were no studies if the surface topography direction (dominant direction) influences high-frequency measurement noise detection and, respectively, reduction processes. Falsely applied methods, e.g., commonly used (available in the commercial software) filters, can cause huge errors in surface topography parameters computing. Moreover, analysis of surface topography, especially when roughness evaluation [49] is required, with the application of frequency or direction methods, desire a very reliable and thorough analysis, therefore should be presented in detail. Even experienced users, when not guided on how to use common methods, can falsely estimate the properties of properly made parts and reject its suitability to further processing.

2. Materials and Methods

2.1. Studied Details

Various types of surfaces were studied, as follows: cylinder liners after the plateau-honing process, plateau-honed cylinder liners with additionally burnished oil pockets (dimples) with various sizes (width ranging from 70 to 800 μm and depth between 7 and 100 μm), turned, ground, milled or, respectively, laser-textured (with 60- or 120-angle texturing processes).

According to the preliminary analysis, all of the studied details were provided with an areal form removal process [50]. Types of procedures for extraction of long-frequency components from the raw measured data were widely studied in previous papers [51], so, correspondingly, were not considered in this research. From the action above, studied surfaces were, in general, flat that did not contain form and, correspondingly, waviness. However, the accuracy of form and waviness removal on the results obtained was not currently studied. If the surface contained non-measured points [52] or other, like individual peaks (spikes [16]), errors, were extracted and removed from the data.

More than 10 (usually between 10 and 15) surfaces from each type of topographies were measured and studied to provide some repeatability in the analysis of the results, but only some of them were presented in detail. Moreover, all of the analyses were improved with modelled data as well and, subsequently, compared with those measured to find some general proposals.

2.2. Measurement Process

Studied details were measured by stylus or optical techniques. The stylus instrument was Talyscan 150 having a nominal tip radius of 2 μm, with a height resolution equal to 10 nm, the measured area 5 by 5 mm with, respectively, 1000 × 1000 measured points. The sampling interval was 5 μm. The measurement velocity was 1 mm/s and, appropriately, its influence was not considered that it was not the preliminary of the research provided, it issue was extensively studied in previous studies.

Secondly, the non-contact measurement equipment was a white light interferometer, Talysurf CCI Lite. Its height resolution was 0.01 nm. The measured area was 3.35 by 3.35 mm with, proportionately, 1024 × 1024 measured points. The spacing was 3.27 μm and, suitably, the effect of sampling or spacing on values of an areal surface texture parameters were not studied in this paper.

2.3. Applied Methods and Procedures

2.3.1. Regular Methods and Procedures

Studied surfaces were analysed by various methods. All of the procedures applied for data processing are commonly used by metrologists and, simultaneously, available in the commercial software (measuring equipment) dedicated to the analysis of the results of surface topography measurements. One of the main purposes of the studies performed was to present some valuable guidance on how to use those approaches with minimisations of the errors in a measured data processing that even precise measuring equipment may not provide relevant results when obtained data are not processed accurately.

Very popular and, correspondingly, available in measuring equipment are PSD and ACF methods, corresponding to the Fourier Transform characterisation, as an outstanding example of frequency-based analysis. Frequency characteristics with spectrum performance, e.g., the FS method, can be an excellent complement for the two above techniques. Another method, an angular spectrum gives quantitative information about the character of surfaces (isotropic or anisotropic) and surface directionality [53]. Therefore, the texture direction (TD) graph can be applied in the processes of surface topography characterisation when a high-frequency noise is studied.

Generally available filters, proposed for a definition of S-F surface [54] (received by definition of S-surface and F-surface) with software equipment performance, are Gaussian (GF), with its robust (RGF) modification, regular isotropic spline (SF), median denoising (MDF) or, continuing, fast Fourier transform (FFTF) algorithms. Those, widely used filters, can be compared by supporting them with proposed techniques for minimisation of errors in surface roughness parameters calculations.

2.3.2. Proposed Approaches for Surface Topography Characterisation

Even surface topography is often analysed with an areal (3D) performance that many issues concerning material contact, in general, is also in areal meaning, much valuable information can be received when selected profiles are studied. The 2D characterisation can be valuable for the definition of L-surface or, respectively, S-surface, both with properly defined bandwidth filtration. The distortion of some features, e.g., dimples [55], can be also more visible when a 2D analysis is provided.

Comparing the 2D analysis against the 3D, it should be considered when PSD, ACF, FS or, consequently, the isometric (profile) view is applied simultaneously. Some multithreaded aspects were also found ad recommended previously. When an areal PSD, ACF and FS method was proposed, differences between a surface containing high-frequency noise and a surface where that type of measurement error was not defined were negligible or, correspondingly, did not exist. Considering areal studies, which can give a more suitable response to the tribological performance that, consequently, the contact is areal, in general. Nevertheless, some analysis must be based on the selected and extracted features, that can be provided with profile (2D) performance.

The process of definition or, generally, detection of high-frequency measurement errors from the results of surface topography measurement, was already proposed with profile characterisation. Except for the number and size of surface topography features, e.g., valleys, the 2D analysis could give a crucial response to the occurrence of the high-frequency error. Even though the PSD calculation was provided in all directions, the 3D characteristic was not convenient and the required frequencies were not visible on the graph. From that point of view, the 3D PSD characterisation was useless and an areal tribological characterisation was limited to the measurement errors caused by the environment vibrations.

Some similar limitations were observed for ACF analysis. In Figure 1 areal ACF graphs were applied for the characterisation of a milled surface. Differences from ACF calculated for surface containing high-frequency errors (j, k) and the same surface but without this type of noise (d, e) were negligible. For this type of analysis, extraction of the centre (located in the middle of the analysed 3D data) profile gave more unambiguous results that extreme (middle) values increase more rapidly when a high-frequency noise is found (f and l).

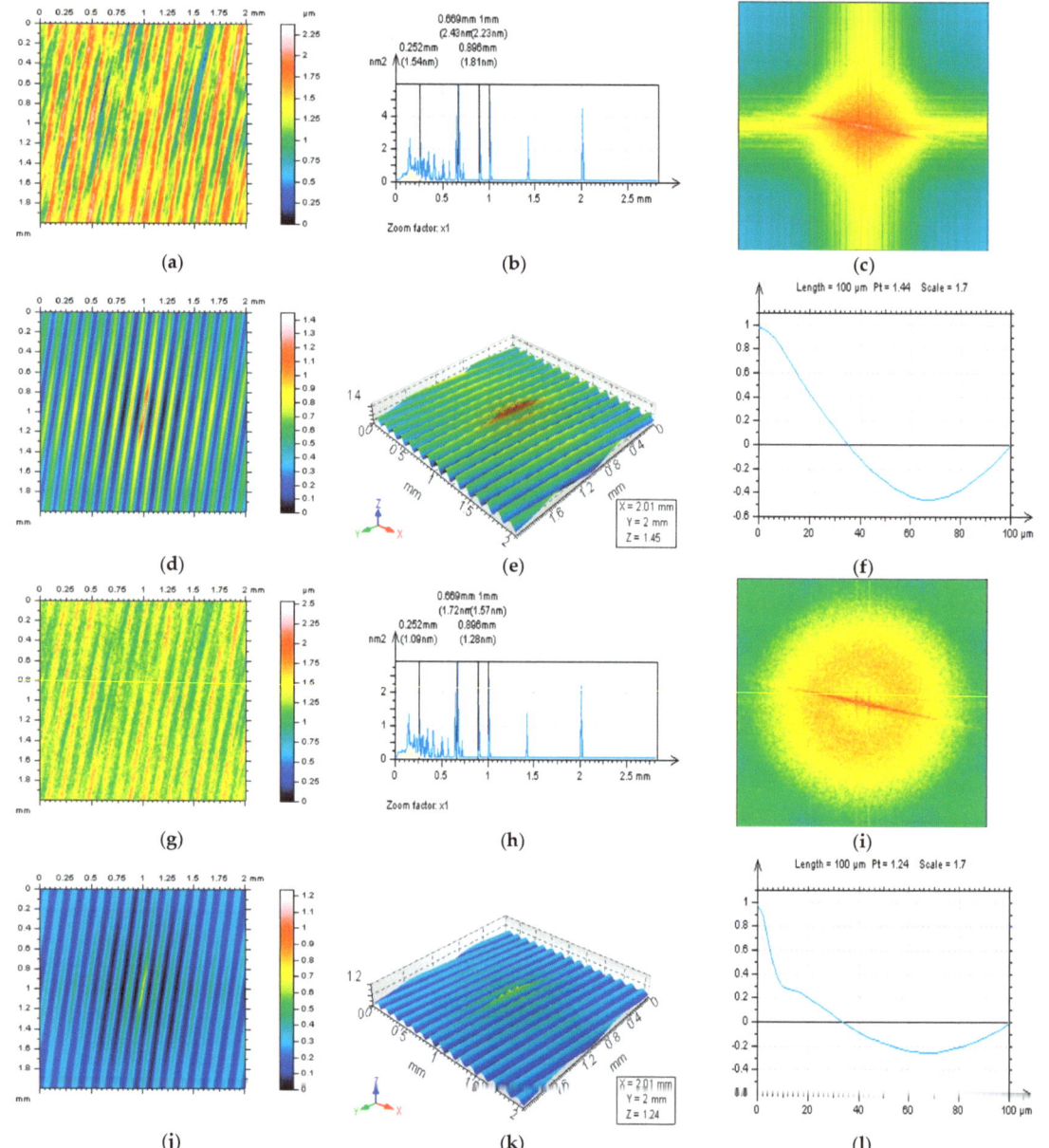

Figure 1. Contour map plots (**a**,**g**), PSDs (**b**,**h**), FS (**c**,**i**) and map plots (**d**,**j**), isometric view (**e**,**k**) and centre–profiles (**f**,**l**) of ACFs received from raw measured data of milled surface (**a**–**f**) and the same surface with a high-frequency noise (**g**–**l**).

The 3D analysis can be valuable only when FS or eye-view studies are performed, nonetheless, this technique may require experienced users. Moreover, the PSD, ACF and FS methods did not always give clear results, e.g., for plateau-honed surface topography (Figure 2), an areal analysis was useless instead of a profile (Figure 3). Therefore, for some cases, the profile (2D) studies are proposed instead of an areal (3D).

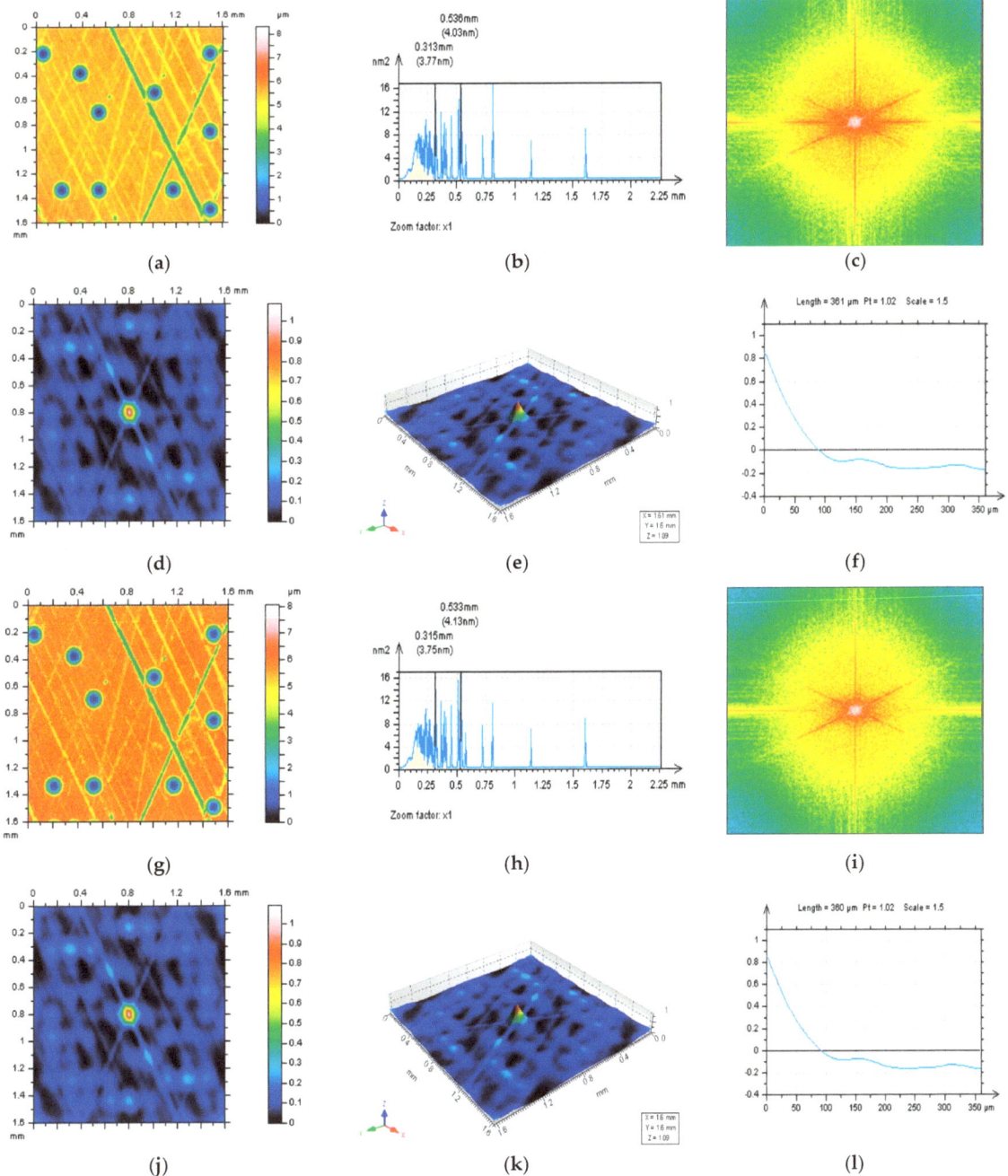

Figure 2. Contour map plots (**a**,**g**), PSDs (**b**,**h**), FS (**c**,**i**) and map plots (**d**,**j**), isometric view (**e**,**k**) and centre–profiles (**f**,**l**) of ACFs received from raw measured data of plateau–honed cylinder liner surface containing oil pockets (**a**–**f**) and the same surface with a high-frequency noise (**g**–**l**).

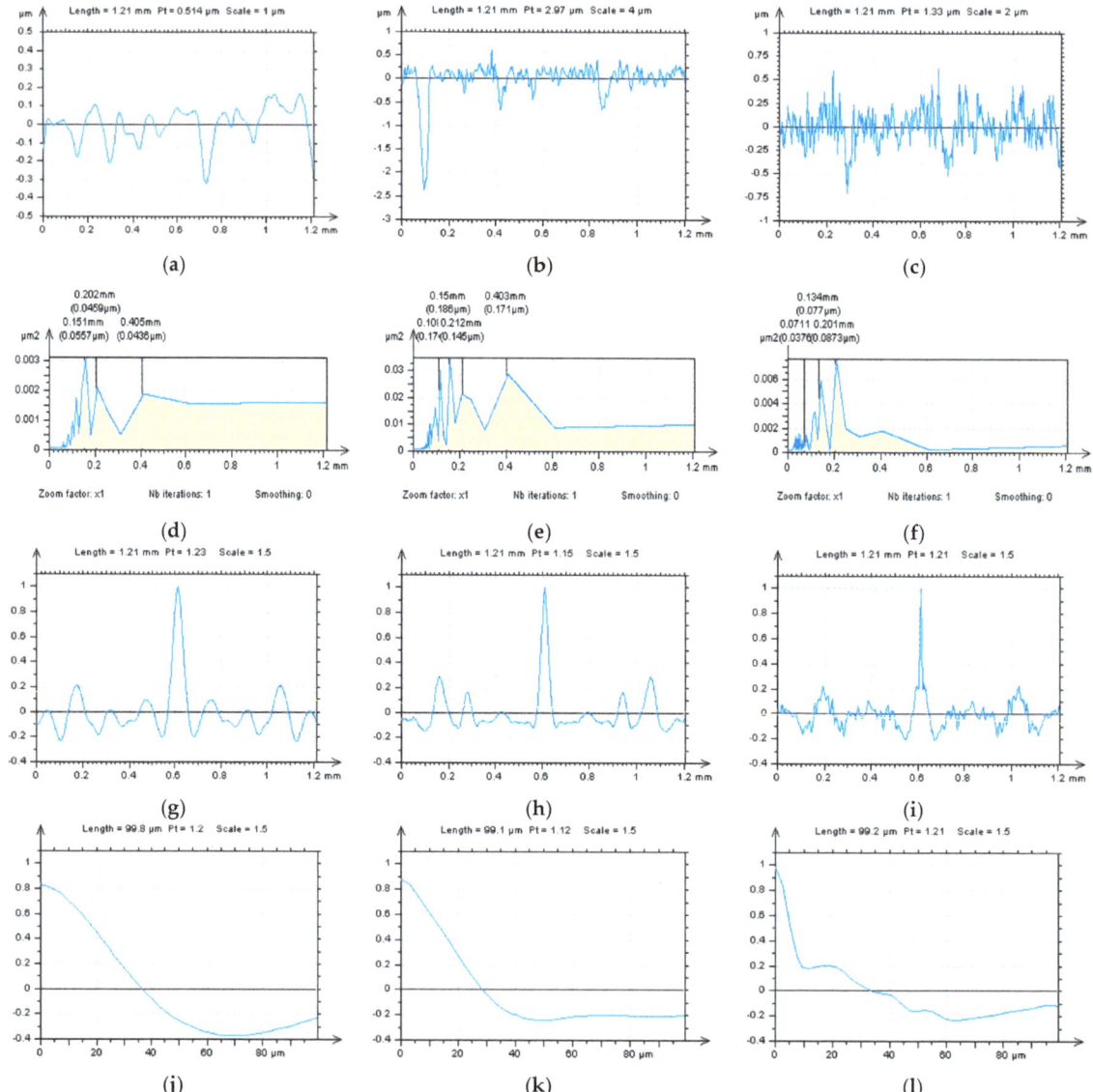

Figure 3. Extracted from a plateau-honed cylinder liner surface with additionally burnished dimples profiles (**a**–**c**): raw measured data (left column), containing high–frequency noise with a dimple (middle) and free-of-feature (right column), their PSDs (**d**–**f**), ACFs (**g**–**i**) and centre–part of ACFs (**j**–**l**) respectively.

Not only analysis of the PSD and ACF graphs can be valuable in processes of surface topography data studies. It was provided in previous studies that the direction of the profile extraction can affect the accuracy of the high-frequency measurement noise detection [56]. Moreover, the direction can also have a considerable influence on the validation of noise removal algorithms that, when using a noise surface characterisation, TD graph can provide relevant results that can be exceedingly valuable in the analysis of surface texture isotropy. The process of detection of high-frequency noise can be improved with an analysis of the high-frequency noise surface (HFNS). The HFNS is a surface received when applying an algorithm for a high-frequency noise separation, e.g., regular digital fileting methods. In practice, the HFNS is a surface removed from the raw measured data when a noise-removal procedure is applied. Previously, when defining the properties of HFNS, it was assumed that it should consist only of the frequencies that noise surfaces are defined, in this case, of the frequencies in the high domain. This property can be described by the PSD analysis. Moreover, the effect of the shape of the centre part of the ACF can be also crucial. According to the further studies it is suggested to deeply analyse the directional properties of the HFNS with a TD graph consideration. If HFNS is defined properly, containing only the high-frequency components should be isotropic as well. This can be verified by calculation and presenting the TD graph.

In previous, provided in already presented results, it was assumed that PSD and ACF characterisation of selected (e.g., those from the centre part of the analysed detail) extracted profiles can give some valuable information about the high-frequency noise occurrence. However, even though the high frequencies are visible in the PSD and ACF graphs, the HFNS can not be isotropic which was shown with plenty of TD graph characterisations. This property can be especially visible when a surface with a determined direction is comprehensively considered. In Figure 4 ground surface is presented. Each type of filtering (GF, SF or MDF) defined an HFNS with no isotropic performance that the TD graph indicates no isotropic textures. This technique can indicate that in the HFNS other than noise features can be found. This is not expected that some required surface features may be removed and surface quality falsely estimated, the values of surface topography parameters can be calculated erroneously as well. Therefore, from the above property, it is suggested to use all of the available techniques, like PSD, ACF, FS and TD methods, in the process of detection of high-frequency measurement errors occurrence and, respectively, to evaluate the surface roughness parameters with minimisation of high-frequency errors influence. One of the most important tasks to be resolved in the presented studies is to select the value of filter bandwidth. The cut-off value was studied in many previous papers, nevertheless, there is extremely difficult to remove the measurement noise without any distortion of the surface texture features, like valleys, dimples, and treatment traces in general. Very popular in surface topography analysis are plenty of Gaussian filters [57]. Many current studies are performed by comparing plenty of methods and approaches to the ISO-defined general Gaussian filtering technique. Some alternatives can be proposed with Fast Fourier Transform and its filtering methods (FFTF). In practice, it is extremely difficult to propose a relevant cut-off value when selecting an appropriate filter for noise removal that, even optimized, a filter can always affect other frequencies (features) of an analysed detail. Removing some of the features can be especially disadvantageous that in the process of control, properly made parts can be classified as lacks and rejected.

Therefore, in the presented studies, the technique based on the analysis of FS is suggested to select the most appropriate cut-off value when a digital filter is used. In Figure 5 FS technique was presented for HFNS received by FFTF filtering of a plateau-honed cylinder liner raw measured surface data. The FS-based technique (FSBT) proposes to select the bandwidth of the filter with the filter selector range. When the range is too big (a) and exceeds the FFT graph features with known direction, the filter can not remove the entire noise data. However, on the other hand, when a range is too small (c) other features can be also separated from measured data. This can be especially visible when an isometric view of HFNS is studied (d, f) or, respectively, TD graphs (j, l) are considered. Application of the filter range equal to the size (length) of the features in the FFT graph (b) turns out to be essential in an accurate definition of HFNS, which is isotropic (TD), is primarily composed of the high-frequencies (PSD) and, correspondingly, has no traces of the treatment trace features (isometric view analysis). The FSBT can be especially valuable for topographies with a determined direction but can be not convincing when isotropic surfaces are studied.

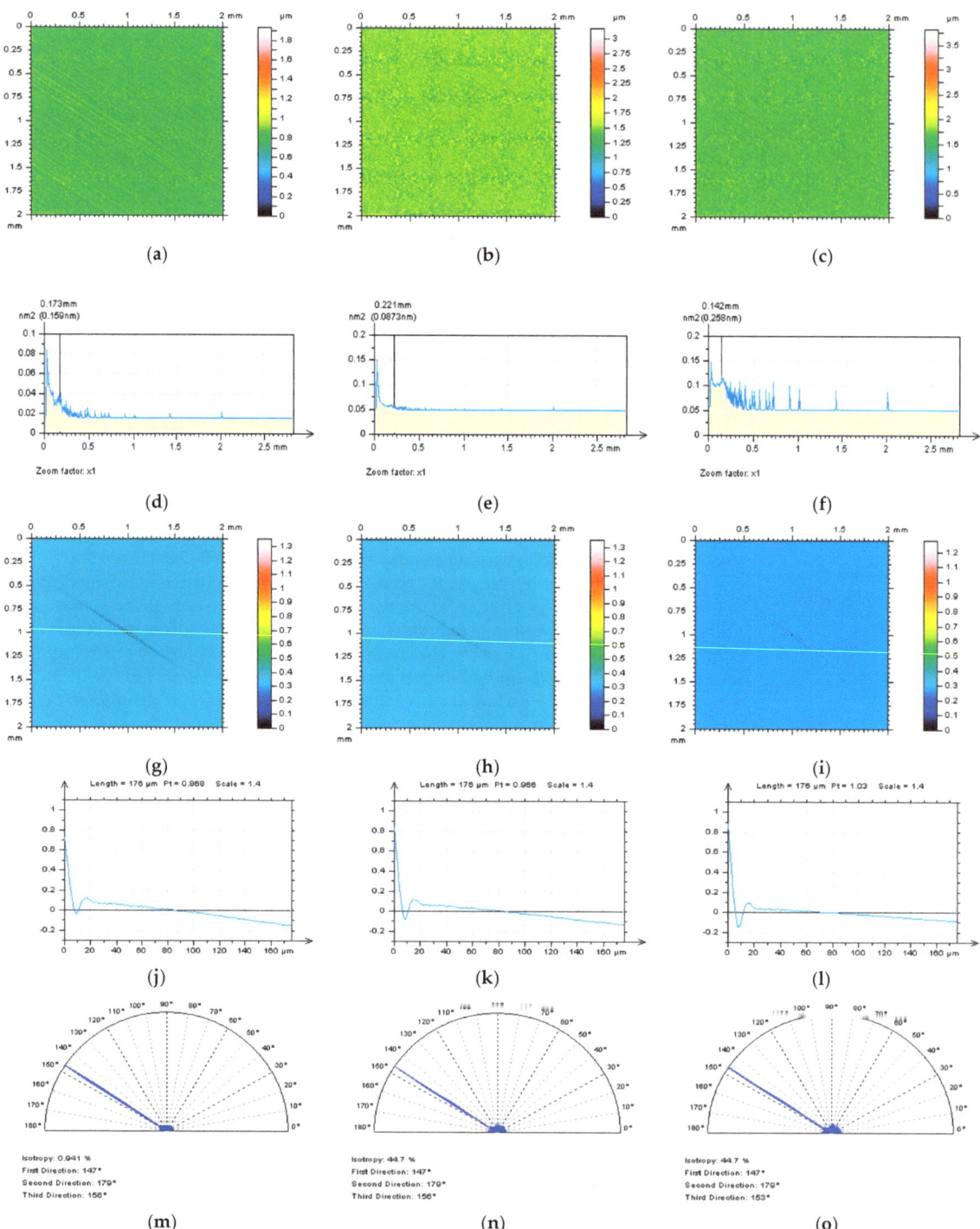

Figure 4. Contour map plots (**a**–**c**) of HFNS received from ground surface after application of GF (left column), SF (middle) and MDF (right column), their PSDs (**d**–**f**), ACFs (**g**–**i**), extraction of ACFs based on direction of the treatment traces (**j**–**l**) and TD graphs (**m**–**o**), respectively, cut-off = 0.025 mm.

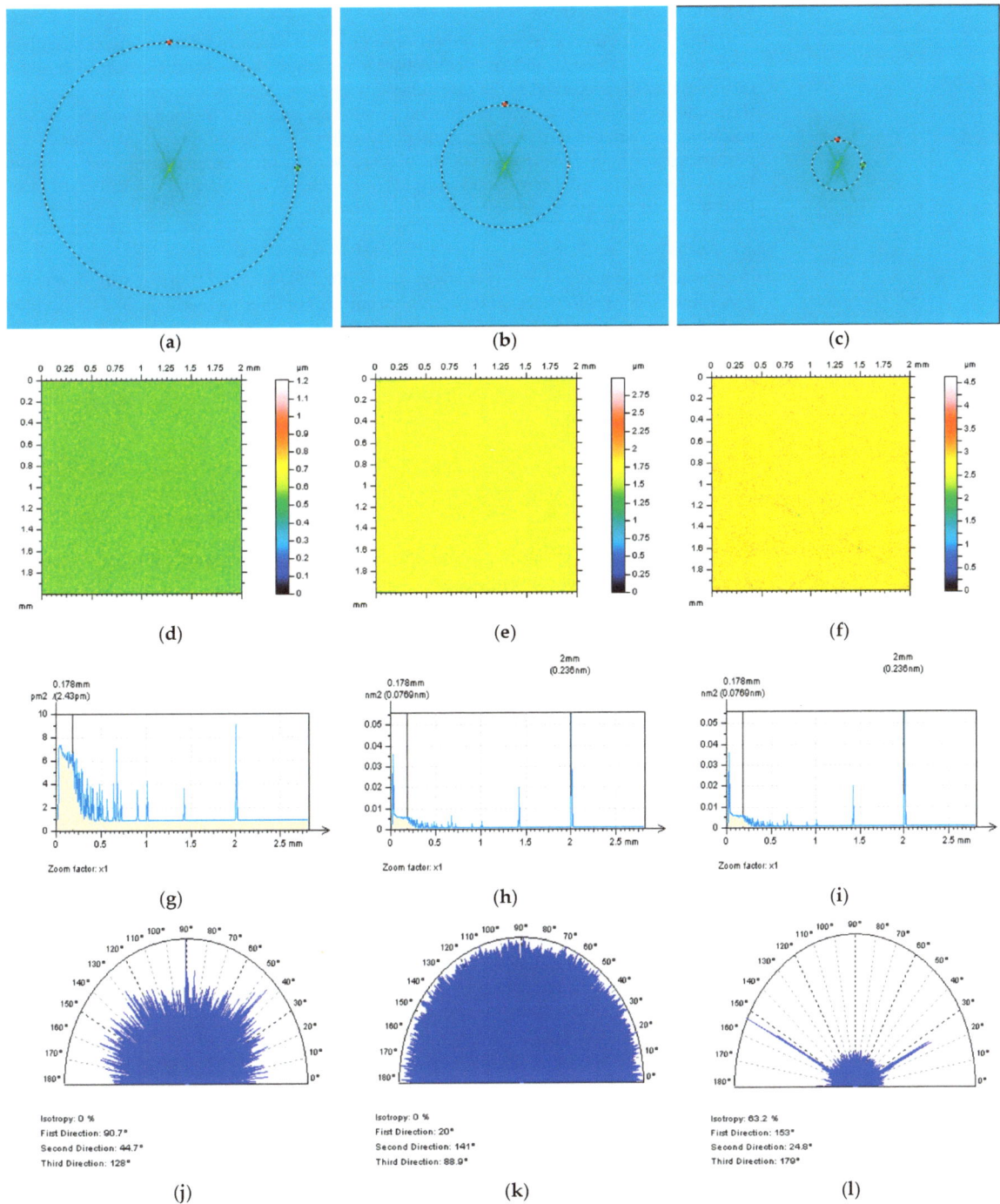

Figure 5. Spectrum edition (**a**–**c**), HFNS (**d**–**f**), their PSDs (**g**–**i**) and TDs (**j**–**l**) from a plateau-honed cylinder liner surface texture received after application of FFTF method.

3. Results and Discussion

Processes of data analysis were proposed for raw measured data received straightly from the measurement and, respectively, for modified data. Modification of the raw measured data is presented as the raw data with rotational actions to improve the validity of the methods proposed. As was mentioned previously, presented methods can be especially valuable for surfaces with a directional performance. Therefore, proposed studies were performed for cylinder liners after the plateau-honing process, plateau-honed cylinder liners with additionally burnished dimples (oil pockets), turned, ground, milled or laser-textured with various angles (60 or 120 in particular).

3.1. Analysis of Raw Measured Data with Proposed Directional Technique

From the performed analysis it was found that PSD areal (3D) characterisation could give some responses for selected types of both surfaces or filtering methods, like GF (Figures 6b and 7b), SF (Figures 6h and 7h) and FFTF (Figures 6n and 7n) for turned or plateau honed cylinder liner with additionally burnished oil pockets surfaces or, respectively, SF (Figure 8h) and FFTF (Figure 8n) for laser-textured topographies. However, when GF or SF was used for the definition of the HFNS from the results of plateau-honed cylinder liner topographies (Figure 7a,d) there were features visible on the isometric views so, simultaneously, those techniques may be not suitable that removing pattern other than the noise in required (high) frequencies. Similar to the above, the application of the RGF technique caused a separation of treatment trace features from the surface topography of a plateau-honed cylinder liner with additionally created valleys (Figure 7d) or laser-textured details (Figure 8d).

For reduction of the influence of high-frequency measurement errors, the FSBT method with application of the FFTF filtration technique was proposed. From the analysis of the 3D PSD graphs, it was found that the high-frequency components were dominant for all of the analysed types of surfaces (Figures 6n, 7n and 8n) when an FFTF method was applied and, correspondingly, there were no non-noise features in the HFNS (Figures 6m, 7m and 8m). Analysing a high-frequency error occurrence can be also improved with an application of the direction-based method with TD graph studies. Generally, when the surface does not contain any direction features, the TD graph is presented in an isotropic performance for both raw measured surface or HFNS characterisation. For each of the types of considered direction surface, like cylinder liners after the plateau-honing process, plateau-honed cylinder liners with additionally burnished oil pockets (valleys), turned, ground, milled or laser-textured with various angles surfaces, application of the FFTF filtration technique with a selection of the cut-off value according to the FSBT approach gave encouraging results.

The 3D PSD analysis provided dominant frequency in the high domain, the TD graphs presented the HFNS as an isotropic surface (Figures 6o, 7o and 8o) and, respectively, no treatment traces were found in the HFNS isometric view. Fulfilment of all these properties simultaneously can minimise the errors in a high-frequency measurement noise definition and, correspondingly, reduction by a digital filtration. Roughly chosen bandwidth can not be convenient when surface feature densities are different. Applications of various filtration techniques, like those plenties of Gaussian filters and their modifications, may not be relevant when not required features are also removed from the raw measured data. Therefore, the FSBT method seems to be fairly advantageous for a regular user.

The best method to validate the FSBT technique's suitability is to analyse the HFNS with a PSD, ACF, FS and TD characterisation simultaneously, the more characteristics are fulfilled the more a high-frequency measurement noise procedure can be classified as sufficiently precise. Application of the FSBT scheme was not considered for isotropic surfaces, nevertheless, an isotropic performance of the HFNS can be also valuable and may be considered in further studies.

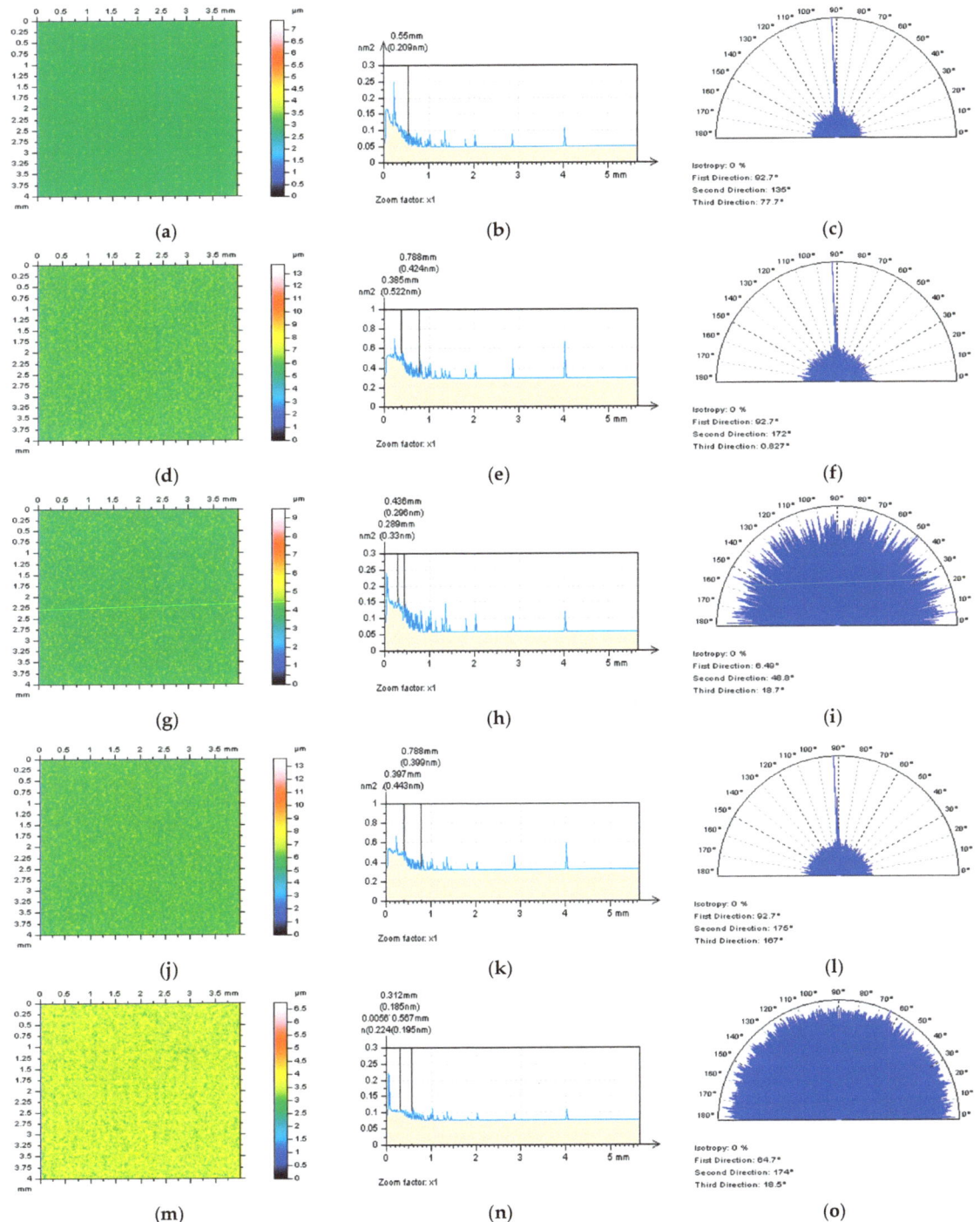

Figure 6. Analysis of a HFNS (left column), their PSDs (middle) and TDs (right column) received from a turned piston skirt surface by application of: GF (**a**–**c**), RGF (**d**–**f**), SF (**g**–**i**), MDF (**j**–**l**) and FFTF (**m**–**o**) method, cut–off = 0.025 mm.

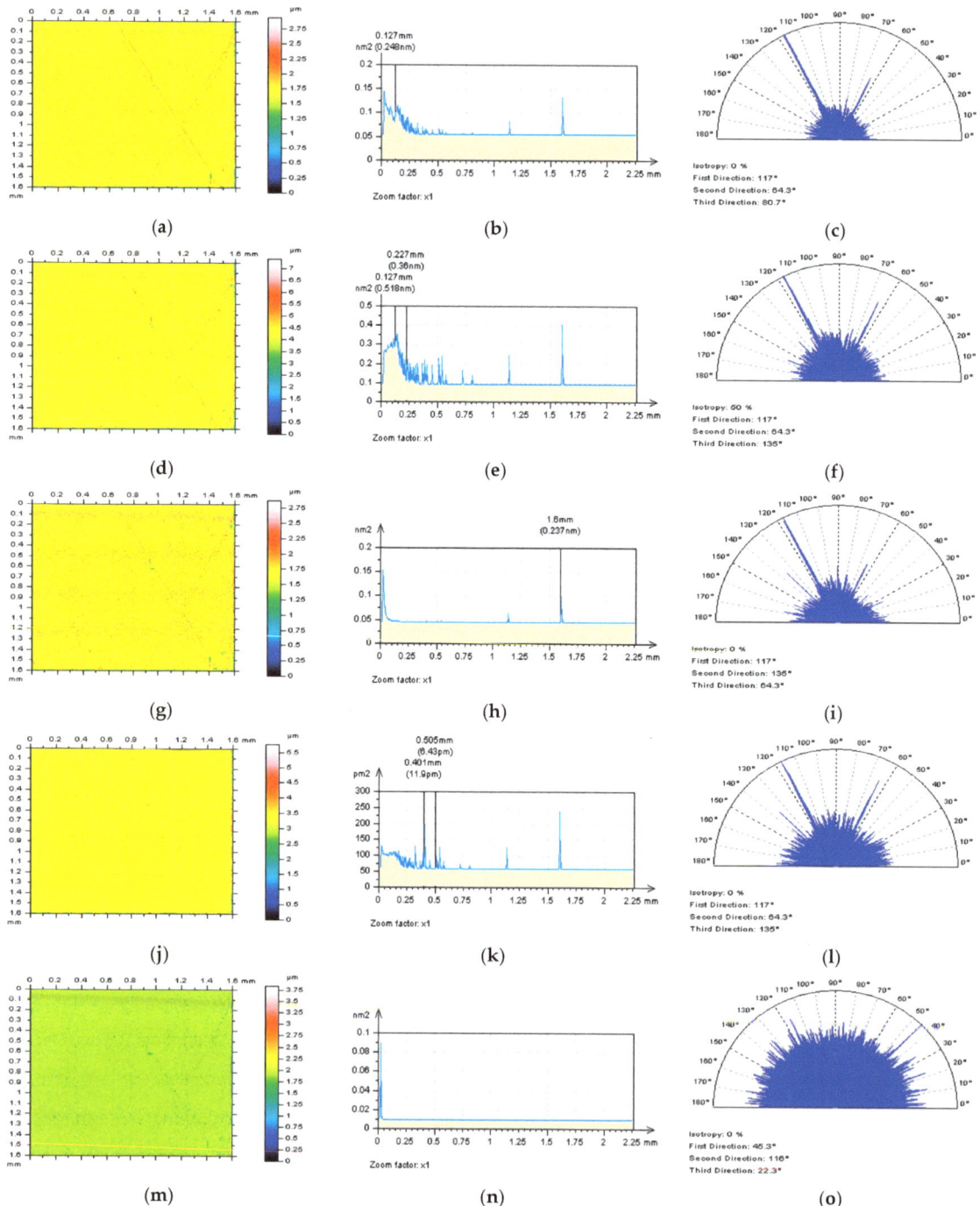

Figure 7. Analysis of a HFNS (left column), their PSDs (middle) and TDs (right column) received from a plateau-honed cylinder surface containing oil pockets by usage of: GF (**a–c**), RGF (**d–f**), SF (**g–i**), MDF (**j–l**) and FFTF (**m–o**) approach, cut–off = 0.015 mm.

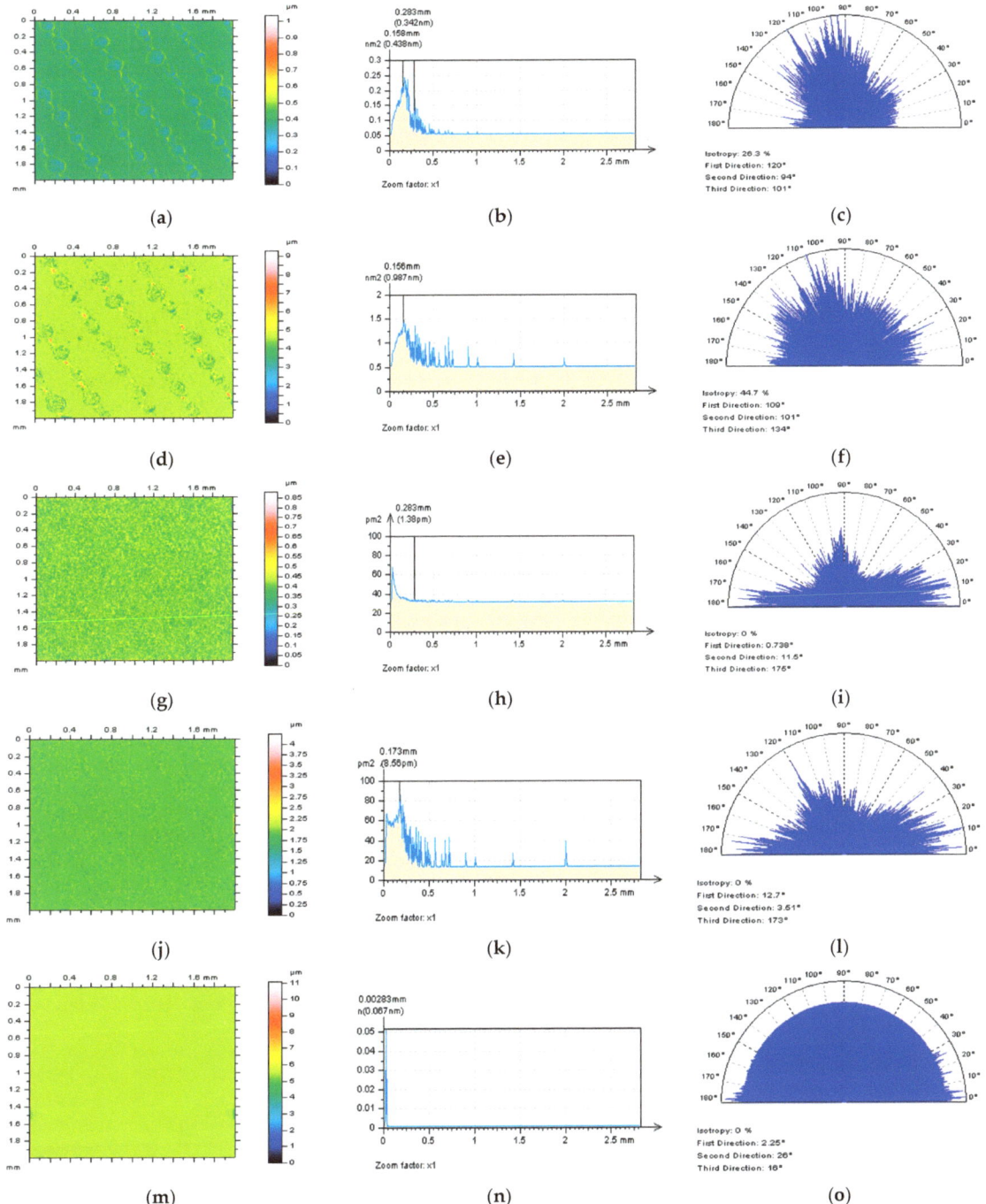

Figure 8. Analysis of a HFNS (left column), their PSDs (middle) and TDs (right column) received from a laser-textured surface by filtering: GF (**a**–**c**), RGF (**d**–**f**), SF (**g**–**i**), MDF (**j**–**l**) and FFTF (**m**–**o**) approach, cut–off = 0.015 mm.

3.2. Validation of Proposed Method with a Data Rotation Process

The proposed HFNS directional method shortly named the FSBT scheme, can be improved with studies performed for raw measured data with the rotation process. For isotropic topographies, the rotation validation was not required that, respectively, data usually have isotropic properties and direction does not influence the actions made. However, when the surface has not had an isotropic characteristic, rotation of the data can be proposed. Therefore, for plateau-honed, ground, milled or laser-textured surfaces, the rotational validation of the HFNS direction method can be applied. For this type of surface texture data, a rotation equal to 30 degrees on the left and right was provided.

Generally, the application of 3D PSD analysis improved that the rotation has no significant influence on the validation of the high-frequency noise detection process. For each type of analysed (directional) surface, differences were not found or they were negligible, even the PSD was calculated as in all directional methods. Usually, the PSD gave some response to the occurrence of the high-frequency noise in the results of surface topography measurements.

Analysis of the isometric view of the HFNS showed that Gaussian filters and their modifications, e.g., GF and RGF, caused a removal of the relevant features from the raw measured surface topography data that, correspondingly, there were found on the HFNS (Figure 9a,d and Figure 10a,d). The occurrence of the non-high-frequency features caused differences in the 3D PSD graphs, some frequencies with bandwidths greater than 0.015 µm were visible (Figure 9b,e and Figure 10b,e). A similar tendency could be found when an MDF was used, the isometric view did not present any non-noise features, nevertheless, from the analysis of the 3D PSD graph, it can be concluded that contain greater than the proposed cut-off (0.015 µm) value features (Figure 9j,k, and Figure 10j,k). In all of those three cases, the isotropy was also lost (Figure 9c,f,l and Figure 10c,f,l). Some encouraging results were obtained when the spline approach (SF) was proposed. Both, PSD and TD graphs gave interesting proposals.

The cut-off value was proposed with an application of the FSBT scheme that, respectively, bandwidth was selected with a technique widely presented in Figure 5. The rest of the algorithms (filters) were proposed with similar cut-off values to correspond to the newly applied approach. In fact, each of the filters can be used with different bandwidths, nonetheless, some general conclusions may be difficult to be suggested in this manner. Therefore, bandwidth characteristic is defined according to the FSBT requirements.

From the studies of TD graphs, it was concluded that the direction (rotation) has a considerable influence on the isotropy of the HFNS when Gaussian filters (GF and RGF) were applied. In some cases, when the angle of the features (e.g., scratches) was changed, the isotropy could be found more than in the rest of the feature directions. Therefore, for these types of filtering methods, the TD and, correspondingly, FSBT scheme, must be improved with an analysis of the isometric view and PSD of the HFNS.

Application of SF and MDF techniques gave more encouraging results that direction and, especially, isotropy received in the TD graph were proportional to the rotation direction. From that matter, those techniques were influenced by the angle of the rotation process and their proportionality can indicate the usefulness of those filters.

The most encouraging results were received when an FFTF method was applied. All of the requirements, containing analysis of the isometric view of HFNS, their PSD, ACF, FS and TD, gave a straight response to the occurrence of the high-frequency measurement errors. When considering PSD, the noise frequency (high-frequency) was entirely dominant. There were no non-noise features on the HFNS and, respectively, the HFNS was isotropic in general. All of the requirements were not met simultaneously when other methods (GF, RGF, SF and MDF) were proposed. From that point of view, considering all of the analysed, commonly-used (available in commercial software) filters, the FFTF method can be the most suitable for definition and, correspondingly, reduction of the high-frequency measurement errors from the results of surface topography measurement of plateau-honed, turned, ground, milled or laser-textures surfaces.

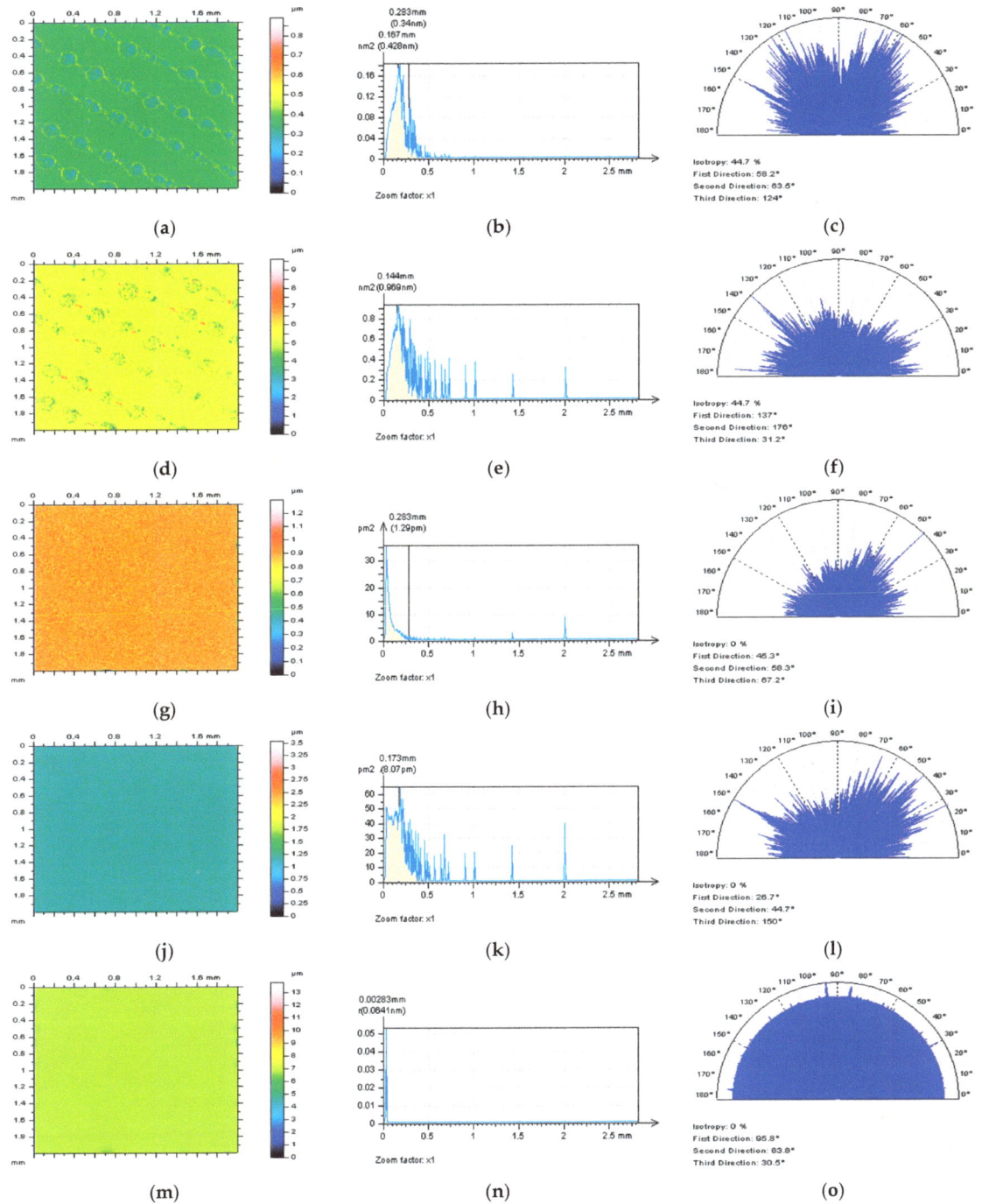

Figure 9. Analysis of a HFNS (left column), their PSDs (middle) and TDs (right column) received from a 30° left rotated laser-textured surface by application of the following filtering method: GF (**a**–**c**), RGF (**d**–**f**), SF (**g**–**i**), MDF (**j**–**l**) and FFTF (**m**–**o**) approach, cut-off = 0.015 mm.

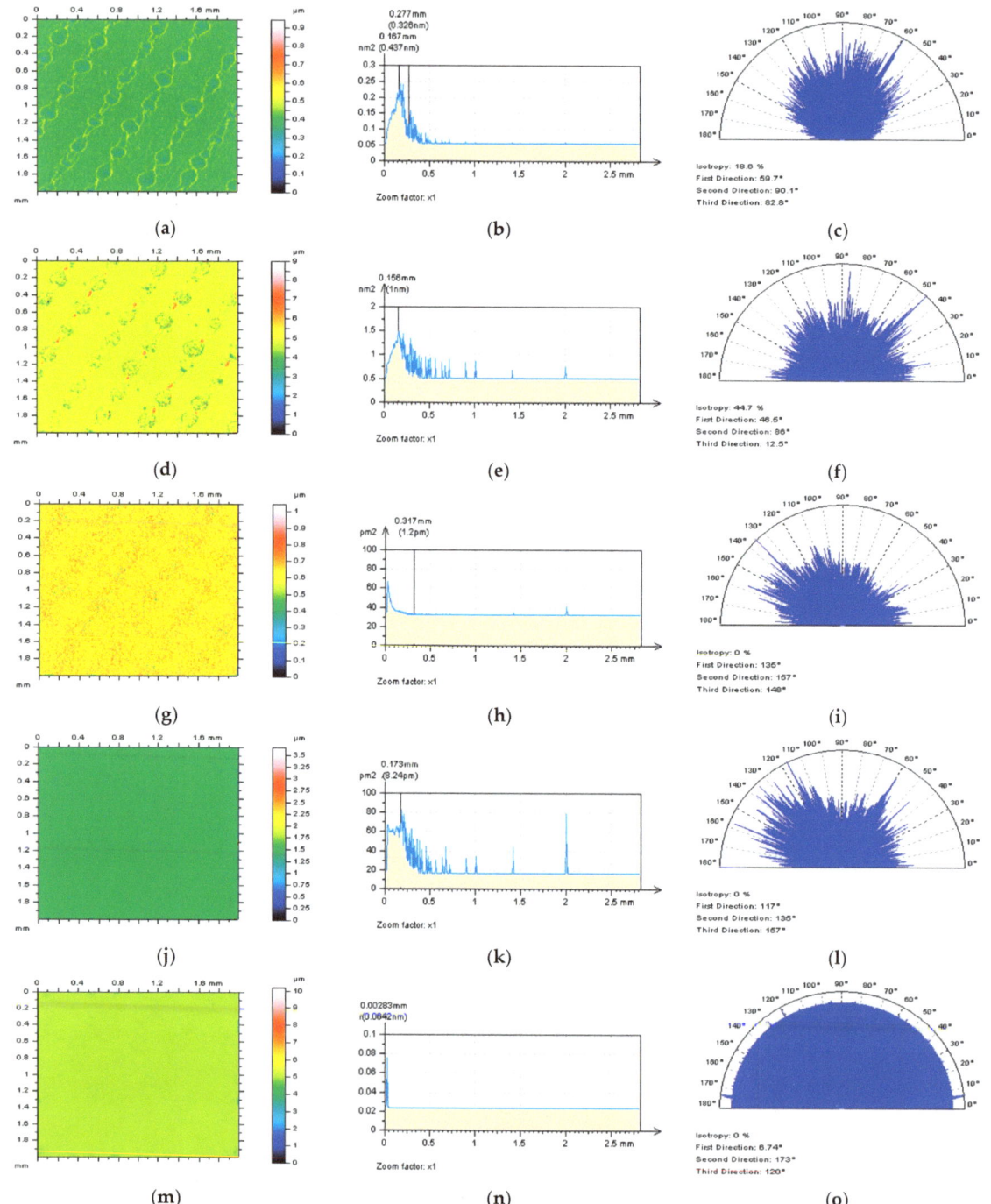

Figure 10. Analysis of a HFNS (left column), their PSDs (middle) and TDs (right column) received from a 30° right rotated laser-textured surface by usage of the following filtering approaches: GF (**a**–**c**), RGF (**d**–**f**), SF (**g**–**i**), MDF (**j**–**l**) and FFTF (**m**–**o**) approach, cut–off = 0.015 mm.

4. The Outlook

There are still many issues that require more sophisticated studies and must be resolved in future. Of those, the most important are:

1. Studies of isotropic surface topography are still an encouraging issue when a high-frequency noise must be defined and reduced. There are no direction features that can be useful for defining the HFNS properties. From that point of view, other, non-feature-based methods must be proposed for this type of surface.
2. The high-frequency measurement errors can be difficult to detect as suggested in paper methods when the amplitude of the noise is relatively small. Some improvement of the proposed method must be included with modifications of the noise amplitude.
3. Furthermore, the effect of the high-frequency measurement noise amplitude on the results of applied techniques was not considered. The influence of the amplitude of noise on the results of filter application and, correspondingly, on the surface topography parameters (from ISO 25178 standard) must be also considered to provide more surface functional advantages.

5. Conclusions

The following conclusions can be proposed:

1. General functions, available in the commercial software, like Power Spectral Density, Autocorrelation Function, Frequency Spectrum or Texture Direction, are very useful in the processes of definition and then reduction of the high-frequency measurement errors. The most important criteria for the selection of relevant procedures is to remove (reduce) the high-frequency noise without any modification of the other feature of the characterized surface data.
2. For a high-frequency measurement noise characterisation, both profile (2D) and an areal (3D) analysis can offer other benefits. The 3D analysis can be valuable only when Frequency Spectrum or eye-view studies are performed, nonetheless, this technique may require experienced users. The Power Spectral Densities, Autocorrelation Function and Frequency Spectrum methods did not always give clear results, e.g., for plateau-honed surface topography, an areal analysis was useless instead of a profile. Therefore, for some cases, the profile (2D) analysis is proposed contrary to an areal (3D).
3. The technique based on the analysis of the Frequency Spectrum is suggested to select the most appropriate cut-off value when a digital filter is used for a high-frequency measurement noise reduction. The Frequency Spectrum-based technique proposes to select the bandwidth of the filter with the filter selector range. When the range is too big and exceeds the Fast Fourier Transform graph features with a known direction, the filter can not separate the entire noise data and, respectively, on the other hand, when a range is too small other features can be also removed from the raw measured data. From that matter, it is difficult to propose one cut-off value for each type of surface that should be dependent on the density and direction of the surface topography features. The main purpose is to increase the value of the cut-off until the non-noise data is detected on the high-frequency noise surface.
4. Considering the studies of Texture Direction graphs, it was found that the rotation (direction) has a significant influence on the isotropy of the high-frequency noise surface when Gaussian filters were applied. When the angle of the surface features (e.g., those after honing process) was changed, the isotropy could be defined more than in the rest of the feature directions so, respectively, for these types of filtering methods, the Texture Direction and Frequency Spectrum Based Technique, must be improved with analysis of the isometric view and Power Spectral Density of the high-frequency noise surface.
5. Generally, the usage of 3D Power Spectral Density characterisation improved that the rotation process has no significant influence on the validation of the process of high-frequency noise detection. For analysed types of directional surfaces, differences were not found or, respectively, they were negligible. Usually, analysis of the Power

Spectral Density graph can give some encouraging response to the high-frequency noise occurrence.
6. The most encouraging results were received when a Fast Fourier Transform Filter method was applied. All of the high-frequency noise surface requirements, containing analysis of the isometric view, their Power Spectral Density, Autocorrelation Function, Frequency Spectrum and Texture Direction, gave an appropriate response to the occurrence of the high-frequency measurement errors. Considering all of the analysed, commonly used or, in the other words, available in commercial software, filters, the Fast Fourier Transform Filter method can be the most suitable for definition (detection and removal) of the high-frequency measurement errors from the results of surface topography measurement of plateau-honed, turned, ground, milled or laser-textures surfaces.

Funding: This research received no external funding.

Institutional Review Board Statement: Not applicable.

Informed Consent Statement: Not applicable.

Data Availability Statement: Data sharing is not applicable to this article.

Conflicts of Interest: The author declares no conflict of interest.

Nomenclature

The following abbreviations (left) are used in the manuscript:

ACF	autocorrelation function
AFM	atomic force microscopy
F-surface	form surface
FFTF	Fast Fourier Transform Filter
FS	Frequency spectrum characterisation
FSBT	Frequency Spectrum-Based Technique
GF	Gaussian filter
HFNS	high-frequency noise surface
L-surface	long-wavelength surface
MDF	median denoising filter
MQCL	Minimum Quantity Cooling Lubrication
PSD	power spectral density
RGF	robust Gaussian filter
S-F surface	is a surface after S- and F- filtering
S-surface	short-wavelength noise surface
SF	spline filter
TD	texture direction graph

References

1. Trzepieciński, T.; Fejkiel, R. On the influence of deformation of deep drawing quality steel sheet on surface topography and friction. *Tribol. Int.* **2017**, *115*, 78–88. [CrossRef]
2. Shao, Y.; Yin, Y.; Du, S.; Xia, T.; Xi, L. Leakage Monitoring in Static Sealing Interface Based on Three Dimensional Surface Topography Indicator. *ASME J. Manuf. Sci. Eng.* **2018**, *140*, 101003. [CrossRef]
3. Morehead, J.; Zou, M. Superhydrophilic surface on Cu substrate to enhance lubricant retention. *J. Adhes. Sci. Technol.* **2014**, *28*, 833–842. [CrossRef]
4. Zheng, M.; Wang, B.; Zhang, W.; Cui, Y.; Zhang, L.; Zhao, S. Analysis and prediction of surface wear resistance of ball-end milling topography. *Surf. Topogr. Metrol. Prop.* **2020**, *8*, 025032. [CrossRef]
5. Dzierwa, A.; Galda, L.; Tupaj, M.; Dudek, K. Investigation of wear resistance of selected materials after slide burnishing process. *Eksploat. Niezawodn.* **2020**, *22*, 432–439. [CrossRef]
6. Szala, M.; Świetlicki, A.; Sofińska-Chmiel, W. Cavitation erosion of electrostatic spray polyester coatings with different surface finish. *Bull. Pol. Acad. Sci. Tech. Sci.* **2021**, *69*, e137519. [CrossRef]
7. Macek, W. Correlation between Fractal Dimension and Areal Surface Parameters for Fracture Analysis after Bending-Torsion Fatigue. *Metals* **2021**, *11*, 1790. [CrossRef]

8. Liewald, M.; Wagner, S.; Becker, D. Influence of Surface Topography on the Tribological Behaviour of Aluminium Alloy 5182 with EDT Surface. *Tribol. Lett.* **2010**, *39*, 135–142. [CrossRef]
9. Leach, R.K.; Evans, C.; He, L.; Davies, A.; Duparré, A.; Henning, A.; Jones, C.W.; O'Connor, D. Open questions in surface topography measurement: A roadmap. *Surf. Topogr. Metrol. Prop.* **2015**, *3*, 013001. [CrossRef]
10. Podulka, P. Selection of Methods of Surface Texture Characterisation for Reduction of the Frequency-Based Errors in the Measurement and Data Analysis Processes. *Sensors* **2022**, *22*, 791. [CrossRef]
11. Pawlus, P.; Wieczorowski, M.; Mathia, T. *The Errors of Stylus Methods in Surface Topography Measurements*; Zapol: Szczecin, Poland, 2014.
12. Pawlus, P. Digitisation of surface topography measurement results. *Measurement* **2007**, *40*, 672–686. [CrossRef]
13. Podulka, P. Bisquare robust polynomial fitting method for dimple distortion minimisation in surface quality analysis. *Surf. Interface Anal.* **2020**, *52*, 875–881. [CrossRef]
14. Podulka, P. The effect of valley depth on areal form removal in surface topography measurements. *Bull. Pol. Acad. Sci. Tech. Sci.* **2019**, *67*, 391–400. [CrossRef]
15. Magdziak, M. Selection of the Best Model of Distribution of Measurement Points in Contact Coordinate Measurements of Free-Form Surfaces of Products. *Sensors* **2019**, *19*, 5346. [CrossRef]
16. Podulka, P.; Pawlus, P.; Dobrzanski, P.; Lenart, A. Spikes removal in surface measurement. *J. Phys. Conf. Ser.* **2014**, *483*, 012025. [CrossRef]
17. De Groot, P.; DiSciacca, J. Definition and evaluation of topography measurement noise in optical instruments. *Opt. Eng.* **2020**, *59*, 064110. [CrossRef]
18. Servin, M.; Estrada, J.C.; Quiroga, J.A.; Mosiño, J.F.; Cywiak, M. Noise in phase shifting interferometry. *Opt. Express* **2009**, *17*, 8789–8794. [CrossRef]
19. Šarbort, M.; Holá, M.; Pavelka, J.; Schovánek, P.; Rerucha, Š.; Oulehla, J.; Fo, T.; Lazar, J. Comparison of three focus sensors for optical topography measurement of rough surfaces. *Opt. Express* **2019**, *27*, 33459. [CrossRef]
20. Podulka, P. Comparisons of envelope morphological filtering methods and various regular algorithms for surface texture analysis. *Metrol. Meas. Syst.* **2020**, *27*, 243–263. [CrossRef]
21. Podulka, P. Suppression of the High-Frequency Errors in Surface Topography Measurements Based on Comparison of Various Spline Filtering Methods. *Materials* **2021**, *14*, 5096. [CrossRef]
22. *ISO 2016 25178-600*; Geometrical Product Specification (GPS)—Surface Texture: Areal Part 600: Metrological Characteristics for Areal-Topography Measuring Methods. International Organization for Standardization: Geneva, Switzerland, 2016.
23. De Groot, P.J. The Meaning and Measure of Vertical Resolution in Optical Surface Topography Measurement. *Appl. Sci.* **2017**, *7*, 54. [CrossRef]
24. Podulka, P. Reduction of Influence of the High-Frequency Noise on the Results of Surface Topography Measurements. *Materials* **2021**, *14*, 333. [CrossRef] [PubMed]
25. Pawlus, P. An analysis of slope of surface topography. *Metrol. Meas. Syst.* **2005**, *12*, 295–313.
26. Santoso, T.; Syam, W.P.; Darukumalli, S.; Leach, R. Development of a compact focus variation microscopy sensor for on-machine surface topography measurement. *Measurement* **2022**, *187*, 110311. [CrossRef]
27. Syam, W.P.; Jianwei, W.; Zhao, B.; Maskery, I.; Elmadih, W.; Leach, R. Design and analysis of strut-based lattice structures for vibration isolation. *Precis. Eng.* **2018**, *52*, 494–506. [CrossRef]
28. Muhamedsalih, H.; Jiang, X.; Gao, F. Vibration compensation of wavelength scanning interferometer for in-process surface inspection. In *Proceedings of the 10th Proceedings of Computing and Engineering Annual Researchers' Conference 2010*; University of Huddersfield: Huddersfield, UK, 2010; pp. 148–153.
29. Syam, W.P. In-process surface topography measurements. In *Advances in Optical Surface Texture Metrology*; Leach, R.K., Ed.; IOP Publishing: Bristol, UK, 2020.
30. Zhang, Z.; Zhang, Y.; Zhu, Y. A new approach to analysis of surface topography. *Precis. Eng.* **2010**, *34*, 807–810. [CrossRef]
31. Alcock, S.G.; Ludbrook, G.D.; Owen, T.; Dockree, R. Using the power spectral density method to characterise the surface topography of optical surfaces. *Proc. SPIE* **2010**, *7801*, 780108. [CrossRef]
32. Elson, J.; Bennett, J. Calculation of the power spectral density from surface profile data. *Appl. Opt.* **1995**, *34*, 201–208. [CrossRef]
33. Podulka, P. Fast Fourier Transform detection and reduction of high-frequency errors from the results of surface topography profile measurements of honed textures. *Eksploat. Niezawodn.* **2021**, *23*, 84–89. [CrossRef]
34. Románszki, L.; Klébert, S.; Héberger, K. Estimating Nanoscale Surface Roughness of Polyethylene Terephthalate Fibers. *ACS Omega* **2020**, *5*, 3670–3677. [CrossRef]
35. Tian, H.; Ribeill, G.; Xu, C.; Reece, C.E.; Kelley, M.J. A novel approach to characterizing the surface topography of niobium superconducting radio frequency (SRF) accelerator cavities. *Appl. Surf. Sci.* **2011**, *257*, 4781–4786. [CrossRef]
36. Jacobs, T.D.B.; Junge, T.; Pastewka, L. Quantitative characterisation of surface topography using spectral analysis. *Surf. Topogr. Metrol. Prop.* **2017**, *5*, 013001. [CrossRef]
37. Duparré, A.; Ferre-Borrull, J.; Gliech, S.; Notni, G.; Steinert, J.; Bennett, J. Surface characterization techniques for determining the root-mean-square roughness and power spectral densities of optical components. *Appl. Opt.* **2002**, *41*, 154–171. [CrossRef] [PubMed]

38. Jiang, Y.; Wang, S.; Qin, H.; Li, B.; Li, Q. Similarity quantification of 3D surface topography measurements. *Measurement* **2021**, *186*, 110207. [CrossRef]
39. Walsh, C.; Leistner, A.; Oreb, B. Power spectral density analysis of optical substrates for gravitational-wave interferometry. *Appl. Opt.* **1999**, *38*, 4790–4801. [CrossRef]
40. Jiang, Z.; Wang, H.; Fei, B. Research into the application of fractal geometry in characterising machined surfaces. *Int. J Mach. Tool Manu.* **2001**, *41*, 2179–2185. [CrossRef]
41. Czifra, Á.; Goda, T.; Garbayo, E. Surface characterisation by parameter-based technique, slicing method and PSD analysis. *Measurement* **2011**, *44*, 906–916. [CrossRef]
42. Krolczyk, G.M.; Maruda, R.W.; Nieslony, P.; Wieczorowski, M. Surface morphology analysis of Duplex Stainless Steel (DSS) in Clean Production using the Power Spectral Density. *Measurement* **2016**, *94*, 464–470. [CrossRef]
43. Raoufi, D. Fractal analyses of ITO thin films: A study based on power spectral density. *Phys. B Condens. Matter* **2010**, *405*, 451–455. [CrossRef]
44. Xu, C.; Tian, H.; Reece, C.E.; Kelley, M.J. Enhanced characterization of niobium surface topography. *Phys. Rev. Accel. Beams* **2011**, *14*, 123501. [CrossRef]
45. Sun, J.; Song, Z.; Heb, G.; Sang, Y. An improved signal determination method on machined surface topography. *Precis. Eng.* **2018**, *51*, 338–347. [CrossRef]
46. Zuo, X.; Peng, M.; Zhou, Y. Influence of noise on the fractal dimension of measured surface topography. *Measurement* **2020**, *152*, 107311. [CrossRef]
47. Lin, T.Y.; Blunt, L.; Stout, K.J. Determination of proper frequency bandwidth for 3D topography measurement using spectral analysis. Part I: Isotropic surfaces. *Wear* **1993**, *166*, 221–232. [CrossRef]
48. Podulka, P. Proposal of frequency-based decomposition approach for minimization of errors in surface texture parameter calculation. *Surf. Interface Anal.* **2020**, *52*, 882–889. [CrossRef]
49. Baofeng, H.; Haibo, Z.; Siyuan, D.; Ruizhao, Y.; Zhaoyao, S. A review of digital filtering in evaluation of surface roughness. *Metrol. Meas. Syst.* **2021**, *28*, 217–253. [CrossRef]
50. Podulka, P. Selection of reference plane by the least squares fitting methods. *Adv. Sci. Technol. Res. J.* **2016**, *10*, 164–175. [CrossRef]
51. Podulka, P. The Effect of Surface Topography Feature Size Density and Distribution on the Results of a Data Processing and Parameters Calculation with a Comparison of Regular Methods. *Materials* **2021**, *14*, 4077. [CrossRef]
52. Pawlus, P.; Reizer, R.; Wieczorowski, M. Problem on non-measured points in surface texture measurements. *Metrol. Meas. Syst.* **2017**, *24*, 525–536. [CrossRef]
53. Pawlus, P.; Reizer, R.; Wieczorowski, M. Functional Importance of Surface Texture Parameters. *Materials* **2021**, *14*, 5326. [CrossRef]
54. *ISO 25178-3:2012*; Geometrical Product Specifications (GPS)—Surface Texture: Areal—Part 3: Specification Operators. International Organization for Standardization: Geneva, Switzerland, 2012.
55. Podulka, P. Edge-area form removal of two-process surfaces with valley excluding method approach. *Matec. Web. Conf.* **2019**, *252*, 05020. [CrossRef]
56. Podulka, P. Improved Procedures for Feature-Based Suppression of Surface Texture High-Frequency Measurement Errors in the Wear Analysis of Cylinder Liner Topographies. *Metals* **2021**, *11*, 143. [CrossRef]
57. *ISO 16610-21:2011*; Geometrical Product Specifications (GPS)—Filtration—Part 21: Linear Profile Filters: Gaussian Filters. International Organization for Standardization: Geneva, Switzerland, 2011.

Resolving Selected Problems in Surface Topography Analysis by Application of the Autocorrelation Function

Przemysław Podulka

Faculty of Mechanical Engineering and Aeronautics, Rzeszow University of Technology, Powstancow Warszawy 8 Street, 35-959 Rzeszów, Poland; p.podulka@prz.edu.pl; Tel.: +48-17-743-2537

Abstract: In this paper, the validity of the application of an autocorrelation function for resolving some surface topography measurement problems was presented. Various types of surfaces were considered: plateau-honed, honed with burnished dimples, ground, turned, milled, laser-textured, or isotropic. They were measured with stylus and non-contact (optical) methods. Extraction of selected features, such as form and waviness (defined as an L-surface) and high-frequency measurement noise (S-surface) from raw measured data, was supported with an autocorrelation function. It was proposed to select the analysis procedures with an application of the autocorrelation function for both profile (2D) and areal (3D) analysis. Moreover, applications of various types of regular (available in the commercial software) analysis methods, such as least-square-fitted polynomial planes, selected Gaussian (regression and robust) functions, median filter, spline approach, and fast Fourier transform scheme, were proposed for the evaluation of surface topography parameters from ISO 25178 standards.

Keywords: surface topography; surface texture; measurement; measurement noise; autocorrelation function

1. Introduction

Analysis of surface topography, which is obtained in the final stage of treatment, can be crucial in many functional applications [1]. The measurement and studies of surface properties, such as wear resistance [2], sealing [3,4], friction [5], lubricant retention [6], energy consumption [7] or eco-friendly factors [8], and many more [9–11], in general, can be received from the description of surface topography. Moreover, in some cases, it is defined as a fingerprint of the manufacturing process [12], especially that can be absolutely essential in the process of control [13].

Considering all of the mentioned and further additional issues, analyzing surface topography can be an encouraging task to be resolved. The whole process of results evaluation can be, even roughly, divided into various, however dependent, sub-processes. Firstly, an appropriate measurement method must be proposed to receive relevant results. Generally, those methods can be divided into stylus [14], proposed with many traditional systems [15], and non-stylus techniques [16], where the optical methods have many advantages [17]. They (stylus and optical techniques) were widely compared in many previous studies [18–20]; nevertheless, the most particularly noticeable improvement was in reducing the time of measurement [21].

Very popular and, respectively, often applied in surface topography analysis are methods with the application of the autocorrelation function (ACF) exhibit possible various, e.g., random and periodic, features buried on the machined surface [22]. Math basics of ACF were presented in [23]. Moreover, the roughness amplitudes with the non-random distribution can be thoroughly evaluated by the ACF method [24]. This function can be used when the accurate prediction of the machined surface topography of abrasive belt flexible grinding is required [25]. Moreover, ACF provides two-point correlated information about the spatial relation and dependence of data, indicating randomness or periodicity and isotropy or anisotropy of surface features can be received with this application [26]. Generally, the surface texture orientation can be properly evaluated as well [27]. Different ACF in combination with statistical parameters can be generated and then compared with the prescribed ground surfaces [28]. In addition, feature extraction can be significantly improved with the function presented [29].

Analyzing image data, e.g., fractal analysis on the surface topography image of thin films [30] can be supported by many advantages of ACF applications, compared with many other approaches [31]. Roughness received by the scanning tunneling microscope (STM) measurement can be represented by the ACF [32]. This method was widely used for plateau-honed [33] or polished [34] topographies for an efficient grain size determination. Generally, surface roughness can be characterized by the widths of ACFs, which can be obtained from the digital processing of surface images [35]. The properties of the ACF can be extremely advantageous, e.g., bi-directionality in the soil surface roughness investigation [36].

Except for the characterization of surface topography, the ACF is often proposed in the process of data generation, e.g., for analyzing mixed lubrication [37]. Furthermore, when investigating the precision of generation, correlation distances can be calculated from the ACF of the generated data [38]. Identification and reconstruction of Bi-Gaussian surfaces can be received with ACF characterization [39]. This is an important function to describe surface roughness and depends on the nature of morphology and, respectively, the powers spectrum can be evaluated from the ACF [40] as well. The function describes the general dependence of the values of the data at one position on the values at another position or, correspondingly, the ACF is used for surface topographic assessment to indicate the randomness and directionality of surface features [41], described as the theoretical directional variogram [42] in some cases.

The advantages of the application of ACF were presented widely, nevertheless, its suitability in the definition of the S-surface (reducing the high-frequency measurement noise) and L-surface (an areal form removal) was not comprehensively studied. The L-surface (S-surface) is obtained by the application of the L-filter (S-filter), a filter eliminating the largest (the smallest) scale elements from the surface [43]. Some of the results presented for a profile (2D) and an areal (3D) roughness evaluation [44] indicated its suitability in the detection of high-frequency errors [45] when describing the S-surface, however, its validation must be improved.

Errors in roughness evaluation can be received from various sources. Firstly, the measuring method can influence the reliability of the results obtained. Optical methods are fast, compared to that stylus [46], nevertheless are fraught with many errors that can radically affect the results of surface topography analysis. One of the types of errors is those in the high-frequency domain. High-frequency measurement noise is a type of noise that can be caused by instability of the mechanics with any influences from the environment or, on the other hand, by internal electrical noise [47,48]. However, in most cases, the high-frequency noise is a result of vibration [49,50].

Considering the analysis of machined parts, such as cylindrical elements of the engine, the surface roughness parameters are calculated after an areal form removal. From that matter, even if both an appropriate and precise measurement technique (device) is applied, properties of the surface can be described erroneously when the data processing is not selected carefully. Considering all of the sophisticated methods, there are many algorithms and, respectively, procedures for the reduction of errors in data processing, their selection requires mindful users [51].

When defining the L-surface, distortions of the results can be found for the areas where deep or wide features are located [52,53]. There were many studies performed for form extraction, taking into consideration the location of the features, especially in the edge area of the analyzed detail [54]. Some modifications of the commonly used in surface characterization, the Gaussian filtration method, can be an alternative solution [55]. The enlargement of dimple distortion was found when features were located near/on the edge of considered data [56]. Much crucial information can be received with a profile exploration [57] that, respectively, in some cases, 2D characterization can allow defining more confidential results than 3D studies [58,59]. From that matter, the effect of feature size, distribution, and density was considered for the definition of the L-surface with an application of general, available in the commercial software, methods [60].

In this paper, the ACF method was applied for validation of methods for an areal form removal and definition of the high-frequency measurement errors. Both surfaces, S-surface and L-surface, can be more precisely obtained when a comprehensive analysis of the ACF is applied. Many advantages of this method were presented in the analysis of the Gaussian surfaces [61–63]. Moreover, the main proposal of the studies presented was to indicate both the usefulness and suitability of the ACF application, proposing its usage with other, also available in the commercial software, function, such as a power spectral density (PSD) [64–68] or thresholding [69,70] methods. The usage of ACF in the definition of the L-surface was not comprehensively studied in previous research. Moreover, many advantages in the calculation of the S-surface (for high-frequency measurement noise reduction) with ACF application must be highlighted as well. It was proposed to validate the methods of definition of L-surface and S-surface with an application of ACF. It was found that the precision of the S-L surface definition can be significantly improved with ACF usage. Errors in the calculation of ISO 25178 surface roughness parameters can be also reduced when the ACF function was applied.

2. Materials and Methods

2.1. Analyzed Surfaces

Various types of surface topographies were considered: plateau-honed, with a cross-hatch pattern equal to 60°, some of them with additionally burnished oil pockets (dimple's average width and average depth were around 0.2 mm and 10 µm, respectively, and the area density of oil pockets was smaller than 20%), turned, ground, and isotropic details. More than 20 measured surfaces of each type were considered to provide some repeatability in the analysis of the results but only a few of them were presented in detail.

In Figure 1, contour map plots (left column), isometric views (middle), and selected ISO 25178 surface topography parameters including some from the Sk family (right column) were presented for each of the types of surfaces. Both types of plateau-honed cylinder liners (Figure 1a,b), turned (Figure 1c) and ground (Figure 1d) details were presented as raw measured data, containing the form, cylindrical shape, and waviness, if existed. For this type of data, both L-surfaces and, respectively, S-surfaces were defined with the ACF improvement. An example of an isotropic surface (Figure 1e) was flat, respectively, an areal form removal process, described as a definition of the L-surface, was not considered in the studies presented. For this type of surface, the ACF method was applied to detect and reduce the effect of the high-frequency errors, obtained as an S-surface, from the results of the measurement process, namely the raw measured data.

The values of the following, located in the ISO 25178 standard [71,72], surface topography parameters were measured, calculated and considered: root mean square height Sq, skewness Ssk, kurtosis Sku, maximum peak height Sp, maximum valley depth Sv, the maximum height of surface Sz and arithmetic mean height Sa from the height parameters; autocorrelation length Sal, texture parameter Str, and texture direction Std from spatial parameters; root mean square gradient Sdq and developed interfacial areal ratio Sdr from hybrid parameters; peak density Spd and arithmetic mean peak curvature Spc from feature parameters; core roughness depth Sk, reduced summit height Spk, reduced valley depth Svk, upper material ratio $Sr1$, and lower material ratio $Sr2$ from functional parameters; surface bearing index Sbi, core fluid retention index Sci and valley fluid retention index Svi from the functional indices [73].

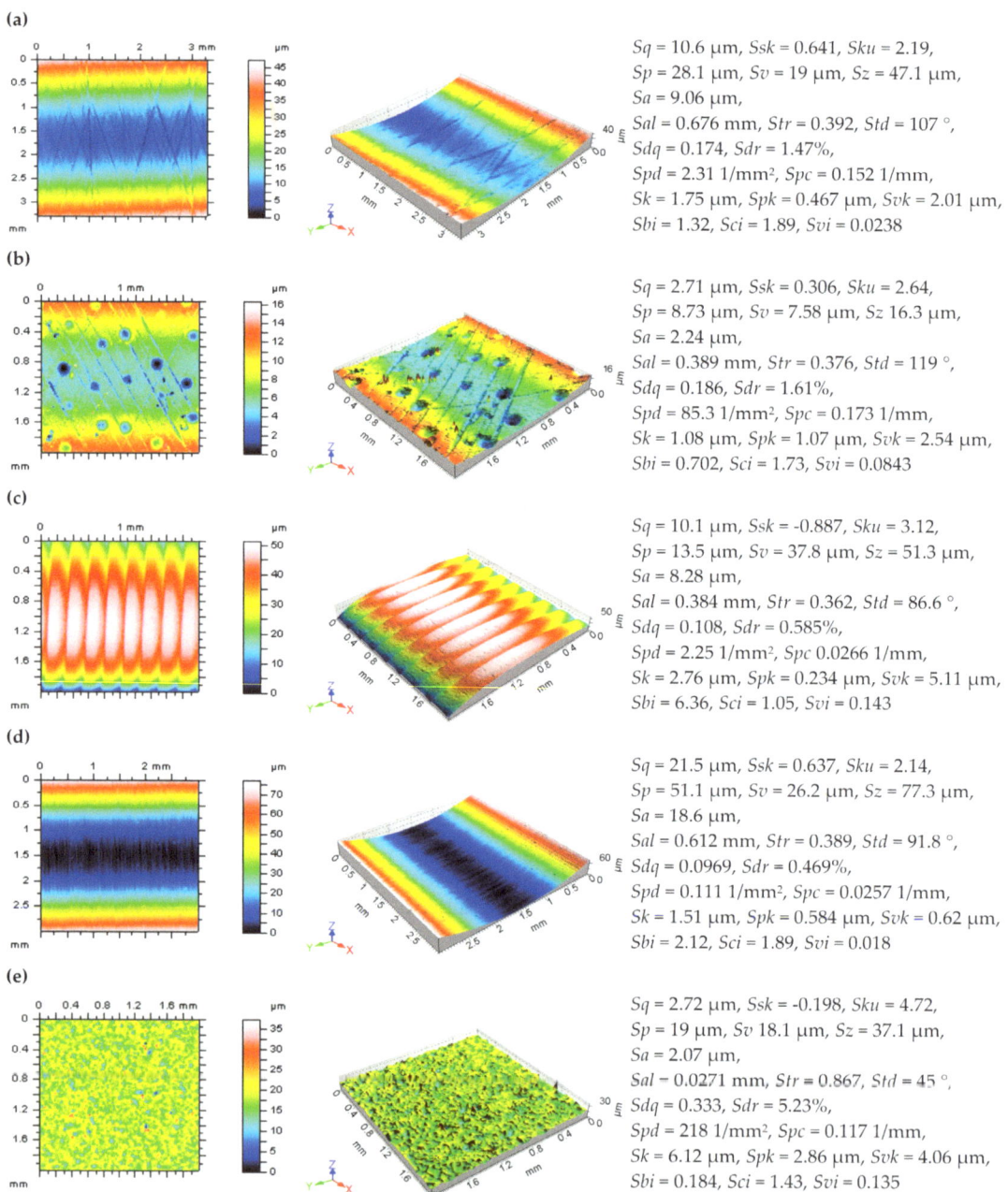

Figure 1. Contour map plots (left), their isometric views (middle) and ISO 25178 parameters (right) of measured surfaces: plateau-honed cylinder liner (**a**), honed cylinder liner with additionally burnished oil pockets (**b**), turned (**c**), ground (**d**) and isotropic (**e**).

2.2. Measurement Process

Analyzed details were measured with different stylus or non-contact (optical) techniques. The Talyscan 150 and Talysurf CCI Lite (produced by Taylor Hobson Ltd, Leicester,

U.K., version 2.8.2.95) devices were applied as the stylus and optical measuring equipment, respectively.

For the stylus instrument: a nominal tip radius of 2 µm, height resolution of 10 nm and the measured area 5 by 5 mm with 1000 × 1000 measured points, were applied. The sampling interval was equal to 5 µm. The measurement speed was 0.75 mm/s. The effect of the measurement velocity was not studied, it was not the subject of the current research but, respectively, was analyzed in the previous studies.

For the non-contact measurement: the height resolution was 0.01 nm, and the measured area was 3.35 by 3.35 mm with the 1024 × 1024 measured points, respectively, obtained. The spacing was 3.27 µm. A Nikon 5×/0.13 TI objective (Nikon GmbH, Düsseldorf, Germany) was used.

In the study presented for both methods, the effect of sampling (stylus) and spacing (non-contact) on the values of 3D surface roughness parameters were not considered.

For both analyses, TalyMap Gold software (version 6.1), copyrighted by Digital Surf (Besançon, France), was used for digital filtering and ISO 25178 surface topography parameters calculation and, respectively, evaluation.

All the surfaces measured by the non-stylus device were carefully studied for the detection of the individual peaks [74] (spikes [75]) errors from the raw measured data. This type of error was observed for the optical measurement technique. Individual peaks were reduced by the thresholding method with a selected material ratio from 0.13% to 99.87% [76].

2.3. Proposals of Improvement of Methods for Surface Topography Analysis by Application of the Autocorrelation Function

Evaluation of surface roughness was provided after an areal form removal (definition of an L-surface) and reduction of high-frequency measurement noise (proposals of an S-surface). For the definition of the L-surface and S-surface, the following methods were proposed: least-squares fitted cylinder (LSFC), the least-square fitted polynomial of 2nd degree (Poly2), regular Gaussian regression filter (GRF), robust Gaussian regression filter (RGRF), regular isotropic spline filter (SF), median denoising filter (MDF) and fast Fourier transform filter (FFTF). All of those methods are available in commercial software (version 6.1). For improvement of the proposed procedures, the following data analysis techniques were additionally used: the autocorrelation function (ACF) and power spectral density (PSD) were received from these sources as well.

The least-square fitting (LSF) is an often-used method for surface roughness evaluation. Considering the ISO 13565-3 standard [77], the LSF method was applied directly for the running-in-wear process analysis of the plateau surface [17]. Furthermore, the LSF method is proposed when defining the ISO 21920–2 parameters needed for evaluating the roughness of plateau structure surfaces [78]. The reference planes defined in this study by L-surface for assessing nominally flat and curved surfaces were recommended with LSF methods [79]. Based on the LSF method and the electrostatic field theory, the linear relationship between the work function (WF) and the surface roughness of the alloys [80] can be found. LSF can be applied in both shapes (cylinder [81]) and polynomial [53] when the definition of the reference plane. Machines such as the coordinate measuring machine (CMM) often employ the LSF algorithms [82]. Furthermore, the LSF approach can be used to calibrate the profile least-square median of the surface texture curve [83].

Widely proposed and, respectively, applied are filters based on the Gaussian function [84]. One of the most commonly used is a regular Gaussian filter (GRF) [85]. Many types of surfaces can be characterized by the Gaussian filters, and apart from many tribological studies, the roughness of granular materials can be evaluated as well [86]. Often proposed modification of the regular Gaussian filter is a robust scheme [87]. The robust Gaussian regression filter (RGRF) has been proposed in many surface characterization problems, including the surface topography analysis of ultra-precision machined surfaces [88].

An alternative to many non-resolved problems with Gaussian filtering methods is a spline. The universal spline filter, combined with the FFT algorithm, can reduce the computation time considerably [89]. When separating the roughness, waviness, and form, the variational problem can be solved by approximating the filtered surface data employing different two-dimensional functions, e.g., B-spline [90]. Nevertheless, compared to the Gaussian methods (e.g., GRF), the spline filter (SF) was often found that generate more errors [91].

Digital data processing, especially of image sources, is often proposed with median filters [92]. The application of a median filter can be valuable in the preprocessing unit that corrects surrounding light levels and filters noise [93]. The median denoising filter (MDF) can be especially valuable in the elimination of the bad points that, respectively, a 3×3 denoising filter is proposed as a spatial operator [94].

As an alternative for ACF characterization, it was provided that PSD can be valuable in the correct determination of surface morphology when scanning force microscopy (SFM) [95] or atomic force microscopy (AFM) [96] imaging analysis of the surface roughness was presented. It is found, additionally, that if PSD roughness is expressed as a polynomial function, the international roughness index (IRI) can be simply calculated [97]. A similar variation of low-frequency roughness of swift heavy ion (SHI) irradiated ultra-thin gold films can be received with PSD application [98]. Fourier techniques, which are the basics of PSD analysis, can characterize the wavefront of optical components in various directions [99]. An angular spectrum, derived from the PSD, can be often employed to research the spatial distribution of spectral energy for the milled surface [100]. The effects of feed and vibrations on surface roughness can be researched with the use of PSD as well [101]. Furthermore, various measuring instruments can be compared for roughness evaluation by using the PSD functions [102]. Comparing, respectively, the fast Fourier transform (FFT) plots and angular spectrum of different types of wear particles can be demonstrated with calculated PSD function [103]. As already mentioned, isotropy and anisotropy [104]. The PSD was proposed to analyze surface roughness profiles in single-point diamond turning [105]. Generally, the profile (2D) PSD advantages were clearly presented previously [106]. Finally, the PSD, considering both profile (2D) and an areal (3D) roughness evaluation, was entirely beneficial in the process of high-frequency measurement error reduction from the raw measured data [44,45,47,58].

3. Results

This section is divided into three parts. Firstly, in Section 3.1., the ACF was applied for improvements of the methods for an areal form removal (definition of L-surface). Secondly, in Section 3.2, the application of ACF in the definition and reduction of high-frequency measurement noise was presented. Moreover, in the end, in Section 3.3, analysis of modeled data was introduced for improving the ACF methods proposed.

3.1. Reducing Errors in an Areal Form Removal

It was found that the detection of errors in the definition of L-surface can be validated by the ACF. In Figure 2, a 3D ACF (right column) was presented for the plateau-honed cylinder liner surface after an areal form removal by various methods. From the view of both isometric surfaces (left column) and 3D ACF, it was found that if the form (shape and waviness) were not entirely removed, it was indicated by arrows. From that matter and, respectively, by minimization of the values of *Sk* and *Spk* parameters [107], the reference plane received by the Poly2 seems to be the best solution from those considered. In Figure 3, extraction of the free-of-dimple method [108] was proposed for validation of an areal form removal approach for plateau-honed cylinder liner surfaces containing additionally burnished oil pockets. Considering both the extracted details (Figure 3c) and its ACF (Figure 3e), the Poly2 seems to provide more encouraging results than GRF and SF schemes.

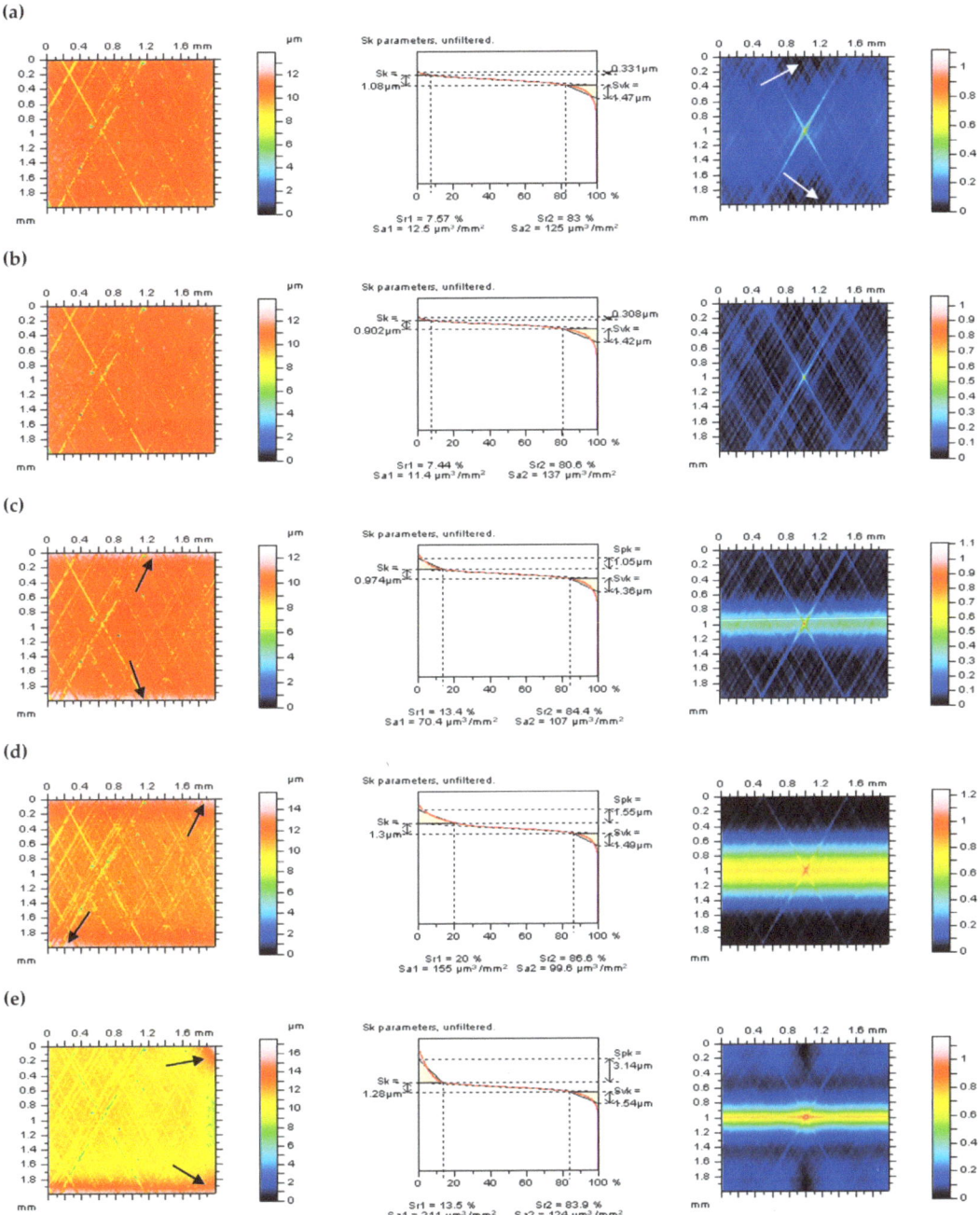

Figure 2. Contour map plots (left column), their material ratio curves (middle), and 3D ACF graphs (right column) of plateau-honed cylinder liner surface after an areal form removal by: LSFC (**a**) Poly2 (**b**), GRF (**c**), RGRF (**d**), and SF (**e**).

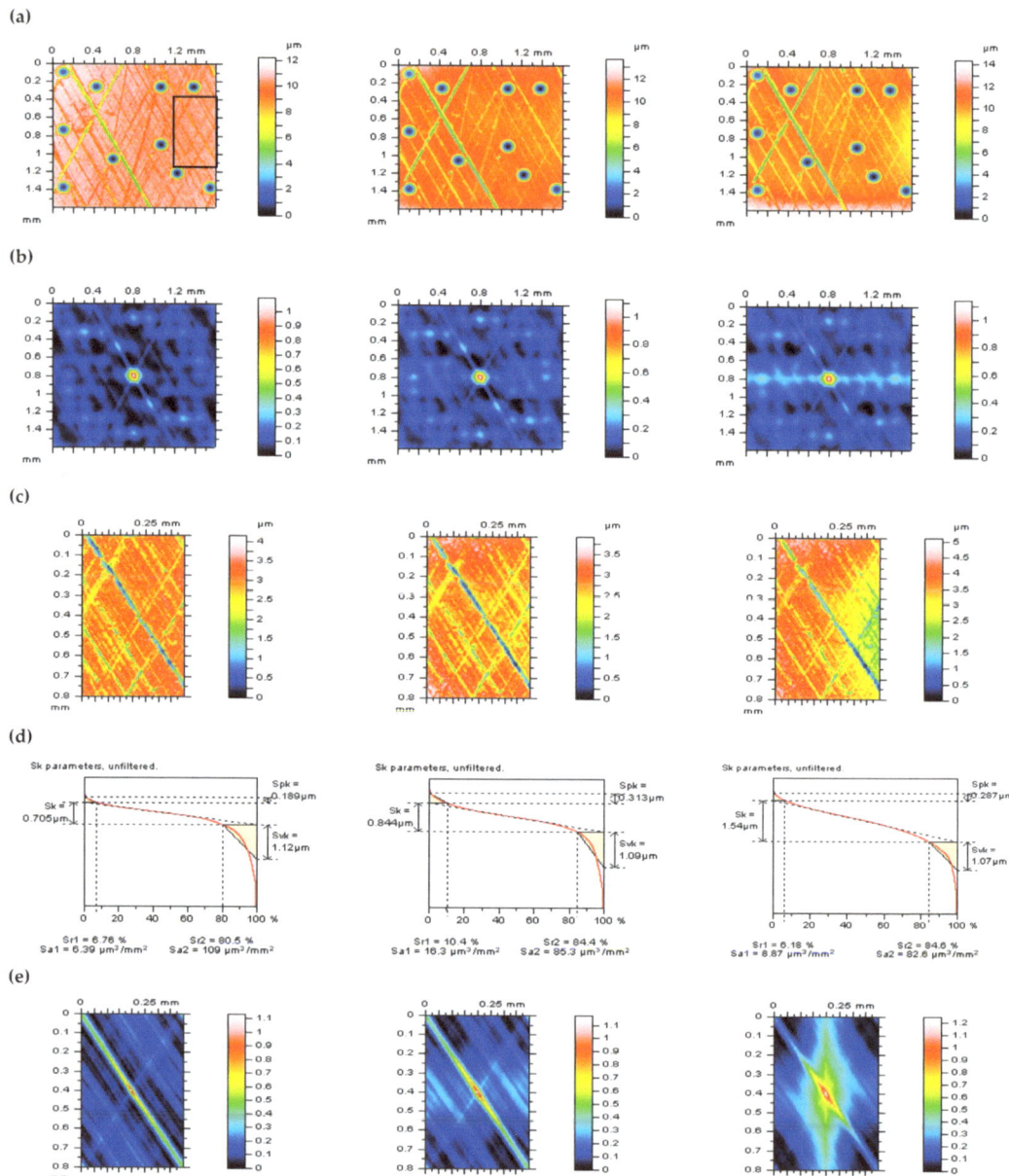

Figure 3. Contour map plots (**a**), their 3D ACF graphs (**b**), enlarged details (**c**) with their material ratio curves (**d**), and 3D ACF (**e**), received from the plateau-honed cylinder liner surface containing additionally burnished dimples after an areal form removal by: Poly2 (left column), GRF (middle) and SF (right column), cut-off = 0.8 mm, the area of an enlarged detail extraction was indicated in Figure 3a.

3.2. Application of an Autocorrelation Function in the Reduction of High-frequency Measurement Errors

When defining an appropriate method for high-frequency measurement noise detection and suppression, the noise surface (NS) [33] was defined. The NS is obtained as the surface created by high-frequency errors. In this case, it is the result of S-filtering. In Figure 4 (right column) were presented NSs created by five various filtering methods of turned piston skirt surface. When analyzing the isometric view of NS, it can be found that some of them contain features that are not located in the high-frequency domain, such as after the application of RGRF or MDF methods (Figure 4b,d). It can indicate that NSs do not contain only high-frequency components. It was found that removal other than noise features from the raw measured data can affect the distortion of surface roughness [12]. Some features can be distorted, such as dimples, oil pockets, holes, scratches, and surface treatment traces in general. This, correspondingly, can significantly affect the values of surface roughness parameters. The more non-noise features are found on the NS, the more distortion in roughness evaluation can be obtained.

From all of the analyzed methods (GRF, RGRF, SF, MDF, and FFTF), considering studies of an isometric view of the S-surface (NS), the application of GRF, SF, and FFTF seem to be particularly suitable; they did not contain non-noise features. To confirm this supposition, an analysis of an areal ACF can be employed. It was found that ACF can be applied for the detection [109] or identification [39] of some surface topography features, especially when various surface textures are considered with fractional approaches [110]. From that matter, in Figure 5, an areal ACFs were presented (left column) for each of the applied filtering procedures. Usage of ACF improved that NS received by RGRF and MDF application caused the removal of non-noise features from the raw data.

Nevertheless, for the GRF scheme, the non-noise features were also defined (Figure 5a) when there were not clearly visible while analyzing the isometric view of the NS. This issue was improved with the thresholding method. The ACF was thresholded with a range equal to 0.13%–99.87% value. This value of thresholding, representing the data truncation, was studied previously [69]. Additionally, for the random surface with Gaussian ordinate distribution height, the difference between material ratios (0.13%–99.87%) was defined as equal to 6 standard deviations of surface amplitude [76]. However, the effect of the thresholding range was not studied in this paper. The accuracy was not considered as it was for the non-noise features indicated.

The occurrence of non-noise features was easily detectable when a thresholded ACF was considered. For results obtained after GRF, RGRF, and MDF application, the non-noise features were found. Contrary to those methods, the usage of SF and FFTF did not define the non-noise features on their NSs (Figure 5c,e). From that procedure, the SF and FFTF approaches seem to cause the smallest errors in the process of high-frequency measurement noise reduction from the raw measured data of turned piston skirt surfaces.

Furthermore, the ACF method was found suitable for both the definition and reduction of high-frequency measurement noise [12,47,58] with profile (2D) and areal (3D) improvements [44]. In Figure 5, additionally, selected profiles (right column) extracted from an areal thresholded ACFs (middle column) were presented. According to the profile characterization, some distortion of the shape of the function was found for the ACF defined with an FFTF method; it was indicated by the arrow in Figure 5e. This exaggeration was found in an areal ACF as well in the center part of the function. The influence of the shape and its distortion, respectively, of the center part of the ACF, was comprehensively studied previously [47], and it was found that its properties can indicate the occurrence of high-frequency measurement errors. In the considered case, it can indicate if the received data still contain noise components.

From the ACF improvement, it can be both validated, firstly, if NS contains non-noise features and, respectively, secondly, if noise data are included in the received surface for which the roughness parameters are calculated.

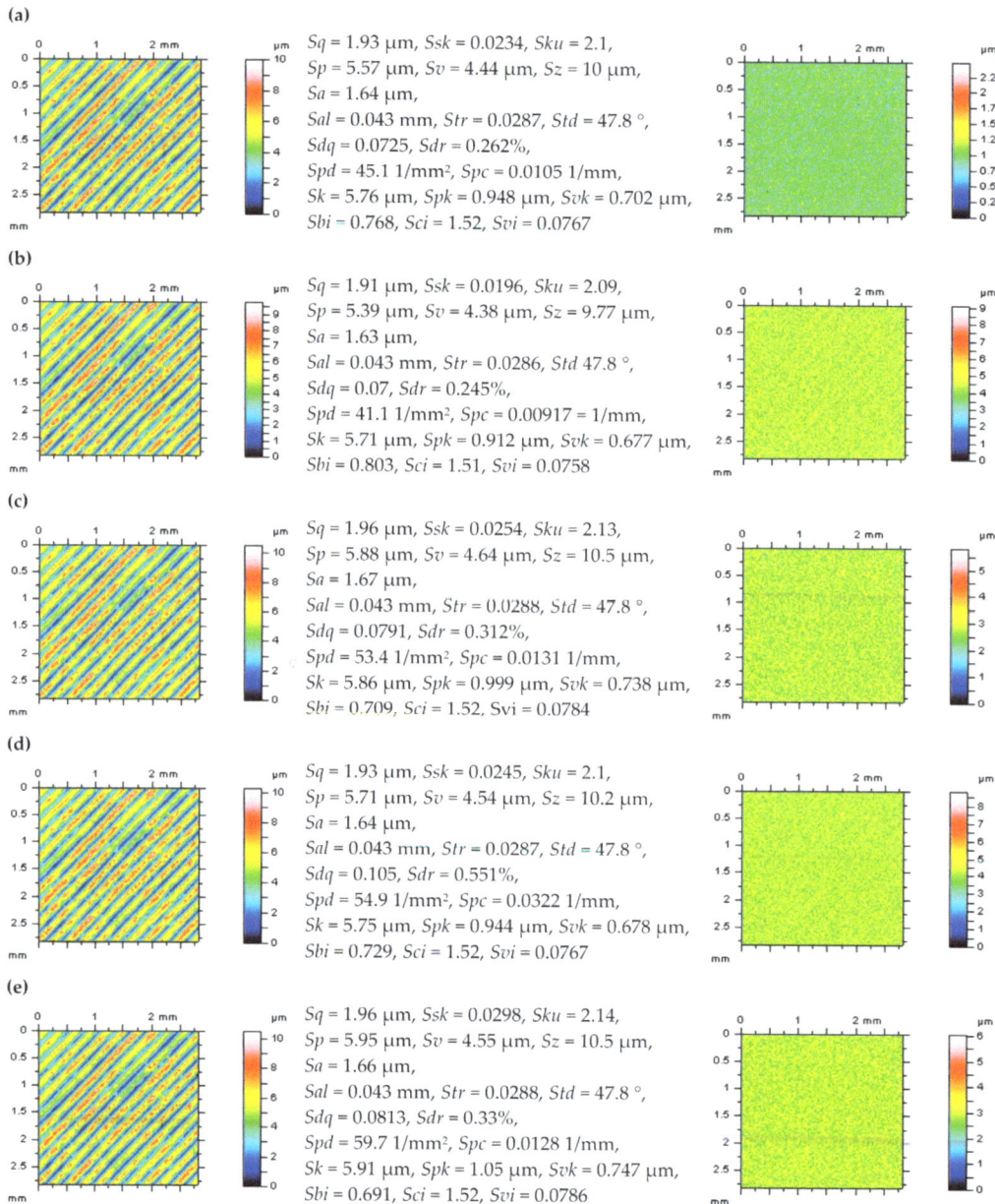

Figure 4. Contour map plots (left column) of the surface after noise removal, their ISO 25178 parameters (middle) and noise surface (right column), respectively, of turned piston skirt surface after application of GF (**a**), RGF (**b**), SF (**c**), MDF (**d**), and FFTF (**e**) method, cut-off = 0.025 mm.

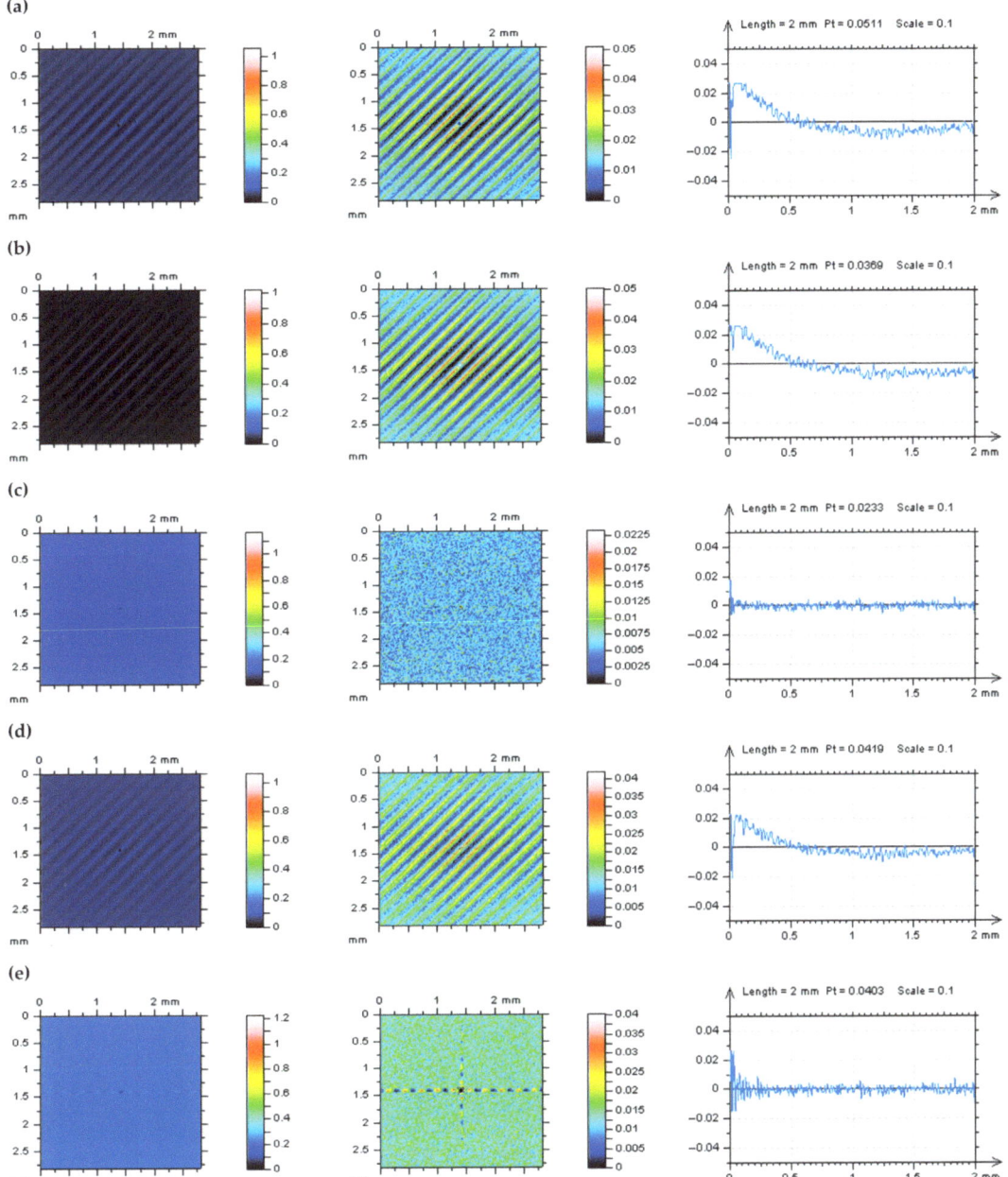

Figure 5. 3D ACF (left column), thresholded (0.31%–99.87%) from previous 3D ACF (middle) and extracted 2D ACF from thresholded ACF (right column) received from turned piston skirt surface after noise removal by GF (**a**), RGF (**b**), SF (**c**), MDF (**d**), and FFT (**e**) approach.

3.3. Improving Proposed Methods with a Modeled Data

For improvement and validation of the results received, surfaces with modeled form (selected L-surfaces) and high-frequency errors (S-surfaces) were considered. Furthermore, both modeled data were then removed by application of various commonly used and available commercial software methods mentioned previously. For an areal form removal, it was improved that least-square fittings of the polynomial plane of 2nd degree (Poly2) caused less distortion of the surface features than various (Gaussian, spline, median or fast Fourier transform) filtering techniques. Moreover, the distortion of modeled surface features was enlarged when they were located near/on the edge of the analyzed detail. Differences in surfaces containing high-frequency noise can be seen in the surface contour map plots (Figure 6a,e). A similar conclusion can be found for profile characteristics. When considering the 3D ACF, differences were difficult to be resolved; nevertheless, some modifications of the center (maximum) value of this function can be resolved. The center part of the ACF was distorted when containing the noise (Figure 6c,g). It was also found that increases in the amplitude of the high-frequency noise caused an enlargement of the distortion of the center part of the ACF [47]. It was also found that properly defined NS should be isotropic. From that matter, its ACF should also exhibit isotropic characteristics. When applying SF and FFTF, it was found that both NS and ACF received the most isotropic properties. From that issue, both mentioned filtering methods can be classified as especially suitable for the suppression of high-frequency noise from the results of surface topography measurements.

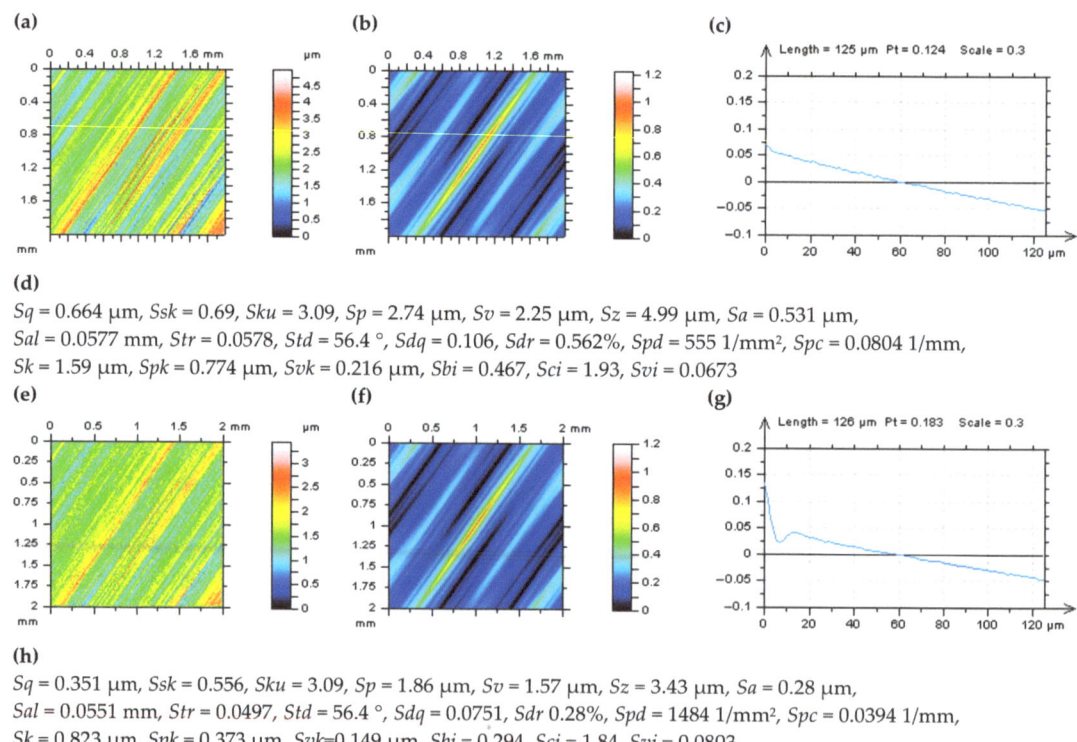

(d)
Sq = 0.664 µm, Ssk = 0.69, Sku = 3.09, Sp = 2.74 µm, Sv = 2.25 µm, Sz = 4.99 µm, Sa = 0.531 µm, Sal = 0.0577 mm, Str = 0.0578, Std = 56.4°, Sdq = 0.106, Sdr = 0.562%, Spd = 555 1/mm², Spc = 0.0804 1/mm, Sk = 1.59 µm, Spk = 0.774 µm, Svk = 0.216 µm, Sbi = 0.467, Sci = 1.93, Svi = 0.0673

(h)
Sq = 0.351 µm, Ssk = 0.556, Sku = 3.09, Sp = 1.86 µm, Sv = 1.57 µm, Sz = 3.43 µm, Sa = 0.28 µm, Sal = 0.0551 mm, Str = 0.0497, Std = 56.4°, Sdq = 0.0751, Sdr 0.28%, Spd = 1484 1/mm², Spc = 0.0394 1/mm, Sk = 0.823 µm, Spk = 0.373 µm, Svk = 0.149 µm, Sbi = 0.294, Sci = 1.84, Svi = 0.0803

Figure 6. Contour map plots (**a,e**), their 3D ACF graphs (**b,f**), extracted 2D ACF profiles (**c,g**) and ISO 25178 roughness parameters (**d,h**) of ground surface: measured (**a–d**) and with added (modeled) high-frequency noise (**e–h**).

4. Conclusions

Based on the results presented, the following conclusions can be raised:

1. The autocorrelation function (ACF) can be exceedingly valuable in the validation of the methods for an areal form removal (definition of L-surface). It was presented that using an ACF with analysis of an isometric view of the measured surface topography can improve the applied method and can be crucial in the reduction of errors in ISO 25178 standard amplitude (height) parameter calculation.
2. When considering the isometric view of the detail after L-surface removal, the distortion of some features (such as burnished dimples) can be important in both measured data and an areal ACF characterization. The higher distortion of data occurs, the greater differences in the isometric view and ACF were obtained. Reduction of errors in both data can, respectively, similarly reduce the distortion in the ISO 25178 roughness parameters calculations.
3. The usage of ACF for details with out-of-feature characteristics can be highly beneficial when the surface contains some deep or wide features, such as dimples. In these cases, omitting the deep features can be necessary for ACF implementation that can affect proposed procedures to be useless. For both analyses, view of surface and ACF graphs, omitting the features is crucial for the definition of S-L-surface.
4. Application of ACF for both profile (2D) and an areal (3D) data can be essential in the process of detection (definition) and reduction (removal) of high-frequency measurement errors from the results of surface roughness evaluation. For both types of data analysis, the center part of the ACF can characterize if the measurement errors in the high-frequency domain were reduced entirely. From that matter, filtering method suitability can also be evaluated.
5. In some cases, the thresholding of the ACF is required. This technique is significant in the definition of S-surface especially. When applying truncation, some required properties of the S-surface (noise surface) can be defined more adequately. This approach, supporting the ACF characterization, can reduce the distortion of surface topography features, so, respectively, the errors in the calculation of the ISO 25178 surface roughness parameters can also be radically reduced.
6. Considering thresholding and out-of-feature methods, they can be applied simultaneously when reducing surface height is crucial in the validation of the filtering technique. The sequence of application of those methods does not affect the analyzed type of surface, but, respectively, the sizes, densities, and distribution of the features located on the studied surface must be thoroughly considered.
7. From all of the studies provided, it was presented that the application of the ACF scheme can significantly reduce the errors in the processes of an areal form removal (definition of L-surface) and suppression of high-frequency measurement noise (indicated as the S-surface). It was proved that even regular and commonly used methods can be especially valuable when applied appropriately.
8. All of those data processing actions are crucial in the process of control of manufactured parts. Reducing errors in surface roughness measurement and data analysis can radically affect the validation of machined parts. Improper definition of S-L surface, especially L-surface, can cause classification of properly made parts as a lack and its rejection. From that matter, all of the actions made on the reduction of errors in roughness evaluation (definition of S-L surface) can be consequently crucial in industrial applications.

5. The Outlook

Despite many issues proposed, there can still be some problems not entirely resolved, as follows:

1. Comprehensive analysis of the ACF for both 3D (areal) and 2D (profile) data must be improved in the edge effect of surface filtering. Many studies provide some general proposals for reducing the effect of edge data on the surface roughness evaluation;

nevertheless, the reduction of errors in ISO 25178 parameters calculation with the usage of ACF was not already presented.
2. Even though least-square methods, such as least-square fitted cylinder elements or polynomial planes of various degrees, are providing encouraging results, digital filtering, e.g., those based on the Gaussian function, is still often used and the results considered. From that matter, more studies on the Gaussian filtering methods must still be proposed.
3. In some cases of analysis with ACF function, the thresholding method is required. Even though a range (0.13%–99.97%) of thresholding techniques was proposed, according to the previously studied cases, those values should be studied considering each type of surface texture. Moreover, the dependences on the surface feature size, density, and location can also be found when the thresholding method is used.
4. The influence of surface directionality on the ACF characterization should be studied as well. It was introduced in some previous studies that direction methods can be exceedingly valuable in the process of high-frequency measurement noise reduction, improving its impact in both the detection and removal of noise data.
5. The effect of the bi-directionality of surface features was not comprehensively studied in both L-surface and S-surface definitions. When reducing errors in high-frequency measurement noise reduction, the S-surface properties were studied when one or even two directions were defined on the analyzed surface. It must be studied widely, and, respectively, some proposals should be unified.

Funding: This research received no external funding.

Institutional Review Board Statement: Not applicable.

Informed Consent Statement: Not applicable.

Data Availability Statement: Data sharing is not applicable to this article.

Conflicts of Interest: The author declares no conflict of interest.

Nomenclature

The following abbreviations and surface roughness parameters are used in the manuscript:
ACF autocorrelation function
AFM atomic force microscopy
CMM coordinate measuring machine
FFT fast Fourier transform
FFTF fast Fourier transform filter
GRF Gaussian regression filter
IRI international roughness index
L-surface long-wavelength surface
LSFC least-square fitted cylinder shape
LSF least-square fitting method
MDF median denoising filter
NS noise surface
Poly2 least-square fitted polynomial plane of 2nd degree
PSD power spectral density
RGRF robust Gaussian regression filter
S-surface short-wavelength noise surface
SF spline filter
SFM scanning force microscopy
SHI swift heavy ion
STM scanning tunneling microscope
Sa arithmetic mean height Sa, µm
Sal autocorrelation length, mm
Sbi surface bearing index

Sci core fluid retention index
Sdq root mean square gradient
Sdr developed interfacial areal ratio, %
Sk core roughness depth, μm
Sku kurtosis
Sp maximum peak height, μm
Spc arithmetic mean peak curvature, 1/mm
Spd peak density, $1/mm^2$
Spk reduced summit height, μm
Sq root mean square height, μm
Sr1 upper material ratio
Sr2 lower material ratio
Ssk skewness
Std texture direction, °
Str texture parameter
Sv maximum valley depth, μm
Svi valley fluid retention index
Svk reduced valley depth, μm
Sz the maximum height of surface, μm

References

1. Panjan, P.; Drnovšek, A. Special Issue: Surface Topography Effects on the Functional Properties of PVD Coatings. *Coatings* **2022**, *12*, 1796. [CrossRef]
2. Zheng, M.; Wang, B.; Zhang, W.; Cui, Y.; Zhang, L.; Zhao, S. Analysis and prediction of surface wear resistance of ball-end milling topography. *Surf. Topogr. Metrol. Prop.* **2020**, *8*, 025032. [CrossRef]
3. Sherrington, I.; Smith, E. The significance of Surface topography in engineering. *Precis. Eng.* **1986**, *8*, 79–87. [CrossRef]
4. Wizner, M.; Jakubiec, W.; Starczak, M. Description of surface topography of sealing rings. *Wear* **2011**, *271*, 571–575. [CrossRef]
5. Shi, R.; Wang, B.; Yan, Z.; Wang, Z.; Dong, L. Effect of Surface Topography Parameters on Friction and Wear of Random Rough Surface. *Materials* **2019**, *12*, 2762. [CrossRef]
6. Jiang, Y.; Suvanto, M.; Pakkanen, T.A. Selective surface modification on lubricant retention. *Surf. Rev. Lett.* **2015**, *23*, 1550097. [CrossRef]
7. Holmberg, K.; Andersson, P.; Erdemir, A. Global energy consumption due to friction in passenger cars. *Tribol. Int.* **2012**, *47*, 221–234. [CrossRef]
8. Awale, A.S.; Vashista, M.; Yusufzai, M.Z.K. Application of eco-friendly lubricants in sustainable grinding of die steel. *Mater. Manuf. Process.* **2021**, *36*, 702–712. [CrossRef]
9. Krolczyk, G.M.; Maruda, R.W.; Krolczyk, J.B.; Nieslony, P.; Wojciechowski, S.; Legutko, S. Parametric and nonparametric description of the surface topography in the dry and MQCL cutting conditions. *Measurement* **2018**, *121*, 225–239. [CrossRef]
10. Królczyk, G.; Kacalak, W.; Wieczorowski, M. 3D Parametric and Nonparametric Description of Surface Topography in Manufacturing Processes. *Materials* **2021**, *14*, 1987. [CrossRef]
11. Cooper, L.F. A role for surface topography in creating and maintaining bone at titanium endosseous implants. *J. Prosthet. Dent.* **2000**, *84*, 522–534. [CrossRef] [PubMed]
12. Podulka, P. Reduction of Influence of the High-Frequency Noise on the Results of Surface Topography Measurements. *Materials* **2021**, *14*, 333. [CrossRef] [PubMed]
13. Matsumura, T.; Iida, F.; Hirose, T.; Yoshino, M. Micro machining for control of wettability with surface topography. *J. Mater. Process. Technol.* **2012**, *212*, 2669–2677. [CrossRef]
14. Ohlsson, R.; Wihlborg, A.; Westberg, H. The accuracy of fast 3D topography measurements. *Int. J. Mach. Tool Manuf.* **2001**, *41*, 1899–1907. [CrossRef]
15. Sherrington, I.; Smith, E.H. Modern measurement techniques in surface metrology: Part I; stylus instruments, electron microscopy and non-optical comparators. *Wear* **1988**, *125*, 271–288. [CrossRef]
16. Leach, R. *Optical Measurement of Surface Topography*; Springer: Berlin, Germany, 2011.
17. Sakakibara, R.; Yoshida, I.; Nagai, S.; Kondo, Y.; Yamashita, K. Surface roughness evaluation method based on roughness parameters in ISO 13565-3 using the least-squares method for running-in wear process analysis of plateau surface. *Tribol. Int.* **2021**, *163*, 107151. [CrossRef]
18. Vorburger, T.V.; Rhee, H.G.; Renegar, T.B.; Song, J.-F.; Zheng, A. Comparison of optical and stylus methods for measurement of surface texture. *Int. J. Adv. Manuf. Technol.* **2007**, *33*, 110–118. [CrossRef]

19. Mahovic Poljacek, S.; Risovic, D.; Furic, K.; Gojo, M. Comparison of fractal and profilometric methods for surface topography characterization. *Appl. Surf. Sci.* **2008**, *254*, 3449–3458. [CrossRef]
20. Thompson, A.; Senin, N.; Giusca, C.; Leach, R. Topography of selectively laser melted surfaces: A comparison of different measurement methods. *CIRP Ann. Manuf. Technol.* **2017**, *66*, 543–546. [CrossRef]
21. Podulka, P. Comparisons of envelope morphological filtering methods and various regular algorithms for surface texture analysis. *Metrol. Meas. Syst.* **2020**, *27*, 243–263. [CrossRef]
22. Aich, U.; Banerjee, S. Characterizing topography of EDM generated surface by time series and autocorrelation function. *Tribol. Int.* **2017**, *111*, 73–90. [CrossRef]
23. Dong, W.P.; Sullivan, P.J.; Stout, K.J. Comprehensive study of parameters for characterising three-dimensional surface topography: III: Parameters for characterising amplitude and some functional properties. *Wear* **1994**, *178*, 29–43. [CrossRef]
24. Fubel, A.; Zech, M.; Leiderer, P.; Klier, J.; Shikin, V. Analysis of roughness of Cs surfaces via evaluation of the autocorrelation function. *Surf. Sci.* **2007**, *601*, 1684–1692. [CrossRef]
25. Zou, L.; Liu, X.; Huang, Y.; Fei, Y. A numerical approach to predict the machined surface topography of abrasive belt flexible grinding. *Int. J. Adv. Manuf. Technol.* **2019**, *104*, 2961–2970. [CrossRef]
26. Järnström, J.; Ihalainen, P.; Backfolk, K.; Peltonen, J. Roughness of pigment coatings and its influence on gloss. *Appl. Surf. Sci.* **2008**, *254*, 5741–5749. [CrossRef]
27. Brown, L.G.; Shvaytser, H. Surface orientation from projective foreshortening of isotropic texture autocorrelation. *IEEE Trans. Pattern Anal.* **1990**, *12*, 584–588. [CrossRef]
28. Liao, D.; Shao, W.; Tang, J.; Li, J. An improved rough surface modeling method based on linear transformation technique. *Tribol. Int.* **2018**, *119*, 786–794. [CrossRef]
29. Brochard, J.; Khoudeir, M.; Augereau, B. Invariant feature extraction for 3D texture analysis using the autocorrelation function. *Pattern Recogn. Lett.* **2001**, *22*, 759–768. [CrossRef]
30. Zhou, W.; Cao, Y.; Zhao, H.; Li, Z.; Feng, P.; Feng, F. Fractal Analysis on Surface Topography of Thin Films: A Review. *Fractal Fract.* **2022**, *6*, 135. [CrossRef]
31. Kulesza, S.; Bramowicz, M. A comparative study of correlation methods for determination of fractal parameters in surface characterization. *Appl. Surf. Sci.* **2014**, *293*, 196–201. [CrossRef]
32. Munoz, R.C.; Vidal, G.; Kremer, G.; Moraga, L.; Arenas, C.; Concha, A. Surface roughness and surface-induced resistivity of gold films on mica: Influence of roughness modelling. *J. Phys. Condens. Matter* **2000**, *12*, 2903. [CrossRef]
33. Podulka, P. Improved Procedures for Feature-Based Suppression of Surface Texture High-Frequency Measurement Errors in the Wear Analysis of Cylinder Liner Topographies. *Metals* **2021**, *11*, 143. [CrossRef]
34. Heilbronner, R.P. The autocorrelation function: An image processing tool for fabric analysis. *Tectonophysics* **1992**, *212*, 351–370. [CrossRef]
35. Dhanasekar, B.; Mohan, N.K.; Bhaduri, B.; Ramamoorthy, B. Evaluation of surface roughness based on monochromatic speckle correlation using image processing. *Precis. Eng.* **2008**, *32*, 196–206. [CrossRef]
36. Dusséaux, R.; Vannier, E. Soil surface roughness modelling with the bidirectional autocorrelation function. *Biosyst. Eng.* **2022**, *220*, 87–102. [CrossRef]
37. Wang, W.-Z.; Chen, H.; Hu, Y.-Z.; Wang, H. Effect of surface roughness parameters on mixed lubrication characteristics. *Tribol. Int.* **2006**, *39*, 522–527. [CrossRef]
38. Uchidate, M.; Shimizu, T.; Iwabuchi, A.; Yanagi, K. Generation of reference data of 3D surface texture using the non-causal 2D AR model. *Wear* **2004**, *257*, 1288–1295. [CrossRef]
39. Hu, S.; Brunetiere, N.; Huang, W.; Liu, X.; Wang, Y. Bi-Gaussian surface identification and reconstruction with revised autocorrelation functions. *Tribol. Int.* **2017**, *110*, 185–194. [CrossRef]
40. Patrikar, R.M. Modeling and simulation of surface roughness. *Appl. Surf. Sci.* **2004**, *228*, 213–220. [CrossRef]
41. Krolczyk, G.; Raos, P.; Legutko, S. Experimental Analysis of Surface Roughness and Surface Texture of Machined and Fused Deposition Modelled Parts. *Teh. Vjesn.* **2014**, *21*, 217–221.
42. Gharechelou, S.; Tateishi, R.; Johnson, B.A. A Simple Method for the Parameterization of Surface Roughness from Microwave Remote Sensing. *Remote Sens.* **2018**, *10*, 1711. [CrossRef]
43. ISO 25178-3:2012; Geometrical Product Specifications (GPS)—Surface Texture: Areal—Part 3: Specification Operators. International Organization for Standardization: Geneva, Switzerland, 2012.
44. Podulka, P. Fast Fourier Transform detection and reduction of high-frequency errors from the results of surface topography profile measurements of honed textures. *Eksploat. Niezawodn.* **2021**, *23*, 84–89. [CrossRef]
45. Podulka, P. Selection of Methods of Surface Texture Characterisation for Reduction of the Frequency-Based Errors in the Measurement and Data Analysis Processes. *Sensors* **2022**, *22*, 791. [CrossRef]
46. Pahk, H.; Stout, K.; Blunt, L. A Comparative Study on the Three-Dimensional Surface Topography for the Polished Surface of Femoral Head. *Int. J. Adv. Manuf. Technol.* **2000**, *16*, 564–570. [CrossRef]
47. Podulka, P. Suppression of the High-Frequency Errors in Surface Topography Measurements Based on Comparison of Various Spline Filtering Methods. *Materials* **2021**, *14*, 5096. [CrossRef]
48. De Groot, P.; DiSciacca, J. Definition and evaluation of topography measurement noise in optical instruments. *Opt. Eng.* **2020**, *59*, 064110. [CrossRef]

49. Zhuo, Y.; Han, Z.; An, D.; Jin, H. Surface topography prediction in peripheral milling of thin-walled parts considering cutting vibration and material removal effect. *Int. J. Mech. Sci.* **2021**, *211*, 106797. [CrossRef]
50. Podulka, P. Advances in Measurement and Data Analysis of Surfaces with Functionalized Coatings. *Coatings* **2022**, *12*, 1331. [CrossRef]
51. Peta, K.; Mendak, M.; Bartkowiak, T. Discharge Energy as a Key Contributing Factor Determining Microgeometry of Aluminum Samples Created by Electrical Discharge Machining. *Crystals* **2021**, *11*, 1371. [CrossRef]
52. Podulka, P. The effect of valley depth on areal form removal in surface topography measurements. *Bull. Pol. Acad. Sci. Tech. Sci.* **2019**, *67*, 391–400. [CrossRef]
53. Podulka, P. Bisquare robust polynomial fitting method for dimple distortion minimisation in surface quality analysis. *Surf. Interface Anal.* **2020**, *52*, 875–881. [CrossRef]
54. ISO 16610-28:2016; Geometrical Product Specifications (GPS). Filtration—Part 28: Profile Filters: End Effects. International Organization for Standardization: Geneva, Switzerland, 2016.
55. Janecki, D. Edge effect elimination in the recursive implementation of Gaussian filters. *Precis. Eng.* **2012**, *36*, 128–136. [CrossRef]
56. Podulka, P. The effect of valley location in two-process surface topography analysis. *Adv. Sci. Technol. Res. J.* **2018**, *12*, 97–102. [CrossRef]
57. Janecki, D. Gaussian filters with profile extrapolation. *Precis. Eng.* **2011**, *35*, 602–606. [CrossRef]
58. Podulka, P. Proposal of frequency-based decomposition approach for minimization of errors in surface texture parameter calculation. *Surf. Interface Anal.* **2020**, *52*, 882–889. [CrossRef]
59. Giusca, C.L.; Leach, R. Calibration of the scales of areal surface topography measuring instruments: Part 3. Resolution. *Meas. Sci. Technol.* **2013**, *24*, 105010. [CrossRef]
60. Podulka, P. The Effect of Surface Topography Feature Size Density and Distribution on the Results of a Data Processing and Parameters Calculation with a Comparison of Regular Methods. *Materials* **2021**, *14*, 4077. [CrossRef]
61. Chen, Y.; Huang, W. Numerical simulation of the geometrical factors affecting surface roughness measurements by AFM. *Meas. Sci. Technol.* **2004**, *15*, 2005–2010. [CrossRef]
62. Pei, J.; Han, X.; Tao, Y.; Feng, S. Mixed elastohydrodynamic lubrication analysis of line contact with Non-Gaussian surface roughness. *Tribol. Int.* **2020**, *151*, 106449. [CrossRef]
63. Fecske, S.K.; Gkagkas, K.; Gachot, C.; Vernes, A. Interdependence of Amplitude Roughness Parameters on Rough Gaussian Surfaces. *Tribol. Lett.* **2020**, *68*, 43. [CrossRef]
64. Jacobs, T.D.B.; Junge, T.; Pastewka, L. Quantitative characterization of surface topography using spectral analysis. *Surf. Topogr. Metrol. Prop.* **2017**, *5*, 013001. [CrossRef]
65. Krolczyk, G.M.; Maruda, R.W.; Nieslony, P.; Wieczorowski, M. Surface morphology analysis of Duplex Stainless Steel (DSS) in Clean Production using the Power Spectral Density. *Measurement* **2016**, *94*, 464–470. [CrossRef]
66. Podulka, P. Proposals of Frequency-Based and Direction Methods to Reduce the Influence of Surface Topography Measurement Errors. *Coatings* **2022**, *12*, 726. [CrossRef]
67. Cai, C.; An, Q.; Ming, W.; Chen, M. Modelling of machined surface topography and anisotropic texture direction considering stochastic tool grinding error and wear in peripheral milling. *J. Mater. Process. Technol.* **2021**, *292*, 117065. [CrossRef]
68. Chen, S.; Zhao, W.; Yan, P.; Qiu, T.; Gu, H.; Jiao, L.; Wang, X. Effect of milling surface topography and texture direction on fatigue behavior of ZK61M magnesium alloy. *Int. J. Fatigue* **2022**, *156*, 106669. [CrossRef]
69. Podulka, P. Thresholding Methods for Reduction in Data Processing Errors in the Laser-Textured Surface Topography Measurements. *Materials* **2022**, *15*, 5137. [CrossRef]
70. Maculotti, G.; Feng, X.; Su, R.; Galetto, M.; Leach, R.K. Residual flatness and scale calibration for a point autofocus surface topography measuring instrument. *Meas. Sci. Technol.* **2019**, *30*, 075005. [CrossRef]
71. ISO 25178-2:2012; Geometrical Product Specifications (GPS)—Surface Texture: Areal—Part 2: Terms, Definitions and Surface Texture Parameters. International Organization for Standardization: Geneva, Switzerland, 2012.
72. Franco, L.A.; Sinatora, A. 3D surface parameters (ISO 25178-2): Actual meaning of Spk and its relationship to Vmp. *Precis. Eng.* **2015**, *40*, 106–111. [CrossRef]
73. Grzesik, W. Prediction of the Functional Performance of Machined Components Based on Surface Topography: State of the Art. *J. Mater. Eng. Perform.* **2016**, *25*, 4460–4468. [CrossRef]
74. Gao, F.; Leach, R.; Petzing, J.; Coupland, J.M. Surface measurement errors using commercial scanning white light interferometers. *Meas. Sci. Technol.* **2008**, *19*, 015303. [CrossRef]
75. Lou, S.; Tang, D.; Zeng, W.; Zhang, T.; Gao, F.; Muhamedsalih, H.; Jiang, X.; Scott, P.J. Application of Clustering Filter for Noise and Outlier Suppression in Optical Measurement of Structured Surfaces. *IEEE Trans. Instrum. Meas.* **2020**, *69*, 6509–6517. [CrossRef]
76. Podulka, P.; Pawlus, P.; Dobrzanski, P.; Lenart, A. Spikes removal in surface measurement. *J. Phys. Conf. Ser.* **2014**, *483*, 012025. [CrossRef]
77. ISO 13565-3:1998; Geometrical Product Specifications (GPS)—Surface Texture: Profile Method; Surfaces Having Stratified Functional Properties—Part 3: Height characterization using the Material Probability Curve. International Organization for Standardization: Geneva, Switzerland, 1998.

78. Nagai, S.; Yoshida, I.; Oshiro, K.; Sakakibara, R. Acceleration of surface roughness evaluation using RANSAC and least squares method for Running-in wear process analysis of plateau surface. *Measurement* **2022**, *203*, 111912. [CrossRef]
79. Dong, W.P.; Mainsah, E.; Stout, K.J. Reference planes for the assessment of surface roughness in three dimensions. *Int. J. Mach. Tools Manuf.* **1995**, *35*, 263–271. [CrossRef]
80. Xue, M.; Peng, S.; Wang, F.; Ou, J.; Li, C.; Li, W. Linear relation between surface roughness and work function of light alloys. *J. Alloys Compd.* **2017**, *692*, 903–907. [CrossRef]
81. Forbes, A.B. *Least Squares Best Fit Geometric Elements*; NLP Report DITC 140/89; National Physical Laboratory: Teddington, UK, 1989.
82. Weber, T.; Motavalli, S.; Fallahi, B.; Cheraghi, S.H. A unified approach to form error evaluation. *Precis. Eng.* **2002**, *26*, 269–278. [CrossRef]
83. Li, Y.; Liu, Y.; Tian, Y.; Wang, Y.; Wang, J. Application of improved fireworks algorithm in grinding surface roughness online monitoring. *J. Manuf. Process.* **2022**, *74*, 400–412. [CrossRef]
84. ISO 16610-21:2011; Geometrical Product Specifications (GPS)—Filtration—Part 21: Linear Profile Filters: Gaussian Filters. International Organization for Standardization: Geneva, Switzerland, 2011.
85. Kondo, Y.; Yoshida, I.; Nakaya, D.; Numada, M.; Koshimizu, H. Verification of Characteristics of Gaussian Filter Series for Surface Roughness in ISO and Proposal of Filter Selection Guidelines. *Nanomanufactur. Metrol.* **2021**, *4*, 97–108. [CrossRef]
86. Li, Y.; Otsubo, M.; Kuwano, R.; Nadimi, S. Quantitative evaluation of surface roughness for granular materials using Gaussian filter method. *Powder Technol.* **2021**, *388*, 251–260. [CrossRef]
87. ISO 16610-31:2016; Geometrical Product Specifications (GPS)—Filtration—Part 31: Robust Profile Filters: Gaussian Regression Filters. International Organization for Standardization: Geneva, Switzerland, 2016.
88. Li, H.; Cheung, C.F.; Jiang, X.Q.; Lee, W.B.; To, S. A novel robust Gaussian filtering method for the characterization of surface generation in ultra-precision machining. *Precis. Eng.* **2006**, *30*, 421–430. [CrossRef]
89. Zhang, H.; Yuan, Y.; Piao, W. A universal spline filter for surface metrology. *Measurement* **2010**, *43*, 1575–1582. [CrossRef]
90. Janecki, D. A two-dimensional isotropic spline filter. *Precis. Eng.* **2013**, *37*, 948–965. [CrossRef]
91. He, B.; Zheng, H.; Ding, S.; Yang, R.; Shi, Z.A. review of digital filtering in evaluation of surface roughness. *Metrol. Meas. Syst.* **2021**, *28*, 217–253. [CrossRef]
92. Zhongxiang, H.; Lei, Z.; Jiaxu, T.; Xuehong, M.; Xiaojun, S. Evaluation of three-dimensional surface roughness parameters based on digital image processing. *Int. J. Adv. Manuf. Technol.* **2009**, *40*, 342–348. [CrossRef]
93. Kim, H.Y.; Shen, Y.F.; Ahn, J.H. Development of a surface roughness measurement system using reflected laser beam. *J. Mater. Process. Technol.* **2002**, *130–131*, 662–667. [CrossRef]
94. Fan, S.; Jiao, L.; Wang, K.; Duan, F. Pool boiling heat transfer of saturated water on rough surfaces with the effect of roughening techniques. *Int. J. Heat Mass Transf.* **2020**, *159*, 120054. [CrossRef]
95. González Martínez, J.F.; Nieto-Carvajal, I.; Abad, J.; Colchero, J. Nanoscale measurement of the power spectral density of surface roughness: How to solve a difficult experimental challenge. *Nanoscale Res. Lett.* **2012**, *7*, 174. [CrossRef]
96. Gong, Y.; Misture, S.T.; Gao, P.; Mellott, N.P. Surface Roughness Measurements Using Power Spectrum Density Analysis with Enhanced Spatial Correlation Length. *J. Phys. Chem. C* **2016**, *120*, 22358–22364. [CrossRef]
97. Sun, L. Simulation of pavement roughness and IRI based on power spectral density. *Math. Comput. Simulat.* **2003**, *61*, 77–88. [CrossRef]
98. Dash, P.; Mallick, P.; Rath, H.; Tripathi, A.; Prakash, J.; Avasthi, D.K.; Mazumder, S.; Varma, S.; Satyam, P.V.; Mishra, N.C. Surface roughness and power spectral density study of SHI irradiated ultra-thin gold films. *Appl. Surf. Sci.* **2009**, *256*, 558–561. [CrossRef]
99. Lawson, J.K.; Wolfe, C.R.; Manes, K.R.; Trertholme, J.B.; Aikens, D.M.; English, R.E.J. Specification of Optical Components Using the Power Spectral Density Function. In *Optical Manufacturing and Testing, Proceedings of the SPIE's 1995 International Symposium on Optical Science, Engineering, and Instrumentation, San Diego, CA, USA, 9–14 July 1995*; SPIE Press: Bellingham, WA, USA, 1995; Volume 2536, pp. 38–50. [CrossRef]
100. Zhang, Q.; Zhang, S. Effects of Feed per Tooth and Radial Depth of Cut on Amplitude Parameters and Power Spectral Density of a Machined Surface. *Materials* **2020**, *13*, 1323. [CrossRef]
101. Khan, G.S.; Sarepaka, R.G.V.; Chattopadhyay, K.D.; Jain, P.K.; Bajpai, R.P. Characterization of nanoscale roughness in single point diamond turned optical surfaces using power spectral density analysis. *Indian J. Eng. Mater. Sci.* **2004**, *11*, 25–30.
102. Duparré, A.; Ferre-Borrull, J.; Gliech, S.; Notni, G.; Steinert, J.; Bennett, J.M. Surface Characterization Techniques for Determining the Root-Mean-Square Roughness and Power Spectral Densities of Optical Components. *Appl. Opt.* **2002**, *41*, 154–41171. [CrossRef] [PubMed]
103. Peng, Z.; Kirk, T.B. Two-dimensional fast Fourier transform and power spectrum for wear particle analysis. *Tribol. Int.* **1997**, *30*, 583–590. [CrossRef]
104. Wu, J.P.; Kirk, T.B.; Peng, Z.; Miller, K.; Zheng, M.H. Utilization of two-dimensional fast Fourier transform and power spectral analysis for assessment of early degeneration of articular cartilage. *J. Musculoskelet. Res.* **2005**, *9*, 119–131. [CrossRef]
105. Cheung, C.F.; Lee, W.B. Characterisation of nanosurface generation in single-point diamond turning. *Int. J. Mach. Tools Manuf.* **2001**, *41*, 851–875. [CrossRef]
106. Elson, J.; Bennett, J. Calculation of the power spectral density from surface profile data. *Appl. Opt.* **1995**, *34*, 201–208. [CrossRef] [PubMed]

107. Podulka, P.; Dobrzański, P.; Pawlus, P.; Lenart, A. The effect of reference plane on values of areal surface topography parameters from cylindrical elements. *Metrol. Meas. Syst.* **2014**, *21*, 247–256. [CrossRef]
108. Podulka, P. Edge-area form removal of two-process surfaces with valley excluding method approach. *Matec. Web. Conf.* **2019**, *252*, 05020. [CrossRef]
109. Sayles, R.S.; Thomas, T.R. The spatial representation of surface roughness by means of the structure function: A practical alternative to correlation. *Wear* **1977**, *42*, 263–276. [CrossRef]
110. Gogolewski, D. Fractional spline wavelets within the surface texture analysis. *Measurement* **2021**, *179*, 109435. [CrossRef]

Disclaimer/Publisher's Note: The statements, opinions and data contained in all publications are solely those of the individual author(s) and contributor(s) and not of MDPI and/or the editor(s). MDPI and/or the editor(s) disclaim responsibility for any injury to people or property resulting from any ideas, methods, instructions or products referred to in the content.

Article

Machine Learning-Driven Optimization of Micro-Textured Surfaces for Enhanced Tribological Performance: A Comparative Analysis of Predictive Models

Zhenghui Ge [1], Qifan Hu [1], Rui Wang [1], Haolin Fei [2,*], Yongwei Zhu [1] and Ziwei Wang [2]

[1] College of Mechanical Engineering, Yangzhou University, Yangzhou 225009, China; zhge@yzu.edu.cn (Z.G.); qifanhu2024@163.com (Q.H.); ywzhu@yzu.edu.cn (Y.Z.)
[2] Department of Engineering, Engineering Building, Lancaster University, Lancaster LA1 4YW, UK; z.wang82@lancaster.ac.uk
* Correspondence: h.fei1@lancaster.ac.uk

Abstract: Micro-textured surfaces show promise in improving tribological properties, but predicting their performance remains challenging due to complex relationships between surface features and frictional behavior. This study evaluates five algorithms—linear regression, decision tree, gradient boosting, support vector machine, and neural network—for their ability to predict load-carrying capacity and friction force based on texture parameters including depth, side length, surface ratio, and shape. The neural network model demonstrated superior performance, achieving the lowest MAE (24.01) and highest R-squared value (0.99) for friction force prediction. The results highlight the potential of machine learning techniques to enhance the understanding and prediction of friction-reducing micro-textures, contributing to the development of more efficient and durable tribological systems in industrial applications.

Keywords: micro-texture surface; tribological properties; machine learning algorithms; prediction

1. Introduction

In industrial fields such as transportation, energy, aerospace, etc., wear and friction are unavoidable due to the surface interactions between moving parts of machines. Wear and friction not only bring energy and economic consumption, but also increase the maintenance cost of mechanical equipment and even reduce the service life of mechanical equipment. With the development of modern industry, the demand for improving the service life and performance of mechanical equipment by reducing wear is becoming increasingly strong [1–3]. Improving the friction performance of moving parts in mechanical equipment has also gradually become the focus of attention and research.

It is generally believed that a smoother surface will bring better tribological properties. However, many studies have shown that uneven surfaces with micro-texture may have better tribological properties [4,5]. Liu et al. [6] found that micro-textured surfaces have better friction performance. Compared to polished surfaces, the average friction coefficient of micro-pit surfaces can be reduced by as much as 33.1% in the stable stage. Wan et al. [7] found that the friction coefficient of texture surfaces was more stable than smooth surfaces, and the minimum friction coefficient of lubricating texture surfaces was 0.25 of non-lubricated surfaces. Tong et al. [8] found that the micro-textured surface can reduce the maximum fatigue and friction wear of the smooth surface by 38.4%, with the same friction conditions and selected test parameters. Feng et al. [9] prepared micro-textures on the coating surface, and the results showed that micro-textured surfaces have lower friction coefficients and wear rates than untextured surfaces, and the wear performance also increases with the increase of micro-textured density. Xu et al. [10] prepared a bionic microstructure similar to shark skin on the metal surface. The experimental results showed that the average friction coefficient and wear depth of the biomimetic microstructure were reduced by 41.52% and 79.07%, respectively, at 25 N, and by 50.77% and 80%, respectively, at 35 N.

Currently, the development of micro-textured surfaces with different characteristic parameters has become one of the favorite topics of researchers. He et al. [11] analyzed the tribological properties of different micro-textured surfaces such as circular pits, square pits, and ring grooves by simulation methods. The research results indicated that micro-textured circular pits have a better friction reduction effect. Zhao et al. [12] found that an appropriate increase in the diameter and depth of micro-textures can increase the average oil film thickness between the surfaces of friction pairs, and effectively improve lubrication characteristics. Yang et al. [13] designed orthogonal experiments of the cutting process and micro-texture parameters, which proved that micro-texture has anti-wear and friction-reducing effects. Moreover, they also found that the evaluation index can be improved by changing the distribution state of the micro-texture, and the optimum results were obtained when the micro-pit diameter was 62.3429 μm and the micro-pit spacing was 235.6443 μm in their research. Yin et al. [14] analyzed the influence of surface micro-texture with different shapes and densities on the disassembly damage and load-carrying capacity of the surplus fit axis from the aspects of micromorphology, microstructure, and material properties. The results showed that the surface of a circular microstructure with an area density of 20% can effectively reduce the disassembly and assembly damage and improve the load-carrying capacity. Xie et al. [15] discussed the effects of texture density and depth parameters on friction performance and the results showed that as the surface texture density and depth increased, the surface friction coefficient decreased first and then increased later. Zheng et al. [16] studied the effect of micro-textured distribution density on tool wear and obtained micro-textured tools with variable density distribution. Chen et al. [17] conducted a study on the friction properties of a diamond-like carbon (DLC) film on the textured surface of high-pressure dry gas seals. In the absence of lubrication, it was observed that the friction coefficient of DLC films with micro-textures was lower than that of untextured sealing surfaces. Furthermore, the coefficient of friction exhibited a decreasing trend as the diameter of the micro-texture increased, while an inverse relationship was observed between the coefficient of friction and the density of the micro-texture area. These studies indicate that better friction performances can be obtained by optimizing micro-texture morphology and parameters. However, due to the diverse morphology and variable parameters of micro-textures, it is difficult to fully explore the potential of improving friction performances through simple simulation or experimental parameter studies. Therefore, how to obtain optimized micro-texture morphology and parameters is the key and difficulty in further improving friction performances.

Machine learning has emerged as a powerful tool for optimization in various fields, including tribology [18]. In the context of surface texture optimization, machine learning algorithms have been successfully applied to improve wear resistance and reduce friction force. For example, Gachot et al. [19] employed a genetic algorithm to optimize the surface texture of a steel-steel friction pair, resulting in a significant reduction in friction coefficient. Zhu et al. [20] systematically studied the optimal patterns of surface textures generated by machine learning under different conditions, and compared with the reported optimal textures, the friction coefficient of machine-generated texture was reduced to 27.3%~49.7%. However, the application of diverse machine learning algorithms to comprehensively predict and optimize micro-texture parameters for friction reduction remains largely unexplored. For instance, decision trees are more effective for classification tasks, while neural networks are well-suited to complex, nonlinear problems [21]. In the case of surface texture prediction and optimization, the complex morphology and numerous parameters associated with friction-reducing micro-textures make it challenging to obtain the optimal configuration. This complexity arises from the wide variety of texture morphologies and the interdependence of texture parameters, which can significantly influence the tribological performance of the friction pair.

To address the challenges associated with optimizing friction-reducing micro-textures, this study analyzed the available data and evaluated the performance of five representative machine learning techniques including linear regression (LR), decision tree (DT), gradient boosting (GB), support vector machine (SVM), and neural network (NN). The selection of these algorithms was based on the specific features of the dataset, ensuring that the most effective approach was employed for predicting the performance of different micro-texture configurations. By leveraging the predictive capabilities of machine learning, this study aimed to streamline the optimization process and identify the optimal texture parameters that yielded the desired tribological performance. The results demonstrate that neural networks, in particular, can provide accurate predictions of the effects of surface texture and parameters on friction performance, enabling the exploration of a wider range of design possibilities and the development of more effective friction-reducing micro-textures.

2. Data Acquisition

2.1. The Friction Sub-Model and Texture Parameter Setting

The typical feature size of a micro-texture is in the range of tens to hundreds of microns, the oil film thickness is between a few microns and a dozen microns, and the diameter of the friction vice is between tens and hundreds of millimeters. Additionally, the curvature radius of the surface is significantly larger than the film thickness, which allows for the effect of surface curvature to be ignored and the study to be simplified to a planar problem. The micro-texture model, comprising four shapes of square (a), rectangle (b), circle (c), and aperture (d), is presented in Figure 1.

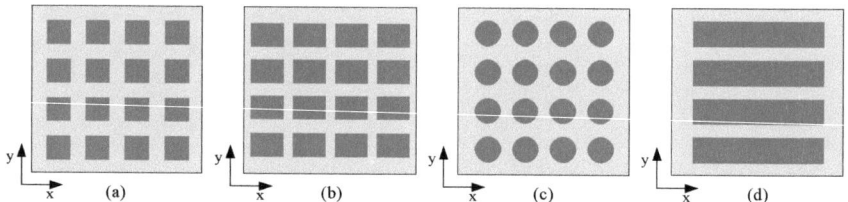

Figure 1. A 2D geometric model with different surface micro-textures: (**a**) square micro-texture; (**b**) rectangular micro-texture; (**c**) circular micro-texture; and (**d**) aperture micro-texture.

A total of 1305 micro-textured watershed models were developed using 3D modeling software. These models were created to represent a range of shapes, edge lengths, area rates, and depths. The configuration of the parameters for each group of micro-textured watersheds is presented in Table 1. The four types of micro-texture possessed the same total area, and therefore the size of other micro-textures could be calculated via Equation (1) based on the size of square micro-texture b.

Table 1. Parameter values used in the simulation model.

Geometry	Size/μm	Area Ratio S/%	Depth h_p/μm
Texture-free	$a \times a$		
Square micro-texture b	25, 50, 100, 150, 200, 300, 400, 500, 600	10%, 20%, 30%, 40%, 50%	1, 2, 3, 5, 7, 8, 9
Rectangular micro-texture	$c \times 1.5c$		
Circular micro-texture	r		
Slit micro-texture	$b \times l$		

$$16b^2 = 24c^2 = 16\pi r^2 = 4bl = Sa^2 \tag{1}$$

where b denotes the length of the square micro-texture, c refers to the short length of the rectangular micro-texture, r indicates the radius of the circular micro-texture, l is the long length of the slit micro-texture, and a represents the length of the substrate. The micro-textures are uniformly distributed on the substrate, and the area ratio S is calculated by the area proportion of micro-textures and the substrate.

2.2. Governing Equations and Boundary Conditions

The micro-texture surfaces were coated with lubricants, resulting in sliding friction occurring in the horizontal direction with a friction pair. A lubrication model was established based on the Navier–Stokes (N-S) equation in order to solve the fluid domain. The following assumptions were considered:

1. The friction pair is rigid and has no deformation;
2. The fluid is incompressible with constant viscosity and density, and the volumetric forces are ignored;
3. The fluid is in laminar and constant mode;
4. Other basic assumptions of the N-S equations [22].

Therefore, the N-S equations and the continuity equation can be expressed as follows:
X direction:

$$\rho\left(u\frac{\partial u}{\partial x}+v\frac{\partial u}{\partial y}+w\frac{\partial u}{\partial z}\right)=-\frac{\partial P}{\partial x}+\eta\left(\frac{\partial^2 u}{\partial x^2}+\frac{\partial^2 u}{\partial y^2}+\frac{\partial^2 u}{\partial z^2}\right) \qquad (2)$$

Y direction:

$$\rho\left(u\frac{\partial v}{\partial x}+v\frac{\partial v}{\partial y}+w\frac{\partial v}{\partial z}\right)=-\frac{\partial P}{\partial x}+\eta\left(\frac{\partial^2 v}{\partial x^2}+\frac{\partial^2 v}{\partial y^2}+\frac{\partial^2 v}{\partial z^2}\right) \qquad (3)$$

Z direction:

$$\rho\left(u\frac{\partial w}{\partial x}+v\frac{\partial w}{\partial y}+w\frac{\partial w}{\partial z}\right)=-\frac{\partial P}{\partial x}+\eta\left(\frac{\partial^2 w}{\partial x^2}+\frac{\partial^2 w}{\partial y^2}+\frac{\partial^2 w}{\partial z^2}\right) \qquad (4)$$

$$\frac{\partial u}{\partial x}+\frac{\partial v}{\partial y}+\frac{\partial w}{\partial z}=0 \qquad (5)$$

where u, v, and w represent the velocity of the fluid along the x, y, and z directions, respectively. ρ represents the density of the lubricant and η represents the dynamic viscosity of the lubricant. P represents the oil film pressure.

The objective of importing the constructed model into Discovery of ANSYS Fluent and setting periodic boundary conditions on its corresponding blocks was to simulate a scenario in which the micro-texture is distributed periodically along the x-axis. The fluid enters the computational domain from the pressure inlet, passes through the domain and exits through the pressure outlet, and then re-enters the pressure inlet boundary of the next computational domain from the pressure outlet in the same way. The boundary conditions of the micro-textured fluid domain are illustrated in Figure 2a. The upper wall was designated as a moving wall with a velocity of U = 5 m/s, the lower wall was designated as a stationary wall, the left wall was designated as a pressure inlet, and the right wall was designated as a pressure outlet. The configuration of the primary parameters of the solution process is illustrated in Figure 2b. The texture models were established in Solidworks and then imported to Ansys Fluent. The automatic meshing function was employed in Fluent with a mesh type of tetrahedral and a mesh growth rate of 1.2. The computational model employed the laminar flow model, which is a standard approach in fluid dynamics. The fluid selected was water, with a density set at 895 kg/m^3 and a dynamic viscosity set at 0.0135145 Pa·s. The pressure with the second order and the momentum with the quick difference were coupled via the SIMPLEC method.

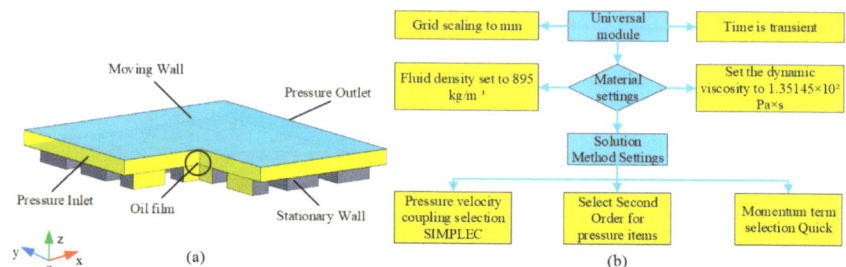

Figure 2. Emulation settings: (**a**) boundary condition settings of the micro-textured watershed; (**b**) key parameter settings for solving mode.

2.3. Simulation Experiment Verification

Texture-free surfaces with lengths of 632.5 µm, 1897.4 µm, and 5060 µm were selected for model validation. The length was calculated based on the square micro-texture model with an area ratio of 10%, and texture lengths of 50, 150, and 400 µm. The validation methodology was employed based on the work conducted by Liu et al. [23], comparing the simulated and theoretical shear stresses of the texture-free surface. The corresponding theoretical equation is shown as Equation (6) and the comparison results are shown in Table 2. The theoretical shear stresses for the three surfaces were all 8446.56 Pa, while the simulation results were 8630.994, 8512.820, and 8480.757 Pa, respectively. The corresponding deviations were 2.18%, 0.78%, and 0.41%. Since the maximum deviation of 2.18% was within the allowable range, this proved that the simulation model was accurate.

$$\tau = \eta \frac{du}{dz} = \eta \frac{U}{h_0} \tag{6}$$

where τ was the shear stress and du/dz was the velocity gradient.

Table 2. Comparison of theoretical and simulated values and errors of shear stress.

Length/µm	Theoretical τ_1/Pa	Simulation τ_2/Pa	Deviation δ/%
632.5		8630.994	2.18%
1897.4	8446.56	8512.820	0.78%
5060		8480.757	0.41%

2.4. Preliminary Analysis of Simulation Data

The load-carrying capacity of the lubricant film on the micro-texture surface was calculated by an area-weighted integral of the positive pressure on the upper film surface, while the friction force was determined by an area-weighted integral of the shear stress on the moving surface along the x-direction of the lubricating film. The corresponding equations can be expressed as follows:

$$F_z = \iint P\, dxdy \tag{7}$$

$$F_x = \iint \tau\, dxdy \tag{8}$$

$$\mu = \frac{F_x}{F_z} \tag{9}$$

where the load-carrying capacity of the oil film on the surface of the micro-texture F_z can be obtained by integrating the area-weighted positive pressure of the moving wall, the friction force Fx can be obtained by integrating the area-weighted shear stress in the x-direction of the lubricating film, τ is the fluid shear stress and μ is the friction coefficient.

A cloud plot of the pressure (top) and shear stress (bottom) distribution of the oil film for each shape texture at a depth of 5 μm, an area ratio of 40%, and a side length of 500 μm is presented in Figure 3. The pressure distribution cloud diagram indicates that the maximum positive pressure value was typically higher than the minimum negative pressure value. Additionally, the inlet pressure of the weaving structure is observed to be smaller than the outlet pressure. This phenomenon can be attributed to the formation of a dispersion wedge gap as a result of the increased distance between the friction surfaces, which occurs as the lubricant flows into the micro-pits. This gap creates a negative pressure in the dispersion region. As the lubricant flows out of the micro-pit, the distance between the surfaces decreases, creating a converging wedge gap that results in a sudden rise in positive pressure in the converging region. This results in an asymmetric pressure direction at the leading and trailing edges of the micro-pit vertically, which in turn generates additional load-bearing effects. The shear stress cloud indicates that the maximum shear stress was observed in the aperture texture configuration, while the minimum was observed in the square texture configuration. Figure 3 demonstrates that the texture has the potential for enhanced friction reduction in the aperture shape.

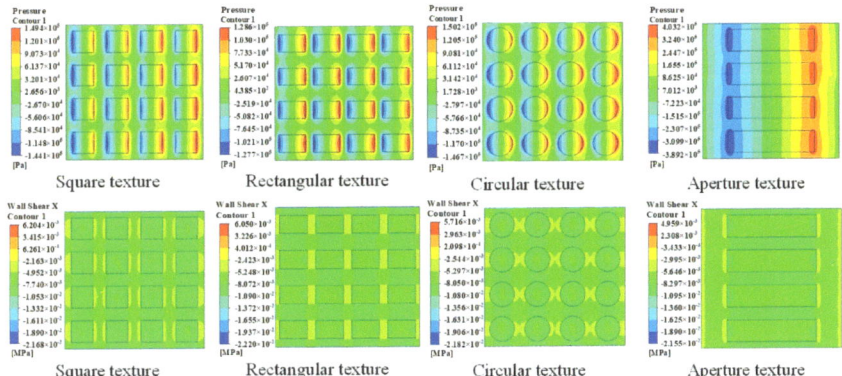

Figure 3. Contour of pressure (**top**) and shear stress (**bottom**) for different texture shapes.

Figures 4 and 5 illustrate the variation of each parameter of the texture as a function of load-carrying capacity and friction coefficient. From Figure 4, it can be observed that the micro-texture structure, when subjected to appropriate parameter selection, exhibited a greater load-carrying capacity than the no-texture structure. However, due to the extensive data exploration conducted in this paper, the multitude of input parameters, and the lack of discernible regularity between the input parameters and the output parameter, it is challenging to draw definitive conclusions.

Figure 5 shows that the friction coefficients of most woven surfaces were smaller than those of their corresponding unwoven surface. Furthermore, the friction coefficients generally exhibited a decreasing trend with the increase in fabric size. However, the influence of individual shape, area ratio, and depth parameters on friction coefficients was not clearly discernible. In conclusion, it was challenging to fully harness the potential of friction performance through the use of basic simulation or experimental parametric studies. This is because it was difficult to study the extent to which micro-texture parameters affected load-carrying capacity and friction coefficients by manual methods, and it was also difficult to predict the optimum texture parameters. Consequently, in this paper, a range of machine learning algorithms was employed to evaluate and predict the friction reduction performance of surface micro-texture parameters.

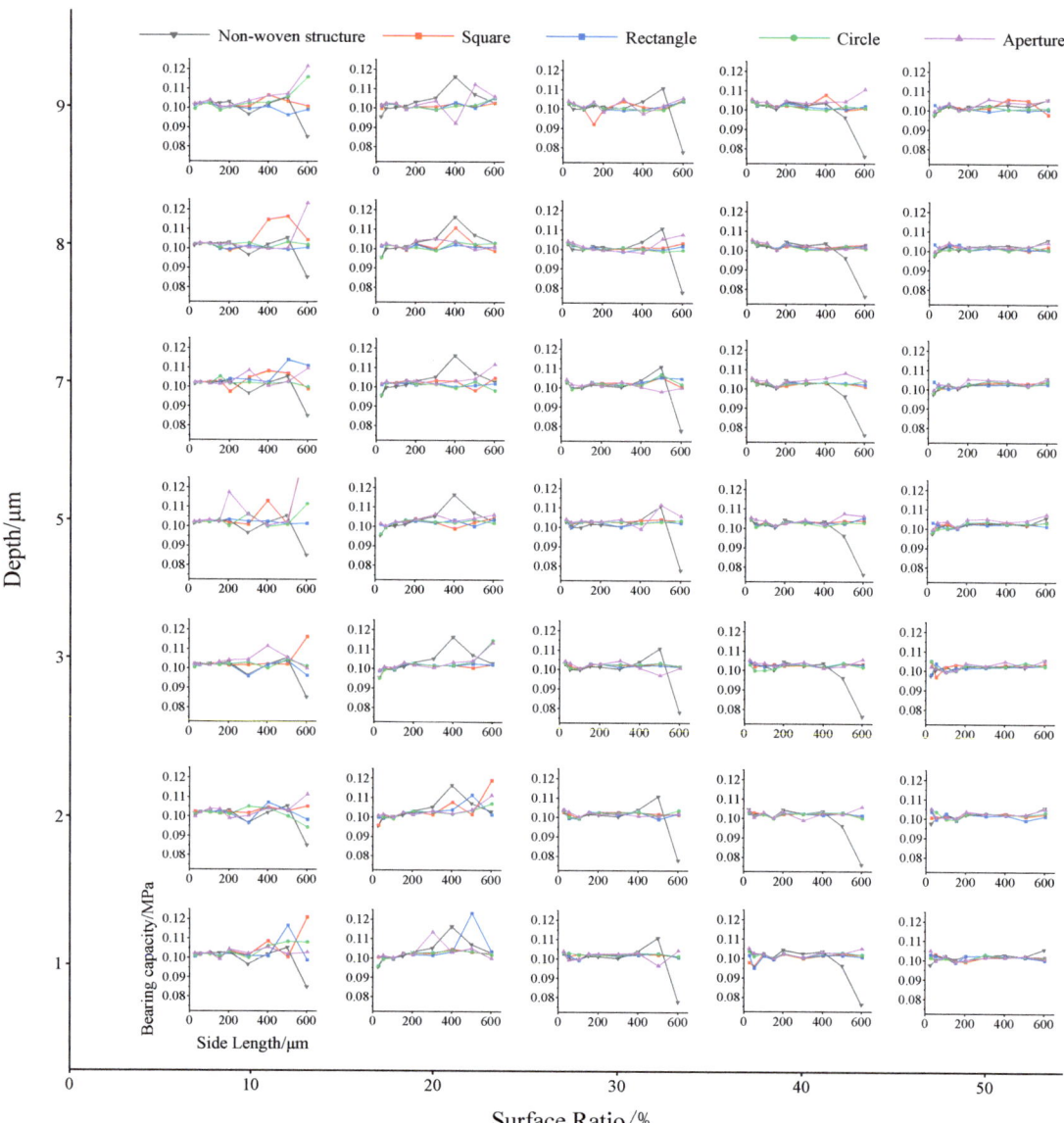

Figure 4. Changes in load-carrying capacity under different texture parameters.

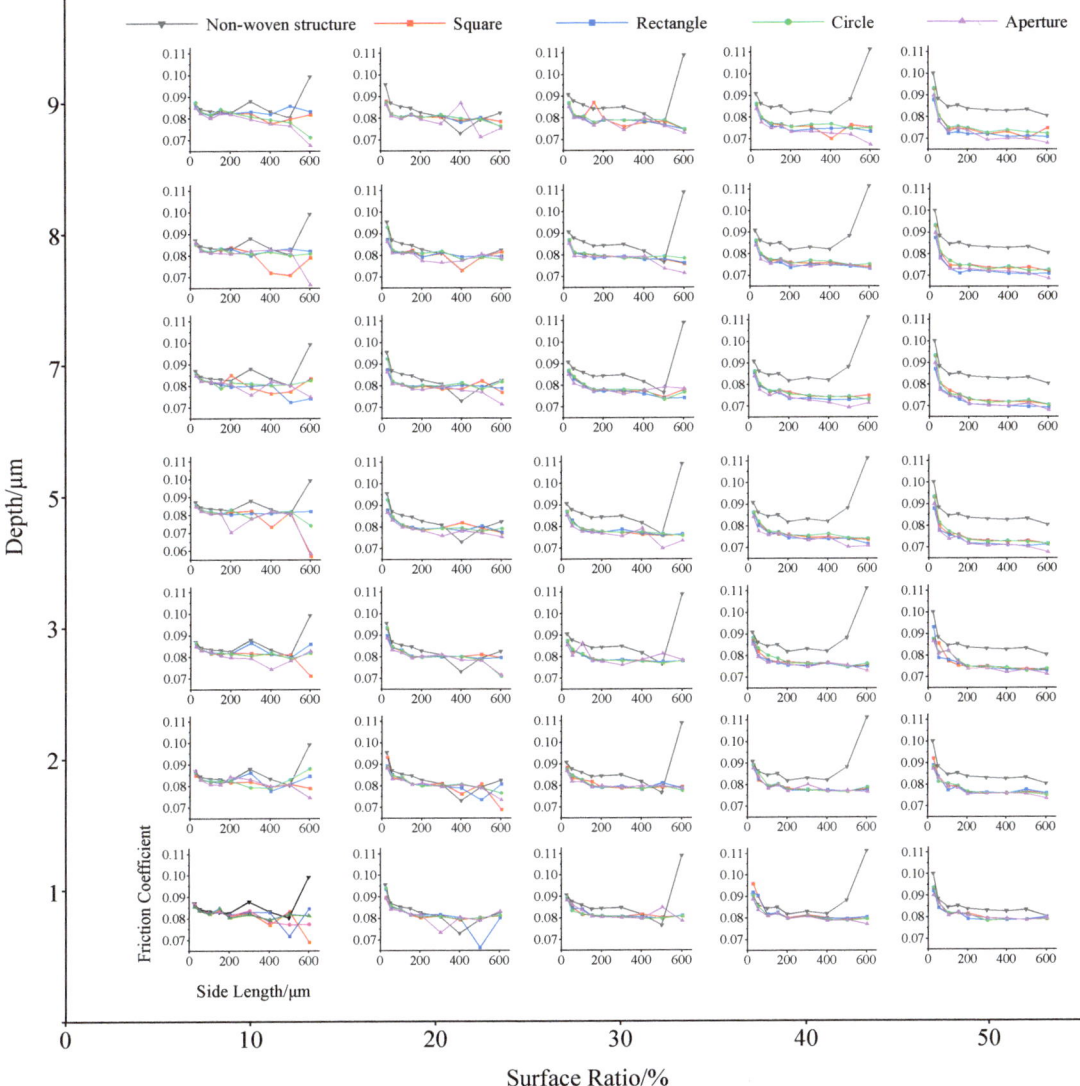

Figure 5. Changes in friction coefficient under different texture parameters.

3. Preliminary Analyses of Simulation Results

3.1. Dataset Construction

This paper focuses on examining the mechanical properties of friction pair parts, emphasizing their responsiveness to various structural characteristics. To construct a comprehensive dataset for subsequent analysis and data-driven model training, simulations were conducted involving parts with diverse parameters such as depth, side length, surface ratio, and shape. The associated key performance metrics, encompassing load-carrying capacity, friction force, friction coefficient, and friction reduction rate, were systematically acquired through simulations involving a range of part configurations. Afterwards, a thorough reorganization was undertaken, accompanied by a discerning data reduction strategy aimed at preserving essential features while mitigating redundancy. This pivotal step in data preparation is essential in optimizing the performance of data-driven models by reducing dataset dimensionality. Parameters exhibiting linear correlations were systematically identified and excluded, resulting in a refined dataset comprising non-linearly correlated input and output pairs. The resulting dataset, characterized by a reduction in input parameters (depth, side length, surface ratio, and shape) and corresponding output parameters (load-carrying capacity and friction force), laid the groundwork for subsequent analyses. The shape feature was encoded for algorithmic processing and analysis, categorically represented as "no texture", "squares", "rectangles", "circles", and "crevices", with numerical values ranging from 0 to 4.

3.2. Exploratory Data Analysis (EDA) Insights

To gain deeper insights into the dataset prior to algorithmic design, exploratory data analysis (EDA) was employed in this study. EDA serves as a commonplace tool for revealing intricacies within datasets. In the context of this study, EDA was utilized to elucidate the nuanced relationships between structural features and critical mechanical behaviors, specifically load-carrying capacity and friction-related properties. This analytical approach enhanced our understanding of the dataset's underlying structure, distributions, and patterns, providing a solid foundation for more informed algorithmic design and the interpretation of results.

We first provide a comprehensive view of the collected data with numerical features summary in Table 3, which includes key statistical measures such as mean, standard deviation, minimum, 25th percentile, median, 75th percentile, and maximum values. Subsequently, we conduct an in-depth analysis of the correlation coefficients existing between each variable within the dataset. The confusion matrix presented in Figure 6 provides insights into the correlations and relationships among the variables D (depth), SL (side length), SR (surface ratio), S (shape), LC (load-carrying capacity), and FF (friction force). The heat map visualizes the pairwise correlations between variables, with darker shades indicating stronger positive (blue) or negative (red) correlations, while lighter shades signify weaker correlations. The matrix reveals several notable patterns and findings that shed light on the intricate interplay between these parameters in the context of the studied system. Each cell within the matrix denotes the correlation strength between two variables, with values ranging from −1 to 1. A value of 1 signifies a robust positive correlation, −1 indicates a pronounced negative correlation, and 0 signifies an absence of correlation.

Table 3. Numerical features summary of the data collected.

	Depth	Side Length	Surface Ratio	Shape	Load-Carrying Capacity	Friction Force	Friction Coefficient
Mean	5.25	258.33	0.38	2.37	102,143.30	8070.26	0.08
Std	2.83	193.05	0.11	1.23	2643.22	486.82	0.01
Min	1.00	25.00	0.10	0.00	76,146.62	7104.28	0.07
25%	2.00	100.00	0.30	1.00	100,955.90	7733.94	0.08
50%	5.00	200.00	0.40	2.00	102,322.10	8020.86	0.08
75%	8.00	400.00	0.50	3.00	103,121.60	8297.02	0.08
Max	9.00	600.00	0.50	4.00	123,444.90	9762.90	0.11

Regarding the load-carrying capacity, the results show the main positive contribution of SL, with a correlation coefficient of 0.2, indicating an increase in SL was generally associated with an improvement in LC. This relationship is intuitively logical, as a larger contact area resulting from increased SL can distribute applied loads more effectively, thereby enhancing load-bearing performance. The secondary factors were SR and S, with correlation coefficients of −0.09 and 0.09, respectively, indicating they possessed the same ability to influence LC in opposite directions. The weakest factor was D, as evidenced by near-zero correlation coefficients (−0.03). Therefore, the main effect of the micro-texture features on the LC can be sequenced as SL (0.2) > SR (−0.09) = S (0.09) > D (−0.03). Similarly, the results show the main negative correlation between SL and FF, with a correlation coefficient of −0.51, indicating that an increase in SL was generally accompanied by a reduction in FF. Other correlation coefficients include a D of −0.23, SR of −0.41, and S of −0.11, indicating that increasing depth, surface ratio, or altering shape potentially led to reductions in friction force. Therefore, the main effect of the micro-texture features on the FF can be sequenced as SL (−0.51) > SR (−0.41) > D (−0.23) > S (−0.11).

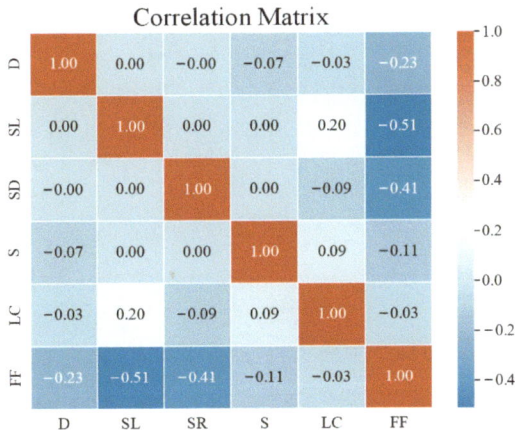

Figure 6. Correlation matrix heat map of the input and output variables.

The confusion matrix also revealed that D, SL, SR, and S had minimal correlations with each other, and the correlation between LC and FF was weak, as evidenced by the near-zero values in their respective cells. This suggests a level of independence among these variables, and they could potentially be tuned independently to optimize the desired performance characteristics. Therefore, the SL and S can be employed as the main optimization parameters in future studies since they can enhance the LC and reduce the FF simultaneously. Moreover, the lack of a dominant correlation in the matrix indicates that no single input feature significantly dominated or influenced specific mechanical properties over others. Given this relationship, employing methods that consider the combined impact of multiple features, such as multivariate regression or machine learning algorithms could be advantageous. These approaches can effectively capture the intricate interplay between various structural characteristics and mechanical properties, providing a more comprehensive understanding of the system.

3.3. Performance Metrics Visualization

In Figure 7, a pair plot is provided offering a comprehensive visual representation of the relationships and distributions among various design parameters and performance metrics for different textile structures. The plot encompasses five distinct shapes: non-woven structure (NWS), square (S), rectangle (R), circle (C), and aperture (A). The analysis reveals clear separations and clustering of data points based on shape, with non-woven structure and aperture exhibiting unique distributions compared to the more regular geometric shapes. This finding highlights the significant impact of structural arrangement and manufacturing process on the behavior and characteristics of these textile structures. One notable observation is the clear separation and clustering of data points based on the shape parameter. The non-woven structure exhibited a distinct distribution compared to the other shapes, indicating its unique properties and performance characteristics. This separation suggests that the manufacturing process and structural arrangement of woven structures have a significant impact on their behavior, setting them apart from the traditional non-woven structure shapes. Within the woven structures, the square, rectangle, and circle structures displayed similar trends and clustering patterns, implying a level of commonality in their performance. However, the aperture shape stood out with its distinct distribution, particularly in terms of load-carrying capacity and friction force. This suggests that the presence of apertures introduces additional complexity and variability in the mechanical properties of the structure.

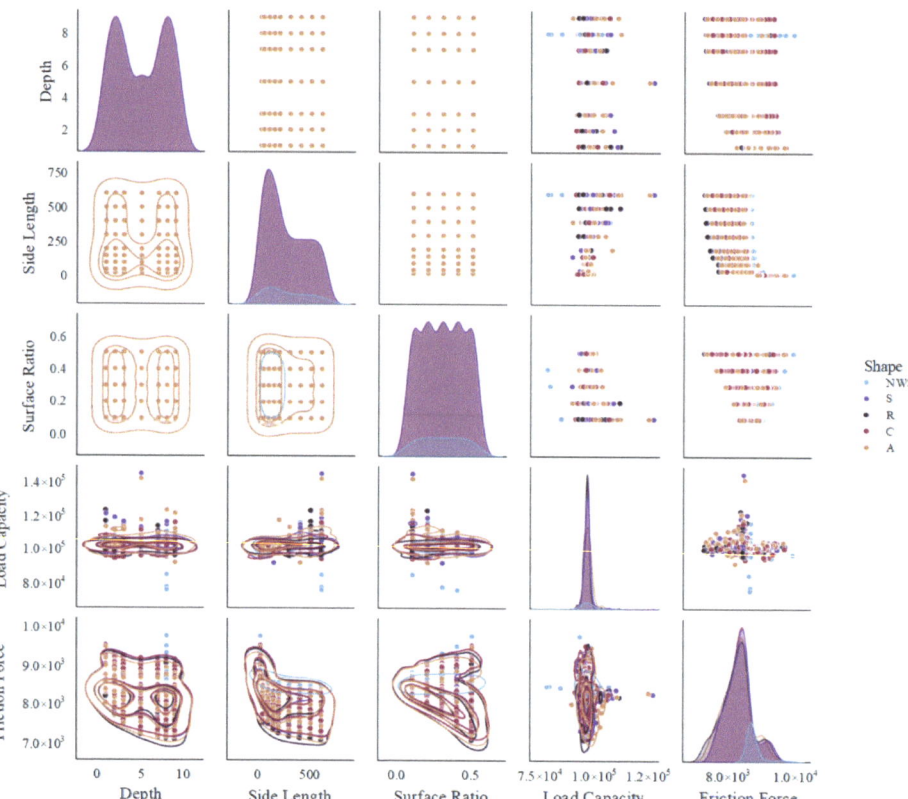

Figure 7. Pair plots depicting the distinct characteristics and performance variations across different textile structure shapes: non-woven structure (NWS), square (S), rectangle (R), circle (C), and aperture (A).

Examining the relationships between design parameters and performance metrics reveals interesting patterns. Both depth and side length showed a positive correlation with load-carrying capacity, indicating that increasing the depth and the side length of the structure generally enhanced its load-bearing capabilities. However, the relationship between depth and friction force was less pronounced, suggesting that depth alone may not be a strong predictor of frictional behavior. Conversely, the surface ratio parameter showed a negative correlation with load-carrying capacity, suggesting that higher surface ratios may compromise the structure's ability to support loads effectively. Interestingly, the friction force metric displayed a wide range of values across different shapes and design parameters. The geometric shapes tended to have lower friction force values compared to the non-woven structure, indicating its inherent suitability to sliding and potential for decreased surface interactions. Among the geometric shapes, the aperture structure exhibited the highest variability in friction force, highlighting the complex interplay between aperture design and frictional properties. The pair plot also reveals the presence of outliers and extreme values, particularly in the load-carrying capacity and friction force metrics. These outliers suggest the existence of specific design configurations or manufacturing conditions that result in exceptional performance characteristics. Further investigation into these outliers could yield valuable insights into optimizing the design and production process for enhanced mechanical properties.

The violin plots presented in Figure 8 provide a comprehensive visualization of the distribution and statistical properties of both friction force and load-carrying capacity across various textile structure shapes. The region of the Y-axis corresponding to the black line segment indicates that 75% of the values fell within that region. The Y-axis value corresponding to the white dot field indicates the mean value of load-carrying capacity and friction for each shape parameter. It can be observed from Figure 8a that the woven structure shapes (square, rectangle, circle, and aperture) generally exhibited lower friction force values compared to the non-woven structure, which displayed a compact and dense distribution with a higher median friction force value. In contrast, the woven structure shapes demonstrated broader distributions of friction force values, generally shifted towards lower magnitudes. This suggests that the woven structures, particularly the geometric shapes, possessed inherent characteristics that contributed to reduced frictional properties. The observation of a wider range of frictional behaviors within each category of woven shapes indicates the potential for further optimization of these structures in order to minimize friction forces. These findings highlight the advantageous tribological properties of woven textures, particularly their ability to reduce surface interactions and promote easier sliding, which is desirable in many applications requiring low friction. The aperture and rectangle shape demonstrated the most pronounced lower tail, implying its potential for achieving reduced frictional properties. On the other hand, the load-carrying capacity violin plot revealed distinct differences in the load-bearing capabilities of the various shapes, as shown in Figure 8b. The non-woven structure exhibited the lowest median load-carrying capacity and a relatively narrow distribution, suggesting limited load-bearing performance compared to the woven structures. Among the woven structures, the square and rectangle shapes exhibited similar distributions and the highest median load capacities, which suggests a higher propensity for load-bearing performance, likely due to their structural stability and efficient load distribution. The circle and aperture shapes also showed relatively high load capacities, with slightly lower median values compared to the square and rectangle shapes. Interestingly, the aperture and square shape exhibited a more extended upper tail in the load-carrying capacity distribution, indicating the presence of samples with exceptionally high load-bearing capabilities. This finding highlights the potential of aperture and square designs in achieving enhanced load-carrying capacity, possibly due to their ability to distribute loads effectively through the aperture structure.

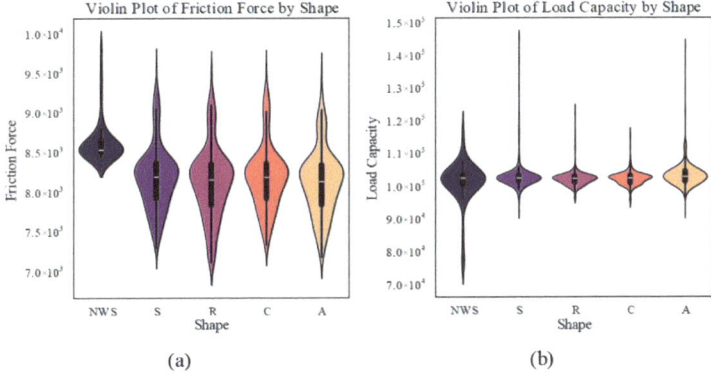

Figure 8. Violin plot of (**a**) friction force and (**b**) load-carrying capacity by shape.

4. Simulation Data Analysis and Insights

4.1. Machine Learning Algorithms

The complex relationships between input and output parameters in modern manufacturing present challenges for traditional fixed-formula methods, which often fail to capture these interactions fully. This limitation has led to the development of data-driven approaches that utilize statistical techniques to learn directly from data, providing more robust solutions for process optimization and prediction. Recently, machine learning (ML) algorithms have become essential in data-driven manufacturing analysis, effectively addressing tasks such as parameter optimization and accurate prediction. The objective of this study was to develop an optimal model capable of accurately characterizing mechanical properties, including force friction and load-carrying capacity, based on limited surface or shape features, without the need for extensive modeling or experiments. Therefore, a regression model was required to predict these properties based on constrained input features.

ML encompasses various algorithms, broadly classified into categories such as linear models, decision trees, ensemble models, kernelized models, and neural networks. The selection of an algorithm depends on data characteristics, interpretability, and specific task requirements. This study selected one representative algorithm from each category based on applicability and successful implementations. For linear models, LR stood out as a pragmatic choice due to its simplicity, interpretability, and effectiveness in capturing linear relationships between features and target variables [24]. DT algorithms are favored for decision-based methods, offering intuitive decision rules and the ability to handle both numerical and categorical data effectively [25]. Ensemble methods, exemplified by the gradient boosting algorithm (XGBoost), are preferred for their capacity to combine multiple weak learners to produce a strong predictor, thereby enhancing overall performance and generalization [26]. Kernelized methods, such as SVMs, excel in capturing complex nonlinear relationships in data by mapping them into a higher-dimensional feature space [27]. Finally, we chose a feedforward neural network for neural network-based methods. Other architectures such as recurrent neural networks (RNN), commonly applied to tasks involving temporal data [28], and convolutional neural networks (CNN), better suited for spatial data [29], possess intricate structures that often demand a large number of parameters. However, given the dataset's limited size, such complexity may not have been optimal. Despite techniques like fine-tuning [30] that can alleviate the need for extensive data, the absence of pre-trained models tailored to the specific dataset rendered overly complex network architectures unsuitable for the task.

This research specifically investigated the applicability of five prominent ML algorithms: LR, DT, GB, SVM, and NN. Each algorithm brings unique strengths and weaknesses to the table, offering varying degrees of effectiveness in addressing the specific regression problem [31]. The implementation details for each algorithm were as follows: the LR model used ordinary least squares regression with a learning rate of 0.01. The DT model was configured with a maximum depth of 10 and a minimum sample per leaf of 5. The GB implementation utilized 100 estimators with a learning rate of 0.1 and a maximum tree depth of 6. The SVM employed a radial basis function kernel with C = 1.0 and epsilon = 0.1. The NN architecture consisted of four layers, including input layer (4 Neurons), hidden layer 1 (64 Neurons), hidden layer 2 (32 Neurons) and output layer (1 Neurons), using ReLU activation functions and Adam optimizer with a learning rate of 0.001, trained for 200 epochs with batch size 32. Therefore, this study aimed to comprehensively evaluate the performance of these algorithms and identify the model that exhibited the highest degree of accuracy in capturing the intricate interplay between the chosen input features and the target outputs.

4.2. Data Preprocessing

In preparing the data for analysis, we addressed the challenge posed by the substantial variance in the dimensions of the input and output features. Given the diverse preprocessing needs of each algorithm under consideration, the approach was tailored to align with these requirements, ensuring optimal model performance. For linear models, sensitivity to feature scale is a notable concern sensitivity. To ensure that all input features contributed equally to the models' learning processes, especially critical for models sensitive to the scale of input variables like linear regression, feature scaling was applied, which is given as:

$$\text{Scaled Feature}_i = \frac{X_i - \mu}{\sigma} \qquad (10)$$

where X_i represents the original value of the feature being scaled, σ and μ represent the standard deviation and the mean of the feature across all samples, respectively. By applying this transformation, each feature is centered around zero with a unit standard deviation, allowing for a harmonized scale across all inputs. In addition, for categorical variables, one-hot encoding transforms these variables into a binary vector representation, which is crucial when including categorical data in machine learning models without implying any ordinal relationship between categories. This process is encapsulated by:

$$\text{One-Hot}_i^n = \begin{cases} 1, \wedge if n = i \\ 0, \wedge otherwise \end{cases} \qquad (11)$$

where n represents the index within the binary vector corresponding to the specific category present in an observation, and i denotes the actual category of the observation from the original dataset. In the one-hot encoded vector, a '1' is placed at position n to indicate the presence of category i, with all other positions set to '0'. This binary vector effectively communicates the presence or absence of categorical features to the model without introducing numeric biases associated with ordinal encoding, ensuring that the model accurately interprets these as distinct, non-ordered categories. Contrastingly, DT inherently manages categorical data without necessitating one-hot encoding. This capability stems from their method of node splitting, which effectively handles both continuous and categorical variables by evaluating whether a feature surpasses a specific threshold. GB models display an indifference to the scale of input features, thus obviating the need for feature scaling or one-hot encoding for categorical variables. However, recognizing the distinct nature of the output targets, we opted to construct two separate models for their prediction, thereby enhancing model specificity and accuracy. For SVMs, both feature and target scaling, alongside one-hot encoding for categorical variables, are essential preprocessing steps. These adjustments are crucial for SVMs due to their sensitivity to the scale of input data, which significantly influences their prediction accuracy. Regarding NN, we experimented with two approaches to model architecture: a dual-output head configuration and separate models for each output target. This experimentation aimed to ascertain the most effective strategy for the prediction tasks, taking into consideration the unique characteristics of neural network models and their flexibility in handling complex data relationships.

5. Results and Discussion

In this study, three evaluation criteria were employed: the mean absolute error (MAE), root mean square error (RMSE), and R-squared (R^2) were employed as evaluation criteria. These metrics are commonly employed for the evaluation of regression models. The mean absolute error (MAE) is a measure of the average magnitude of errors, calculated as the mean of the absolute differences between predicted and true values. This makes the MAE resistant to outliers. In contrast, RMSE penalizes larger errors more heavily, rendering it sensitive to outliers and thus suitable for scenarios where larger errors are more critical. R^2 quantifies the proportion of variance in the dependent variable explained by the independent variables, with higher values indicating a superior model fit. These three metrics were selected for their robustness and interpretability in evaluating regression model performance [32].

The results presented in Tables 4 and 5 present a comparative analysis of the performance of LR, DT, GB, SVM, and NN in predicting a target variable. The optimal values of the corresponding evaluation indicators under different methods are presented in the table in black. In order to facilitate a more intuitive comparison and analysis of the performance of the predictor variables, this paper plotted (Figure 9) them corresponding to Tables 4 and 5. This allowed the prediction of the load-carrying capacity using different machine learning methods to be observed. The normalized mean absolute error (MAE) under the NN method and the normalized root mean square error (RMSE) under the SVM method had the smallest values, with values of 0.50 and 0.94, respectively. In contrast, the decision tree method had the largest values. The normalized MAE and normalized RMSE under the decision tree method exhibited the greatest values, with respective values of 0.71 and 1.46. The value of R-square was not close to 1 in any of the five methods, which may be attributed to the fact that the load-carrying capacity of the liquid oil film was primarily influenced by the viscosity of the lubricant, the film thickness, the rate of formation of the lubricant film, and the surface roughness, and was less correlated with the various parameters of the input texture. A quantitative comparison of the principal assessment metrics for predicting friction utilizing the contrasting methodologies revealed that the neural network model exhibited the most optimal overall performance among the algorithms subjected to this analysis. It achieved the lowest MAE (24.01) and the lowest normalized MAE (0.06), indicating its superior accuracy in predicting the target variable. Moreover, the NN model also attained the lowest normalized RMSE (0.11), which can be attributed to the NN's ability to capture complex nonlinear relationships through its multilayer architecture and effective feature learning capabilities. In terms of the R-square metric, which measures the proportion of variance in the target variable explained by the model, the NN model achieved an impressive value of 0.99. This suggests that the NN model captured almost all the variability in the target variable, providing a highly accurate and reliable prediction. The GB model also demonstrated strong performance, with the second-lowest normalized MAE (0.11), as well as the second-lowest normalized RMSE (0.15). The DT model showed moderate performance and its R-square value of 0.89 suggests that the DT model captured a substantial amount of the variability in the target variable, although not as effectively as the NN and GB models. On the other hand, the LR and SVM models exhibited relatively poor performance compared to the other algorithms, with RMSE of 0.71 and 0.98, indicating a higher level of prediction errors and limited effectiveness in capturing the underlying patterns in the data. The superior performance of the neural network model in this study highlights its capability of modeling complex, nonlinear relationships between input parameters and output performance metrics. This prediction model allows further identification of the optimal combination of texture morphology and parameters that yield the most significant improvements in wear resistance and friction reduction. In conclusion, the NN method was selected for a more precise evaluation of the prediction of load-carrying capacity and friction.

Table 4. Evaluation metrics comparison for load-carrying capacity prediction with different methods including normalized MAE, RMSE, and R square.

	MAE	Normalized MAE	RMSE	Normalized RMSE	R-Square
LR	1626.06	0.55	2819.51	0.95	0.10
DT	2097.43	0.71	4243.13	1.46	−1.14
GB	1568.70	0.53	3031.53	1.02	−0.04
SVM	1563.31	0.53	2803.57	0.94	0.11
NN	1495.61	0.50	2986.95	1.01	−0.01

Table 5. Evaluation metrics comparison for friction force prediction with different methods including normalized MAE, RMSE, and R-square.

	MAE	Normalized MAE	RMSE	Normalized RMSE	R-Square
LR	209.08	0.53	282.81	0.71	0.49
DT	77.97	0.20	131.90	0.33	0.89
GB	41.83	0.11	59.16	0.15	0.98
SVM	248.05	0.62	390.05	0.98	0.04
NN	24.01	0.06	47.58	0.11	0.99

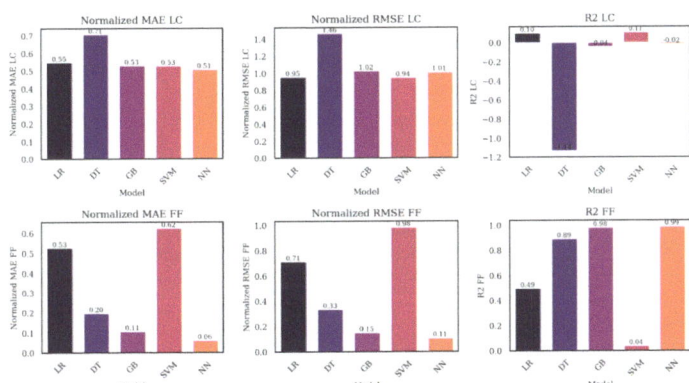

Figure 9. Comparison of normalized MAE regression results across decision trees, gradient boost, linear, NN, and SVM.

This paper further validated the performance and reliability of NN by presenting residual plots in Figure 10. These plots visualize the differences between the true values and the predicted values, enabling an assessment of the model's predictive accuracy and potential biases. In the study of the frictional properties of the surface of the friction partner, the friction force is typically the most important factor. This is because micro-textiles are designed to improve the frictional properties, reduce the friction force, increase the lubrication effect of the surface, and thus reduce the energy consumption and wear. It can be seen from the friction force scatter plot (Figure 10b) that the predicted values and true values are evenly distributed around the y = x line. This indicates that the NN model did not exhibit systematic bias in its predictions, as it did not consistently overestimate or underestimate the friction force. The absence of any discernible pattern in the friction force prediction residuals suggests that the model effectively captured the underlying relationship between the input features and friction force. Furthermore, the magnitude of the residuals remained relatively small compared to the range of friction force values, as shown in Figure 10d. This indicates that the neural network model was capable of providing accurate predictions with minimal error, thereby offering some guidance on the selection of micro-texture parameters for engineering applications. The tight clustering of residuals near zero indicates a high level of precision in the model's predictions, instilling confidence in its ability to estimate load-carrying capacity accurately. However, it is worth noting that the scatter plot of load-carrying capacity. Figure 10a does not demonstrate an even distribution along diagnosis and their residuals shown in Figure 10c exhibit a wider spread compared to the friction force residuals. This suggests that the NN model's predictions of load-carrying capacity had higher variability or uncertainty compared to its predictions of friction force. Nevertheless, the overall magnitude of the load-carrying capacity residuals remained small, with the majority falling within the range of −2–2, indicating that the model still provided reliable and accurate predictions. The normalized residuals plot provides further validation of the NN model's performance.

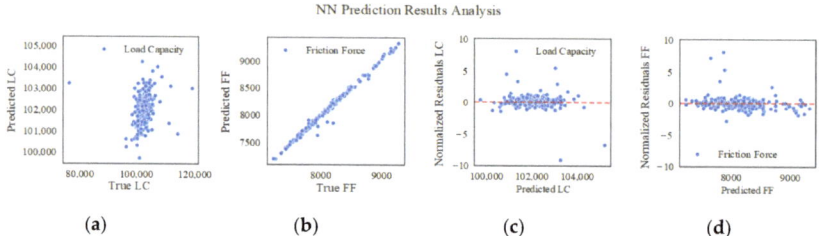

Figure 10. Scatter plots contrasting predicted: (**a**) load-carrying capacity values; and (**b**) friction force values with true values. Residual plots: (**c**) load-carrying capacity values; and (**d**) friction force values.

In short, the computational cost and accuracy of the ML models varied significantly, influencing their suitability for real-world applications. NN achieved the highest accuracy, capturing complex nonlinear relationships with $R^2 = 0.99$ for friction force, while it incurred higher computational costs. In contrast, models like DT and GB struck a balance between accuracy and computational efficiency, making them viable for scenarios where real-time processing is prioritized. LR serves as a baseline but lacks predictive power for load-carrying capacity and friction force, while computationally inexpensive.

6. Conclusions and Perspective

In this study, we proposed a machine learning-based approach to predict and analyze the tribological performance of micro-textured surfaces, with a focus on load-carrying capacity and friction force. Five different ML models were employed to systematically investigate the influence of texture parameters, including depth, side length, surface ratio, and shape, on performance metrics. Among these models, the NN exhibited superior predictive performance, achieving exceptional accuracy in capturing the complex and nonlinear relationships between texture parameters and tribological performance. Specifically, the NN attained an R^2 value of 0.99 and the lowest MAE of 24.01 for friction force prediction, significantly outperforming the other ML models. This approach facilitates rapid exploration of a broad design space, enabling efficient evaluation of texture configurations without relying on extensive physical experimentation. The findings underscore the potential of machine learning, particularly neural networks, as a powerful tool for optimizing tribological performance, providing actionable insights that may be overlooked by conventional analysis methods.

Despite its achievements, this study has several limitations that must be acknowledged. First, it relied entirely on simulation data, which may not fully capture the complexity of real-world conditions. Second, the dataset used was limited in scope, omitting critical factors such as viscosity, temperature, and others that could significantly impact the accurate modeling of tribological systems. Additionally, the computational cost of the neural network model poses a significant barrier to broader industrial adoption, particularly in real-time processing environments. Future research should aim to expand the dataset by including additional relevant features, integrate predictive models with automated design optimization algorithms, and develop lightweight or hybrid machine learning models to mitigate computational challenges. Crucially, experimental validation is necessary to confirm the model's predictions and ensure their reliability in practical applications. Addressing these challenges will help evolve this approach into a comprehensive framework for tackling complex tribological problems, thereby contributing to more efficient and sustainable surface engineering solutions.

Author Contributions: Z.G. (investigation, writing—review and editing), Q.H. (simulation, writing—original draft), R.W. (data processing, writing—original draft), H.F. (methodology, writing—original draft), Y.Z. (conceptualization, review and editing), Z.W. (supervision, review and editing). All authors have read and agreed to the published version of the manuscript.

Funding: This work was financially supported by the Jiangsu Agricultural Science and Technology Innovation Fund (CX(21)3154) and the National Natural Science Foundation of China (52175438).

Institutional Review Board Statement: Not applicable.

Informed Consent Statement: Not applicable.

Data Availability Statement: Data will be made available on request.

Conflicts of Interest: The authors declare no conflict of interest.

References

1. Du, C.; Yu, T.; Zhang, L.; Shen, R.; Wu, Z.; Li, X.; He, X.; Feng, Y.; Wang, D. Robust and universal macroscale superlubricity with natural phytic acid solutions. *Tribol. Int.* **2023**, *183*, 108387. [CrossRef]
2. Huang, J.; Guan, Y.; Ramakrishna, S. Tribological behavior of femtosecond laser-textured leaded brass. *Tribol. Int.* **2021**, *162*, 107115. [CrossRef]
3. Li, S.; Yang, X.; Kang, Y.; Li, Z.; Li, H. Progress on Current-Carry Friction and Wear: An Overview from Measurements to Mechanism. *Coatings* **2022**, *12*, 1345. [CrossRef]
4. Zhang, H.; Dai, S.; Liu, Y.; Zhu, Y.; Xu, Y.; Li, B.; Dong, G. Fishbone-like micro-textured surface for unidirectional spreading of droplets and lubricity improvement. *Tribol. Int.* **2024**, *198*, 109932. [CrossRef]
5. Wang, P.; Liang, H.; Jiang, L.; Qian, L. Effect of nanoscale surface roughness on sliding friction and wear in mixed lubrication. *Wear* **2023**, *530–531*, 204995. [CrossRef]
6. Liu, X.; Zhang, J.; Li, L. Surface Integrity and Friction Performance of Brass H62 Textured by One-Dimensional Ultrasonic Vibration-Assisted Turning. *Micromachines* **2021**, *12*, 1398. [CrossRef]
7. Wan, Q.; Gao, P.; Zhang, Z. Friction and wear performance of lubricated micro-textured surface formed by laser processing. *Surf. Eng.* **2021**, *37*, 1523–1531. [CrossRef]
8. Tong, X.; Shen, J.; Su, S. Properties of variable distribution density of micro-textures on a cemented carbide surface. *J. Mater. Res. Technol.* **2021**, *15*, 1547–1561. [CrossRef]
9. Feng, X.; Wang, R.; Wei, G.; Zheng, Y.; Hu, H.; Yang, L.; Zhang, K.; Zhou, H. Effect of a micro-textured surface with deposited MoS_2-Ti film on long-term wear performance in vacuum. *Surf. Coat. Technol.* **2022**, *445*, 128722. [CrossRef]
10. Xu, L.; Shi, X.; Xue, Y.; Zhang, K.; Huang, Q.; Wu, C.; Ma, J.; Shu, J. Improving Tribological Performance of 42CrMo under Dry Sliding Conditions by Combining Rhombic-Textured Surfaces with Sn–Ag–Cu Solid Lubricant and MXene-Ti3C2TX. *J. Mater. Eng. Perform.* **2023**, *32*, 1275–1291. [CrossRef]
11. He, C.; Yang, S.; Zheng, M. Analysis of synergistic friction reduction effect on micro-textured cemented carbide surface by laser processing. *Opt. Laser Technol.* **2022**, *155*, 108343. [CrossRef]
12. Zhao, J.; Li, Z.; Zhang, H.; Zhu, R. Prediction of Contact and Lubrication Characteristics of Micro-textured Surface Under Thermal Line Contact EHL. *Front. Mech. Eng.* **2021**, *7*, 672588. [CrossRef]
13. Yang, H.; Yang, S.; Tong, X. Study on the Matching of Surface Texture Parameters and Processing Parameters of Coated Cemented Carbide Tools. *Coatings* **2023**, *13*, 681. [CrossRef]
14. Yin, C.; Huang, H.; Dan, Z.; Liu, Z. Disassembly damage and load-carrying capacity of shaft with surface micro-texture. *Surf. Eng.* **2022**, *38*, 562–570. [CrossRef]
15. Xie, X.; Hua, X.; Li, J.; Cao, X.; Tian, Z.; Peng, R.; Yin, B.; Zhang, P. Synergistic effect of micro-textures and MoS_2 on the tribological properties of PTFE film against GCr15 bearing steel. *J. Mater. Res. Technol.* **2021**, *35*, 2151–2160. [CrossRef]
16. Zheng, M.; He, C.; Yang, S. Optimization of Texture Density Distribution of Carbide Alloy Micro-Textured Ball-End Milling Cutter Based on Stress Field. *Appl. Sci.* **2020**, *10*, 818. [CrossRef]
17. Chen, D.; Ding, X.; Yu, S.; Zhang, W. Friction performance of DLC film textured surface of high pressure dry gas sealing ring. *J. Braz. Soc. Mech. Sci.* **2019**, *41*, 161. [CrossRef]
18. Wu, H.; Wei, P.; Hu, R.; Lin, H.; Du, X.; Zhou, P.; Zhu, C. Study on the relationship between machining errors and transmission accuracy of planetary roller screw mechanism using analytical calculations and machine-learning model. *J. Comput. Des. Eng.* **2023**, *10*, 398–413. [CrossRef]
19. Gachot, C.; Rosenkranz, A.; Hsu, S.M.; Costa, H.L. A critical assessment of surface texturing for friction and wear improvement. *Wear* **2017**, *372–373*, 21–41. [CrossRef]
20. Zhu, B.; Zhang, W.; Zhang, W.; Li, H. Generative design of texture for sliding surface based on machine learning. *Tribol. Int.* **2023**, *179*, 108139. [CrossRef]
21. Kotsiantis, S.B.; Zaharakis, I.D.; Pintelas, P.E. Machine learning: A review of classification and combining techniques. *Artif. Intell. Rev.* **2006**, *26*, 159–190. [CrossRef]

22. Jia, J.; Hao, Y.; Ma, B.; Xu, T.; Li, S.; Xu, J.; Zhong, L. Gas–liquid two-phase flow field analysis of two processing teeth spiral incremental cathode for the deep special-shaped hole in ECM. *Int. J. Adv. Manuf. Technol.* **2023**, *127*, 5831–5846. [CrossRef]
23. Liu, W.; Ni, H.; Chen, H.; Wang, P. Numerical simulation and experimental investigation on tribological performance of micro-dimples textured surface under hydrodynamic lubrication. *Int. J. Mech. Sci.* **2019**, *163*, 105095. [CrossRef]
24. Weisberg, S. *Applied Linear Regression*; John Wiley & Sons: Hoboken, NJ, USA, 2005. [CrossRef]
25. Song, Y.; Lu, Y. Decision tree methods: Applications for classification and prediction. *Shanghai Arch. Psychiatry* **2015**, *27*, 130–135. [CrossRef] [PubMed]
26. Chen, T.; Guestrin, C. XGBoost. In Proceedings of the 22nd ACM SIGKDD International Conference on Knowledge Discovery and Data Mining, San Francisco, CA, USA, 13–17 August 2016; ACM: New York, NY, USA, 2016; pp. 785–794. [CrossRef]
27. Suthaharan, S. Support Vector Machine. In *Machine Learning Models and Algorithms for Big Data Classification*; Springer: Boston, MA, USA, 2016; pp. 207–235. [CrossRef]
28. Sherstinsky, A. Fundamentals of Recurrent Neural Network (RNN) and Long Short-Term Memory (LSTM) network. *Physica D* **2020**, *404*, 132306. [CrossRef]
29. Gu, J.; Wang, Z.; Kuen, J.; Ma, L.; Shahroudy, A.; Shuai, B.; Liu, T.; Wang, X.; Wang, G.; Cai, J.; et al. Recent advances in convolutional neural networks. *Pattern Recognit.* **2018**, *77*, 354–377. [CrossRef]
30. Church, K.W.; Chen, Z.; Ma, Y. Emerging trends: A gentle introduction to fine-tuning. *Nat. Lang. Eng.* **2021**, *27*, 763–778. [CrossRef]
31. Huang, J.-C.; Ko, K.-M.; Shu, M.-H.; Hsu, B.-M. Application and comparison of several machine learning algorithms and their integration models in regression problems. *Neural Comput. Appl.* **2020**, *32*, 5461–5469. [CrossRef]
32. Carvalho, D.V.; Pereira, E.M.; Cardoso, J.S. Machine Learning Interpretability: A Survey on Methods and Metrics. *Electronics* **2019**, *8*, 832. [CrossRef]

Disclaimer/Publisher's Note: The statements, opinions and data contained in all publications are solely those of the individual author(s) and contributor(s) and not of MDPI and/or the editor(s). MDPI and/or the editor(s) disclaim responsibility for any injury to people or property resulting from any ideas, methods, instructions or products referred to in the content.

Article

Analysis of Surface Topography Changes during Friction Testing in Cold Metal Forming of DC03 Steel Samples

Tomasz Trzepieciński [1], Krzysztof Szwajka [2,*] and Marek Szewczyk [2]

[1] Department of Manufacturing Processes and Production Engineering, Rzeszow University of Technology, al. Powst. Warszawy 8, 35-959 Rzeszów, Poland; tomtrz@prz.edu.pl

[2] Department of Integrated Design and Tribology Systems, Faculty of Mechanics and Technology, Rzeszow University of Technology, ul. Kwiatkowskiego 4, 37-450 Stalowa Wola, Poland; m.szewczyk@prz.edu.pl

* Correspondence: kszwajka@prz.edu.pl

Abstract: Predicting changes in the surface roughness caused by friction allows the quality of the product and the suitability of the surface for final treatments of varnishing or painting to be assessed. The results of changes in the surface roughness of DC03 steel sheets after friction testing are presented in this paper. Strip drawing tests with a flat die and forced oil pressure lubrication were carried out. The experiments were conducted under various contact pressures and lubricant pressures, and lubrication was carried out using various oils intended for deep-drawing operations. Multilayer perceptrons (MLPs) were used to find relationships between friction process parameters and other parameters (Sa, Ssk and Sku). The following statistical measures of contact force were used as inputs in MLPs: the average value of contact force, standard deviation, kurtosis and skewness. Many analyses were carried out in order to find the best network. It was found that the lubricant pressure and lubricant viscosity most significantly affected the value of the roughness parameter, Sa, of the sheet metal after the friction process. Increasing the lubricant pressure reduced the average roughness parameter (Sa). In contrast, skewness (Ssk) increased with increasing lubrication pressure. The kurtosis (Sku) of the sheet surface after the friction process was the most affected by the value of contact force and lubricant pressure.

Keywords: surface roughness; sheet metal forming; steel sheets; surface topography

1. Introduction

The factors influencing the friction in conventional sheet metal forming using stamping dies include, among others, the topography of the cooperating surfaces, the character of the movement of the tools (static or dynamic) and physico-chemical phenomena in the contact interface and forming temperature [1]. The friction phenomenon occurring in hot and warm forming conditions is more intense than in cold forming conditions, owing to the intensification of the galling phenomenon. At high temperatures, the tribological system is subjected to extreme conditions, and no lubrication can be applied [2]. Physico-chemical phenomena occurring in the contact zone also depend on the materials of the rubbing pair. Large deformation values of the surface asperities and the related increased local temperature in micro-areas cause the phenomenon of adhesion [3,4].

Surface topography is used for the quantitative characterisation of sheet metal surfaces in metal forming [5,6]. Appropriate surface topography determines the formation of oil pockets [7], which reduce friction by creating a pressure lubricant cushion and acting as a lubricant reservoir. Closed lubricant pockets [8] are of the greatest importance in reducing the coefficient of friction. The lubricant contained in these pockets is subjected to high hydrodynamic pressure [9]. The efficiency of transferring the external load through the lubricant depends primarily on the viscosity of the lubricant and the load-bearing capacity of the sheet profile determined by the Abbott–Firestone curve [10].

These contact conditions affect the change in the initial topography of the sheet metal in the sheet metal forming processes [11]. Therefore, surface roughness must be considered as a variable factor during the forming process. It should also be noted that the real contact area, A_{real}, in the micro-scale is smaller than the nominal contact area, A_{nom}; hence, the load is transferred through the deformed summits of the asperities [12].

It is necessary to understand the impact of the surface roughness parameters of the tool and sheet metal on the phenomenon of friction. The kurtosis, Sku; the skewness, Ssk; the Sa and the Sq are the parameters that are used most often to describe the surface roughness of sheets in industrial practice [13]. However, the parameters Sq and Sq are strongly correlated, so analysing only one of these two parameters is sufficient. Sedlacek et al. [14] found that in the case of friction in the lubrication regime, Sku and Ssk are the most suitable for describing the friction occurring during conventional sheet metal forming. Surfaces with more negative values of the skewness parameter and with higher kurtosis values show lower values of the friction coefficient and, consequently, change to a lesser extent during the metal forming process.

The study of changes in surface topography has been the occupation of many authors in recent years. Wu et al. [15] predicted the effect of surface roughing based on the material data of DX56 low-carbon steel sheet metal. They found that surface deformation has a negative effect on the lubrication in metal forming. Azushima and Kudo [16] observed a sheet surface in a friction test and concluded that the flattening of the surface asperities influenced the value of the friction coefficient. Han et al. [17] established a finite element-based model of surface topography based on analysing the effect of forming parameters on the surface topography of T3 pure copper sheet metal. It was concluded that the surface roughness of the sheet metal depended approximately linearly on the initial surface roughness of the workpiece. Klimczak et al. [18], based on the results of analytical studies, recommended the Sq parameter for controlling the sheet metal-formed drawpieces in production. Sigvant et al. [19] numerically investigated the stamping die surface roughness in a VDA239 CR4 GI sheet material. The increase in the blankholder surface roughness resulted in significantly higher restraining of the workpiece. Consequently, the coefficient of friction increased. Sanchez and Hartfield-Wunsch [20] analysed the evolution of surface roughness changes on mill-finished and electron discharge textured (EDTed) aluminium sheets using a draw dead simulator. Average surface roughness parameter, Sa, changes of up to 400% were observed for the mill-finished material. In the case of EDT surfaces, the maximum charge of the roughness parameter, Sa, was approximately 30%. Trzepieciński et al. [21] investigated the evolution of the surface topography due to the deformation of a DC04 steel sheet in a friction test with rounded countersamples. It was found that an increase in the plastic deformation of sheets caused an increase in the value of parameters Rp, Ra and Rt, measured both along the rolling direction of the sheet metal and across it. Zhang et al. [22] developed a finite element (FE) model of the contact of a single roughness peak and flat tool. They concluded that the height of the asperity decreased with increased local pressure, and this effect was much higher than the nominal pressure. Çavuşoğlu and Gürün [23] investigated the surface roughness effect on the formation of the EN AW-3003-H111 sheets using the FE method. The surface roughness influenced the amount of sheet thinning (influence of 56%) in the process of stretching the sheet with a hemispherical punch.

According to this literature review, most authors are focused on the experimental or numerical investigation of the effect of test parameters on the surface topography of deformed sheets. It should be emphasised that the experimental determination of the intricate relations between many friction parameters and the surface roughness of a workpiece is complex. For this reason, in this paper, artificial neural networks (ANNs) were employed to find the relation between parameters of the friction process and the main surface roughness parameters. The advantage of artificial neural networks is that they provide ability to acquire knowledge, even on a limited set of data, as a result of the training process. The mean roughness, Sa, most used in the industrial practice of sheet metal forming, was selected as the main roughness parameter. Based on the results of Sedlacek et al. [14], the skewness, Ssk, and the kurtosis, Sku, were also considered.

2. Methods and Materials

2.1. Test Material

The workpiece material was a DC03 (1.0347) steel sheet with a thickness of 1.2 mm. Mild low-carbon DC03 steel exhibits excellent deep-drawing capability. The chemical constitution of this steel is shown in Table 1. For information purposes, the mechanical properties of this sheet were determined using samples cut along the rolling direction. Uniaxial tensile tests on a universal testing machine were carried out in accordance with standard EN ISO 6892-1 [24]. The average values of the selected mechanical parameters (yield stress $R_{p0.2}$, ultimate tensile strength R_m and elongation A_{50}) based on three measurements are presented in Table 2.

Table 1. Chemical constitution of the DC03 material (wt.%).

C	P (max.)	Mn	S	Al	N	Fe
0.05	0.20	0.20	0.01	0.04	0.003	remainder

Table 2. Basic material parameters of the DC03 sheet.

$R_{p0.2}$, MPa	R_m, MPa	A_{50}, %
203.9	322.7	23.9

2.2. Experimental Methodology

In the experimental studies, the basic and the most used test to determine the value of the friction coefficient of was used. Formally, this friction test is used to determine the friction coefficient in the region of action of the blankholder in metal forming. The test was conducted using a specially designed device (Figure 1a), allowing for innovative pressure-assisted lubrication. The device was mounted in the lower holder of the universal testing machine. Lubricant was supplied to the contact zone through channels (Figure 1b) connected to an Argo-Hytos hydraulic pump (Figure 2).

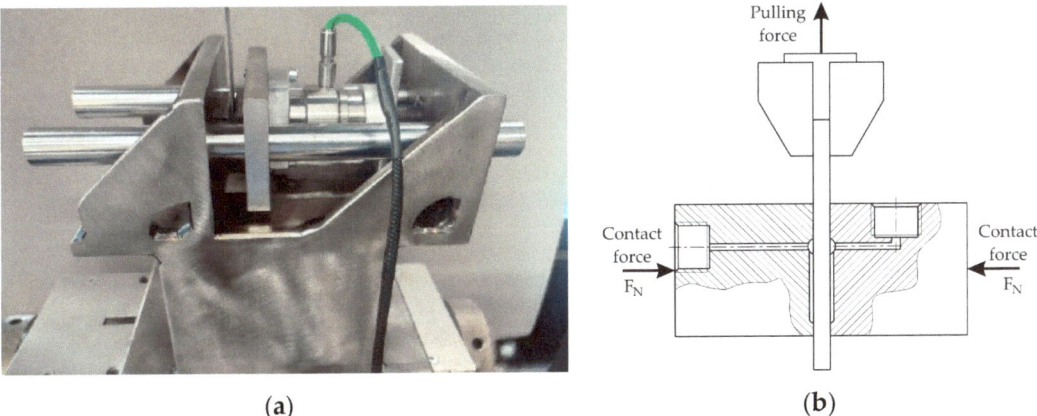

Figure 1. (a) View of the tribotester and (b) the cross-sectional view of the working countersamples.

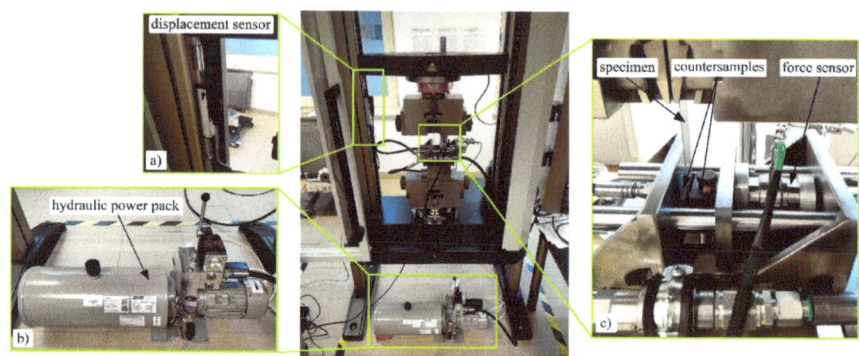

Figure 2. Experimental stand: (**a**) strain sensor, (**b**) hydraulic power pack, (**c**) tribotester [25].

The working part of the device consisted of two countersamples with a flat surface made of 145Cr6 cold work steel (hardness of 197.2 HV). The test involved pulling strips of sheet metal between countersamples pressed together with a specific force. The dimensions of the strips were as follows: width = 25 mm, and length = 130 mm. Various contact force F_N (Figure 1b) values (1000, 2000, 3000 and 4000 N) were used to obtain nominal contact pressures, p_c, of 2, 4, 6 and 8 MPa. Two synthetic oils, intended specifically for lubricating sheet metals in deep-drawing operations—S100 Plus (kinematic viscosity at 20 °C: 360 mm^2/s) and S300 (kinematic viscosity at 20 °C: 1136 mm^2/s)—both produced by Naftochem, were used in the tests. The value of the contact force was set using a clamping screw. The force sensor Kistler type 9345B was used to control the value of the contact force. Oil was delivered to the contact zone conventionally, without pressure-assisted lubrication. In comparison, the lubricants were supplied at pressure, p_l, of 0.6, 1.2 and 1.8 MPa. The upper value of lubricant pressure was selected in such a way as not to lead to oil leakage from the contact zone. So, the value of 1.8 MPa was the maximum oil pressure without leakage under the conditions of the applied contact forces between 1000 and 4000 N. The strip drawing tests were carried out at 20 °C. The effect of the temperature on the friction coefficient determined in strip drawing tests at room temperature is generally ignored, especially during tests with low pressures and sliding velocities [4,26–28]. Some publications point out the increase in the temperature of the surface asperities due to friction [26]; however, measuring the temperature in these asperities is problematic. So, the research assumed that the effect of changing the temperature in the friction interface on the coefficient of friction was negligible, and each strip was tested in stable conditions.

2.3. Surface Roughness

The main three-dimensional surface roughness parameters of the as-received specimens and specimens after friction testing were determined on a surface of 5 × 5 mm using a T8000RC profilometer. The following surface roughness parameters were determined according to standard ISO 25178-2 [29]: Sa, Sq, Ssk, Sp, Sz and Sv.

The surface topography of the workpiece is shown in Figure 3. The surfaces of the countersamples were characterised by the following parameters: Sa = 0.237 µm; Sq = 0.384 µm; Sku = 24.7; Ssk = −2.87; Sp = 4.28 µm; Sz = 10.8 µm; Sv = 6.50 µm.

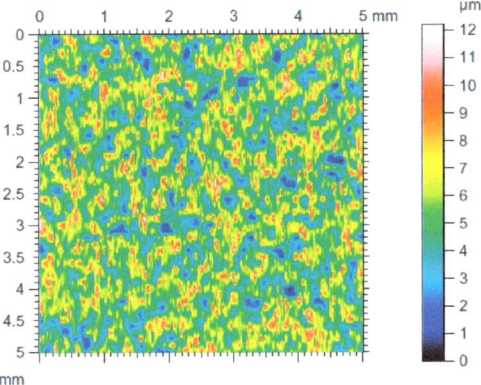

Figure 3. The topography of the DC03 steel sheet.

2.4. Determination of Measures from Contact Force Signal

For the analysis of the registered contact force (F_N) signals, our own application in the LabVIEW environment was developed for the analysis of the values of selected statistical measures of the registered contact force signals. The following statistical measures of contact force, F_N, were analysed: the average value of the contact force (F_{Nmean}), standard deviation (F_{NSD}), kurtosis (F_{Nku}) and skewness (F_{Nsk}). However, before it was possible to determine the values of the adopted statistical measures, it was necessary to select such fragments of the nominal force signal that best represented the value of the contact force, F_N.

The signal fragments from the stabilised waveform were the most suitable for such analysis. They avoided random changes in the value of the contact force signal. Before proceeding to determine the adopted measures of the force signal, the signal offset had to be removed. After removing the offset, the search for the most useful fragment of the signal began. After determining the beginning of the friction process, signal segmentation was carried out, which consisted of dividing the contact force signal (during its analysis) into equal time fragments. Then, from such segments, the average value of the analysed signal was determined. The method of evaluating the stability of the signal (fluctuation) has been presented and described in detail in [30]. Statistical measures from the contact force signal were determined from a fragment of the signal corresponding to the displacement of the sheet in the process over a length of 20 mm (Figure 4).

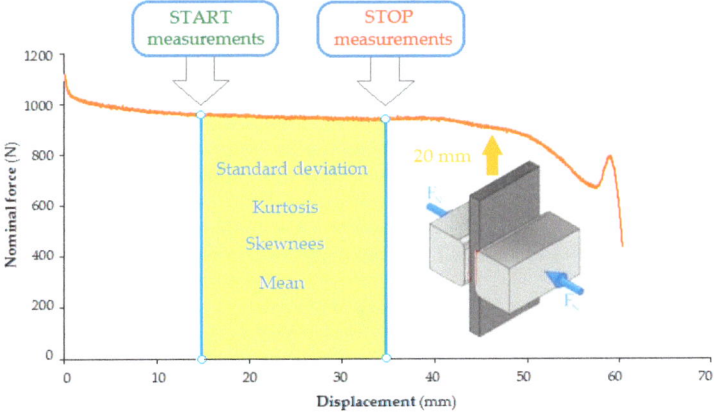

Figure 4. Method for determining statistical measures from the F_N signal.

2.5. Artificial Neural Networks

The analyses with the multilayer perceptrons (MLPs) were carried out based on the results of the friction test. As numerical input parameters, oil pressure and measures of contact force signals listed in Section 2.4 (F_{Nmean}, F_{NSD}, F_{Nku}, F_{Nsk}) were selected. Friction conditions were considered at the network input as categorisation variables: no friction, conventional lubrication (without pressure-assisted lubrication) and pressure-assisted lubrication with oil pressure p_l of 0.6, 1.2 and 1.8 MPa. At a later stage, the significance of these variables was checked by means of a sensitivity analysis. The variation in the input parameters and coefficient of friction, μ (output parameter), is presented in Table 3.

Table 3. The variation in input parameters and coefficient of friction.

Parameter	Range of Variability
F_N, kN	1–4
p_C, MPa	2–8
p_l, MPa	0–1.8
F_{Nmean}, N	960.4–4359.9
F_{NSD}, N	2.13–12.24
F_{Nku}	−1.18–0.53
F_{Nsk}	−0.41–1.01
μ	0.069–0.341
Sa, µm	0.982–1.64
Sku	2.63–3.18
Ssk	−0.0945–0.498

The training set contained 36 experimental training sets (input parameters and a corresponding output value).

Most attention was paid to the multilayer perceptrons (MLPs) and their prediction possibilities in terms of the surface roughness of the specimens after the friction process. Unidirectional MLPs are the most frequently described and the most often used neural network architectures in practical applications. Their dissemination is related to the development of the back propagation training algorithm, which enabled effective training of this type of network in a relatively simple way.

The purpose of this analysis was to find the optimal structure of the MLP and type of activation function while maintaining a good quality of response, i.e., a low number of errors. To model the considered problem using MLPs, the Statistica 13.3 program was used. The minimisation of the network structure mainly involved the size of the input vector—the number of measures and friction parameters analysed by the network. The number of hidden neurons was optimised. The input variables were oil pressure, lubricant viscosity and selected statistical measures of the contact force, F_N (mean value, standard deviation, skewness, kurtosis). On the other hand, the surface roughness parameters, Sa, Ssk and Sku, were used as the output. The input and output parameters used in the selection of the ANN architecture are presented in Figure 5.

In addition, when choosing the appropriate architecture of the ANN, the adopted type of the activation function was also considered (Figure 6).

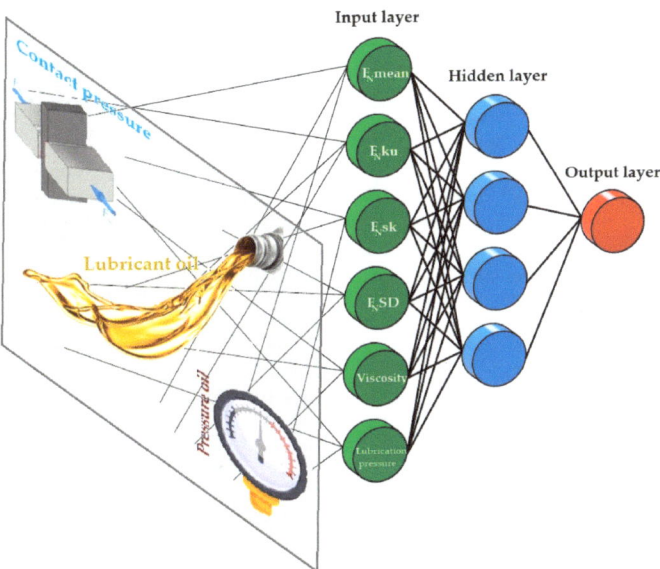

Figure 5. The structure of the ANN adopted in the analyses.

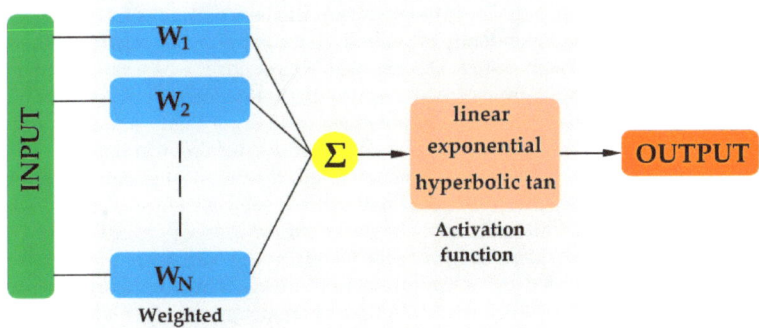

Figure 6. Structure of the MLP.

Three types of the neuron activation function were used for the analysis: linear, exponential and hyperbolic. Their general function of activation functions is to map any real input signal over a limited range, typically 0 to +1 or −1 to +1. In this study, three of the most used activation functions were selected to investigate their effect on the prediction of the surface topography of sheet metal using MLPs.

The operation of the artificial neural network when choosing a linear function consists of directly transferring the value expressing the total activation of the neuron to its output. The problem with a linear activation function is that it cannot be defined within a certain range. It has a range ($-\infty, \infty$), and this makes the ANN sometimes unable to cope with complex problems. Another limitation of ANN application is that the gradient does not depend on the inputs at all. As a result, during backpropagation, the error changes in a constant way. Regardless of the structure of MLP, the last layer always acts as a linear function of the first layer, which makes it difficult to analyse complicated problems.

The exponential and Gaussian-like activation functions are perfect for use in radial basis neural networks [31,32]. However, they are also commonly used in MLPs [33]. The function outputs, y, are from 0 to ∞. The combination of the radial aggregation function and the exponential function with a negative exponent de facto defines the neurons modelling the Gaussian function centred in relation to the weights vector [34]:

$$y = e^{-\alpha} \tag{1}$$

A hyperbolic function (Equation (2)) is an S-shaped (sigmoidal) function which is commonly used in MLPs. The hyperbolic activation function is often approximated by the mathematical function hyperbolic tangent (denoted tanh or tgh). Like the logistic (sigmoid) function, it is an S-shaped curve [34]. Such a function can work perfectly well in numerous non-linear neural networks—especially in multi-layer perceptrons [35,36].

$$y = \frac{1}{1 - e^{-\alpha}} \tag{2}$$

3. Results and Discussion

3.1. MLP Architectures

From the set of experimental data, the test and validation sets of data, whose task was to check the correct operation of the training algorithm, were separated, and the rest of the data were included in the so-called training set (used directly for teaching the network). The experimental set of data was divided into three groups: training data, validation data and test data in proportions of 70%, 15% and 15%, respectively.

Some surface roughness parameters can be correlated in the specific conditions of the surface topography; therefore, it was assumed necessary to consider independent multilayer networks with only one of the roughness parameters, Sa, Ssk or Sku, at the output. Then, the results of the best networks from each group of input parameters were independently evaluated. In the experiments regarding the search for optimal MLP models, the Statistica 13.3 package was used. It was assumed that the network designer build in Statistica software would generate a certain number of networks that would be trained based on the available data from experiments and that only the best three networks would be retained after validation of the training results.

A search was carried out for the most suitable network architecture. The predictive accuracy of neural networks is subject to random effects, and even ANNs with different architectures may have similar predictive properties due to a more favourable initial set of weights. On this basis, analyses were carried out for networks with different architectures and different numbers of neurons. Three networks were selected to predict each of the selected surface roughness parameters of sheet metal. Then, based on the training error values for the validation and training sets, the most advantageous network was selected. The architectures of three networks with the smallest errors for each of the output parameters and each neuron activation function are presented in Tables 4–12 (in the second column), and in the further columns, the values of quality parameters of the MLPs are listed. In these tables, the training, test and validation sets are abbreviated as TR, TE and V, respectively.

Table 4. Quality parameters of the MLP with a linear activation function for the prediction of the Sa parameter.

No.	MLP Architecture	Correlation			Error		
		TR	TE	V	TR	TE	V
1	MLP 10-6-1	0.848561	0.799852	0.456798	0.006260	0.005392	0.007145
2	MLP 10-9-1	0.848783	0.790019	0.457024	0.006246	0.005529	0.007192
3	MLP 10-7-1	0.848783	0.790019	0.457024	0.006246	0.005529	0.007192

Table 5. Quality parameters of the MLP with the exponential activation function for the prediction of the Sa parameter.

No.	MLP Architecture	Correlation			Error		
		TR	TE	V	TR	TE	V
1	MLP 10-5-1	0.964086	0.966114	0.962563	0.001618	0.001729	0.001357
2	MLP 10-10-1	0.969939	0.977829	0.943718	0.001385	0.000794	0.002179
3	MLP 10-4-1	0.965439	0.991373	0.971247	0.001553	0.000574	0.000816

Table 6. Quality parameters of the MLP with the tanh activation function for the prediction of the Sa parameter.

No.	MLP Architecture	Correlation			Error		
		TR	TE	V	TR	TE	V
1	MLP 10-6-1	0.961718	0.961872	0.936167	0.001681	0.001342	0.001421
2	MLP 10-11-1	0.949534	0.989222	0.931313	0.002200	0.001625	0.001367
3	MLP 10-7-1	0.980119	0.990977	0.949637	0.000880	0.001272	0.000859

Table 7. Quality parameters of the MLP with a linear activation function for the prediction of the Ssk parameter.

No.	MLP Architecture	Correlation			Error		
		TR	TE	V	TR	TE	V
1	MLP 10-8-1	0.642631	0.234950	0.754865	0.004041	0.018242	0.007486
2	MLP 10-4-1	0.657397	0.232405	0.787614	0.003797	0.017681	0.005741
3	MLP 10-6-1	0.614863	0.243348	0.756677	0.004418	0.016229	0.005429

Table 8. Quality parameters of the MLP with the exponential activation function for the prediction of the Ssk parameter.

No.	MLP Architecture	Correlation			Error		
		TR	TE	V	TR	TE	V
1	MLP 10-4-1	0.686226	0.262450	0.780858	0.003634	0.017766	0.005997
2	MLP 10-3-1	0.681380	0.255007	0.783729	0.003593	0.017347	0.005152
3	MLP 10-6-1	0.655258	0.298071	0.844212	0.003844	0.016428	0.004866

Table 9. Quality parameters of the MLP with the tanh activation function for the prediction of the Ssk parameter.

No.	MLP Architecture	Correlation			Error		
		TR	TE	V	TR	TE	V
1	MLP 10-7-1	0.632725	0.233232	0.869024	0.003965	0.018361	0.006852
2	MLP 10-8-1	0.613228	0.407401	0.847233	0.004429	0.016867	0.005412
3	MLP 10-9-1	0.659507	0.268853	0.818048	0.003788	0.017803	0.005567

Table 10. Quality parameters of the MLP with a linear activation function for the prediction of the Sku parameter.

No.	MLP Architecture	Correlation			Error		
		TR	TE	V	TR	TE	V
1	MLP 10-8-1	0.891092	0.925749	0.392733	0.002065	0.004704	0.008775
2	MLP 10-3-1	0.891206	0.920480	0.409767	0.002062	0.004854	0.008590
3	MLP 10-4-1	0.889840	0.922791	0.489860	0.002144	0.004079	0.007595

Table 11. Quality parameters of the MLP with the exponential activation function for the prediction of the Sku parameter.

No.	MLP Architecture	Correlation			Error		
		TR	TE	V	TR	TE	V
1	MLP 10-4-1	0.983367	0.940394	0.924044	0.000331	0.003057	0.002618
2	MLP 10-3-1	0.967260	0.902881	0.923602	0.000650	0.003811	0.003660
3	MLP 10-11-1	0.990302	0.969133	0.955919	0.000196	0.001395	0.001561

Table 12. Quality parameters of the MLP with the tanh activation function for the prediction of the Sku parameter.

No.	MLP Architecture	Correlation			Error		
		TR	TE	V	TR	TE	V
1	MLP 10-9-1	0.993754	0.954898	0.837398	0.000129	0.003025	0.002892
2	MLP 10-11-1	0.990686	0.952943	0.839915	0.000195	0.003630	0.003490
3	MLP 10-5-1	0.976751	0.951792	0.880115	0.000470	0.003789	0.002757

3.2. MLP for Prediction of Roughness Parameter Sa

Considering the smallest training errors for both the training and validation sets, the MLP 10-7-1 network (Table 6) was selected for further analysis. For this network, the values of Pearson's correlation R (Equation (3)) for the training, test and validation sets were approximately 0.98, 0.99, and 0.94, respectively. The network training process was carried out until there was no further decrease in the network error value for the test set. Further continuation of the learning process, even with a continuous decrease in error for the test set, led to the excessive correlation of the network response to the training data. The ability of the MLP network to generalise was assessed on the data contained in the test set, the data of which were not used in the learning process. The training process was terminated after 32 epochs (Figure 7).

$$R = \frac{\sum_{i=1}^{n} \left(x_i - \frac{1}{n}\sum_{i=1}^{n} x_i\right)\left(y_i - \frac{1}{n}\sum_{i=1}^{n} y_i\right)}{\sqrt{\sum_{i=1}^{n}\left(x_i - \frac{1}{n}\sum_{i=1}^{n} x_i\right)^2}\sqrt{\sum_{i=1}^{n}\left(y_i - \frac{1}{n}\sum_{i=1}^{n} y_i\right)^2}} \quad (3)$$

where n is the sample size, and x_i and y_i are individual sample points indexed with 'i'.

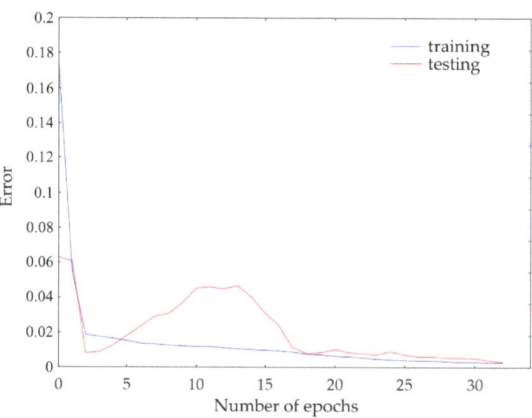

Figure 7. Variation in the network error during the training process.

Variable sensitivity analysis makes it possible to assess the significance of the influence of a specific variable on the value predicted by the output neuron. The basic measure of network sensitivity is the quotient of the error obtained for a data set that does not contain a specific variable and the error obtained with a complete set of variables [34]. The global sensitivity analysis for selected variables in the training set is as follows: oil pressure 57.811, oil viscosity 7.620, F_{NSD} 4.377, F_{Nku} 2.978, F_{Nmean} 2.099 and F_{Nsk} 1.277 (Table 13).

Table 13. Sensitivity analysis of input variables for MLP 10-7-1.

Oil Pressure	Oil Viscosity	F_{NSD}	F_{Nku}	F_{Nmean}	F_{Nsk}
57.811	7.620	4.377	2.978	2.099	1.277

If the error ratio presented in Table 13 is equal to or greater than 1, then this variable has a significant impact on the mean roughness parameter (Sa). The variables that affect the value of the Sa roughness parameter most are oil viscosity and its pressure. Subsequently, the surface roughness parameter, Sa, after the friction process is influenced by statistical measures of force, F_N: standard deviation (F_{NSD}), kurtosis (F_{Nku}), mean value of contact force (F_{Nmean}) and skewness (F_{Nsk}).

The effect of oil viscosity and oil pressure on the value of the Sa parameter should be considered together. The viscosity of the oil determines the durability of the lubricant film and its adhesion to the surfaces of the sheet and the countersample. When lubricating with higher-viscosity oil, higher pressures can be applied in a strip drawing test without the risk of lubricant leaks. In turn, low-viscosity oil is characterized by low flow resistance, and it is easier to break the lubricating film, even at low pressures. Under conditions of increasing contact pressure, the influence of lubricant pressure on changing the surface topography decreases. However, the mechanical cooperation of the surface asperities through flattening and ploughing mechanisms begins to play a dominant role.

Figures 8–11 show a comparison of measured data and responses of the MLP 10-7-1 network for selected configurations of input parameters. The response of the neural network (Figures 8b, 9b, 10b and 11b) perfectly reflects the results of experimental studies (Figures 8a, 9a, 10a and 11a). Under lubricated conditions, if the lubricant pressure increases, the Sa value decreases. This generally applies to negative values of the kurtosis of contact force. For positive values of kurtosis, the Sa value decreases with increasing lubricant pressure. Higher lubricant pressure ensures better penetration of the cavities located in the roughness valleys and, in the case of closed lubricant pockets, an additional reduction in the metallic interaction of the surface roughness asperities.

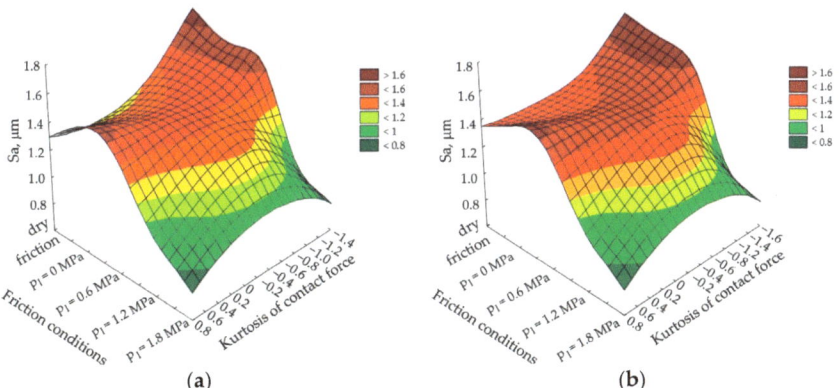

Figure 8. The effect of kurtosis of contact force and oil pressure on the value of Sa parameter: (a) measured data, (b) response surface for predicted data.

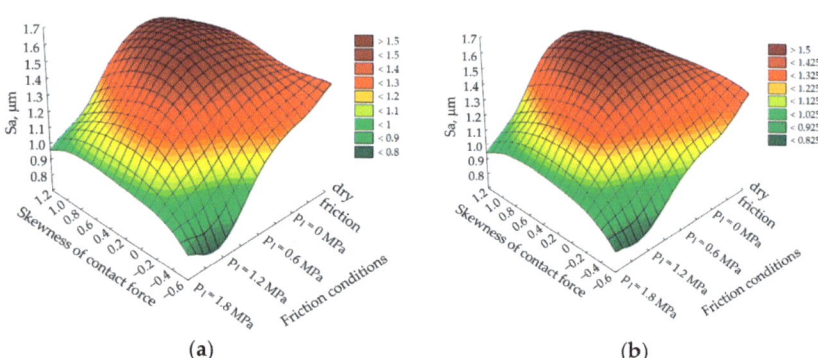

Figure 9. The effect of skewness of contact force and oil pressure on the value of Sa parameter: (**a**) measured data, (**b**) response surface for predicted data.

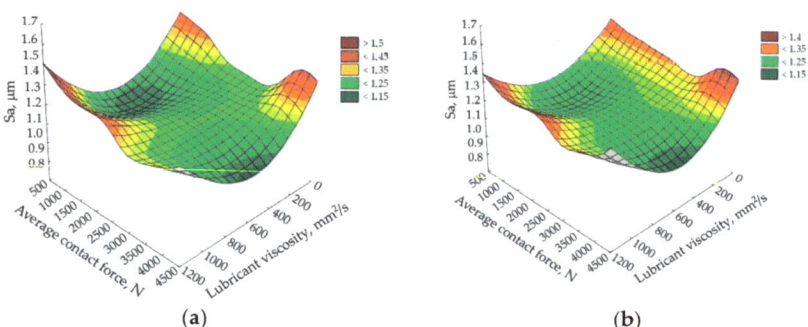

Figure 10. The effect of average contact force and oil viscosity on the value of the Sa parameter: (**a**) measured data, (**b**) response surface for predicted data.

The greatest increase in the value of the Sa parameter is observed for dry friction conditions and during conventional lubrication ($p_l = 0$ MPa). During lubrication with a lubricant pressure of $p_l = 1.8$ MPa, the smallest difference in the Sa parameter of the sheet metal after friction is observed in relation to the as-received sheet metal (Sa = 1.40 µm). Under these conditions, the mean surface roughness Sa increased by approximately 29%–43% (Figure 8). However, in dry friction conditions, the mean surface roughness varied by approximately 8%–21% depending on the kurtosis of contact force. In conventional lubrication conditions ($p_l = 0$ MPa), the mean roughness increased by approximately 2%–14%, depending on the kurtosis of contact force. So, the pressurised lubrication ensures lower average roughness of the workpiece surface after friction compared to the conventional lubrication. In [37], it was observed that friction under pressure lubrication conditions reduces the mean roughness of DC01 steel sheets by approximately 32%. Similarly, in [38], it was found that the Sa parameter of DC04 steel sheets increases with a reduction in normal pressure by approximately 12%–47%, in the range of contact pressures between 3 and 12 MPa. Most published works indicate a reduction in the surface roughness of sheet metal as a result of the surface flattening of asperities [39,40], or an increase in it as a result of the ploughing mechanism [41].

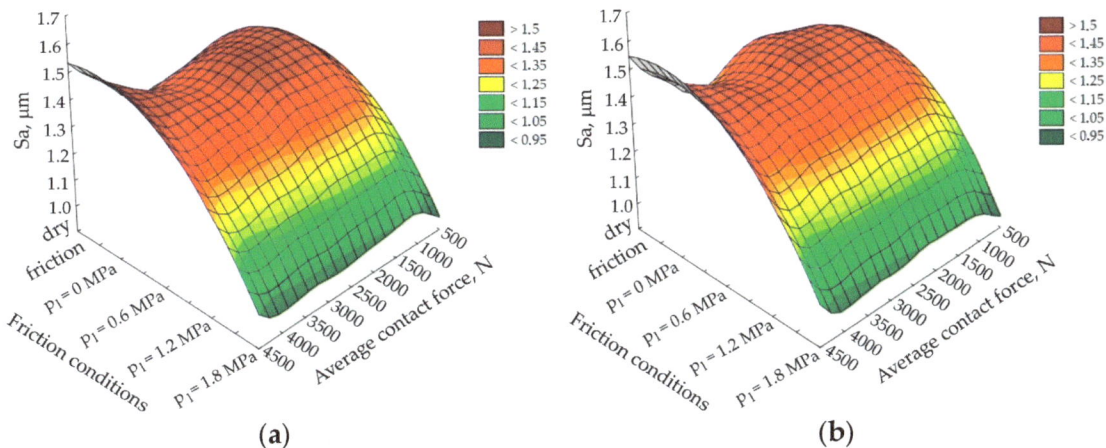

Figure 11. The effect of kurtosis of contact force and lubricant pressure on the value of the Sa parameter: (**a**) measured data, (**b**) response surface for predicted data.

The influence of the skewness of contact force and lubricant pressure on the roughness parameter, Sa, is similar (Figure 9). Increasing the oil pressure reduces the surface roughness of the sheet metal defined by the Sa parameter.

The response surface of the neural network for the influence of the average contact force and lubricant viscosity is represented by two troughs with the minimum values of the Sa parameter (Figure 10). The most unfavourable friction conditions due to the minimisation of the Sa roughness parameter occur at the highest average contact force and in the absence of lubrication (lubricant viscosity of 0 mm^2/s).

In the range of the average contact force between about 1500 and 3500 N, a stabilisation of the Sa roughness parameter value for specific lubrication conditions is visible (Figure 11). Irrespective of the value of the oil pressure, first, the Sa parameter increases; then, its stabilisation occurs in a wide range of changes described above, and finally, after exceeding the value of the average contact force of 3500 N, the Sa parameter increases sharply. This may be justified by the intensification of flattening and ploughing mechanisms under conditions of high contact forces. Then, at high contact pressures, the share of mechanical cooperation of the surface asperities in the total resistance to friction increases.

A scatter plot between observed values and predicted values of roughness parameter Sa is shown in Figure 12. The values present normal distribution, and they are proportionally distributed along the diagonal line. In a normal distribution, the frequency of occurrence of events with the average value of the examined feature is therefore the highest, or, in other words, the probability of such an event occurring is the highest. The frequency of an event occurring (the probability of occurrence) decreases according to the increase in the deviation of the random variable from its expected value.

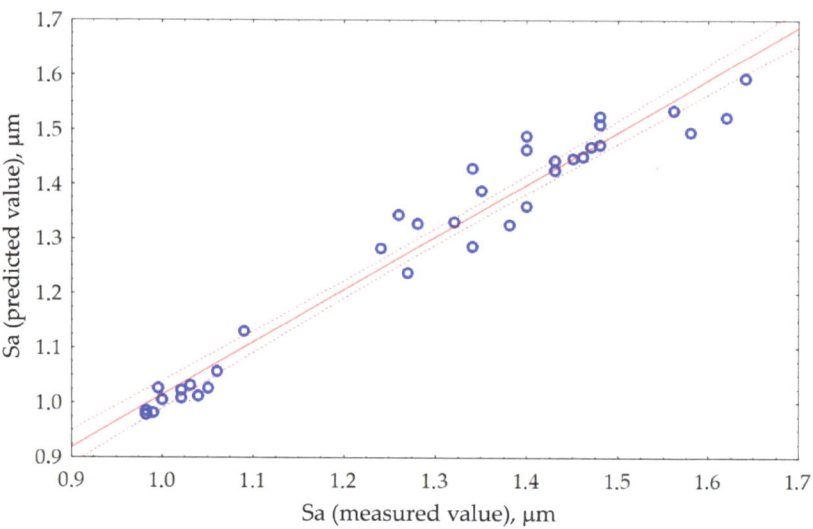

Figure 12. Scatter plot between observed values and predicted values of roughness parameter Sa using MLP 10-7-1.

3.3. MLP for Prediction of Skewness, Ssk

The MLP 10-6-1 neural network (Table 8) predicting the value of skewness, Ssk, of the sheet surface is characterised by worse Pearson's coefficients compared to the network predicting the value of the roughness parameter, Sa. The Pearson's coefficient values for the training, test and validation sets are 0.655, 0.298 and 0.844, respectively. The value of the error ratio for individual input variables (Table 14) is much more even compared to the network predicting the value of the Sa parameter (Table 13). However, none of the quotient values is lower than 1. Oil pressure, average value of contact force and oil viscosity have the greatest informative capacity.

Table 14. Sensitivity analysis of input variables for MLP 10-6-1.

Oil Pressure	F_{Nmean}	Oil Viscosity	F_{NSD}	F_{Nsk}	F_{Nku}
1.402	1.180	1.112	1.015	1.008	1.000

The experimental results and the response surfaces of the MLP 10-6-1 network of the influence of the kurtosis of contact force and the lubrication pressure on the value of the skewness, Ssk, are shown in Figure 13a,b, respectively. The prediction of the neural network is a flattening of the experimental curve. Due to the learning algorithm being stopped after the minimum error value for the test set is reached, the neural network reaches the optimal quality of data generalisation. Skewness, Ssk, increases with increasing lubrication pressure. Most of the sheets tested experimentally are characterised by a positive skewness, Ssk, value (Figure 13). Intense friction leads to rough interaction between the surface asperities, because of which, the surface is flattened.

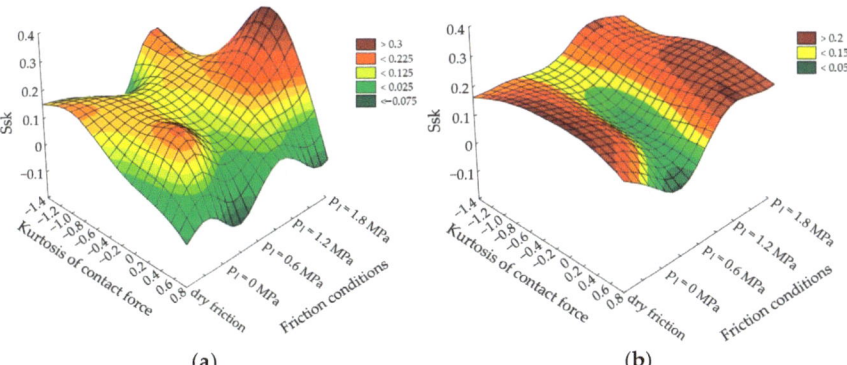

Figure 13. The effect of kurtosis of contact force and oil pressure on the value of the Ssk parameter: (**a**) measured data, (**b**) response surface for predicted data.

The influence of oil pressure and lubricant viscosity on the value of the roughness parameter, Ssk, is represented by the saddle surface. For dry friction (kinematic viscosity 0 mm^2/s), the skewness, Ssk, value shows negative values. It is local in nature. So, the neural network model predicts a saddle surface in a more flattened form, not considering local disturbances of experimental data. In the absence of lubrication, the Ssk value takes relatively large values above 0.3 (Figure 14). Increasing the lubricant pressure under pressure-assisted lubrication slightly increases the skewness value, but this effect is valid for oil with a viscosity of about 600 mm^2/s, corresponding to the saddle point position. A saddle point is a point on a surface with the property that, in any of its surroundings, there are points lying on both sides of the tangent.

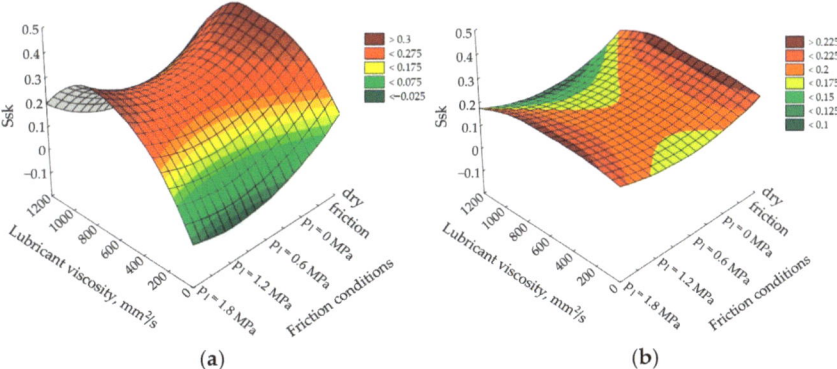

Figure 14. The effect of oil viscosity and lubricant pressure on the value of the Ssk parameter: (**a**) measured data, (**b**) response surface for predicted data.

In general, the shape of the predicted response surface for the network predicting the skewness, Ssk, value deviates much more from the experimental surfaces than the network predicting the value of average roughness Sa (Section 3.2). The reasons should be grounded in different values of the activation function of neurons. The function outputs, y, of the MLP analysed in this section are exponential, and they are in the range of 0 to ∞. The optimal MLP for modelling the average roughness, Sa, contains neurons with hyperbolic activation function with the output values in the range between −1 and +1. So, owing the possibility of including both negative and positive values of the output, the MLP 10-7-1 (Sa) network tends to better match the experimental data than the MLP 10-6-1 (Ssk).

3.4. MLP for Prediction of Kurtosis, Sku

The architecture of the ANN that provides the smallest error values for the training and test sets is the MLP 10-9-1 network containing neurons with the tanh activation function (Table 12). For the training set, the Pearson's correlation is 0.993, and for the test and validation sets, 0.954 and 0.837, respectively. The average value of contact force and oil pressure at almost the same level affects the value of kurtosis, Sku, of the sheet surface. In order of decreasing importance, kurtosis of contact force, standard deviation of contact force, oil viscosity and skewness of contact force were ranked further (Table 15).

Table 15. Sensitivity analysis of input variables for MLP 10-9-1.

F_{Nmean}	Oil Pressure	F_{Nku}	F_{NSD}	Oil Viscosity	F_{Nsk}
13.215	12.277	3.045	2.526	2.152	1.702

The influence of oil viscosity and lubrication pressure on the value of the kurtosis, Sku, is given by a typical saddle curve (Figure 15). In the absence of lubrication and when lubricating with oil with a viscosity of 1.136 mm^2/s, the skewness values are below 3. Under these conditions, the distribution curve is platykurtoic with relatively low valleys and few high peaks. In the second case (Sku > 3), the distribution curve of the profile is leptokurtoic and is characterised by relatively low valleys and many high peaks [42]. The roughness parameter, Sku, increases to values above 3 with the increase in the lubricant pressure. This is a result of the formation of a lubricant 'cushion' and the separation of rubbing surfaces.

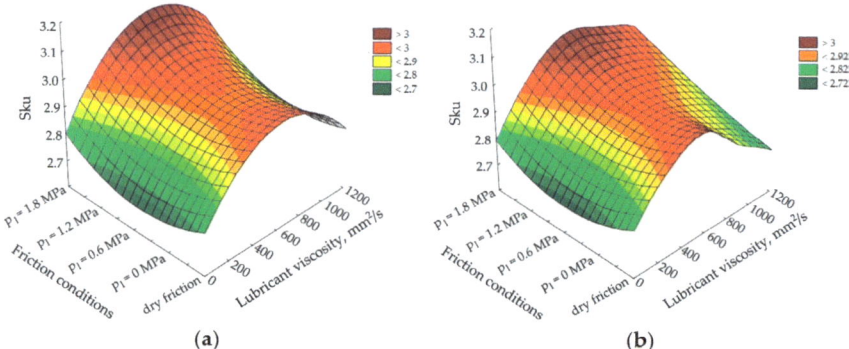

Figure 15. The effect of lubricant pressure and kurtosis of normal force on the value of the Ssk parameter: (**a**) measured data, (**b**) response surface for predicted data.

Increasing the average signal of contact force clearly reduces the value of kurtosis, Sku (Figure 16). This effect is observed for the entire range of oil viscosity changes. Under dry friction conditions (viscosity 0 mm^2/s), the distribution curve of the surface profile is platykurtoic with high peaks or deep scratches, which are formed as a result of the mechanical action of the summits of asperities under lubricant-free conditions.

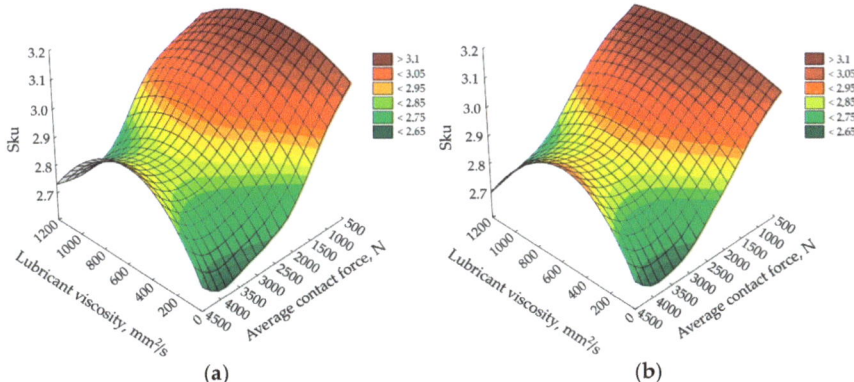

Figure 16. The effect of kurtosis of normal force and oil pressure on the value of the Sku parameter: (**a**) measured data, (**b**) response surface for predicted data.

Changing the friction conditions from dry friction to oil lubrication leads to an increase in the kurtosis, Sku, value (Figure 17). Similarly, under specific friction conditions, an increase in the average signal of contact force causes a decrease in the Sku parameter. As mentioned earlier, the lubrication of the contact interface, under pressure-assisted lubrication conditions, limits the mechanical interaction of the friction surfaces, specifically the interaction of the hard tool with the relatively soft sheet metal. An additional mechanism that is activated under high contact forces is the phenomenon of strain hardening. This phenomenon consists of an increase in the yield stresses that plasticise the material of the surface asperities along with the increasing plastic deformation of the sheet material.

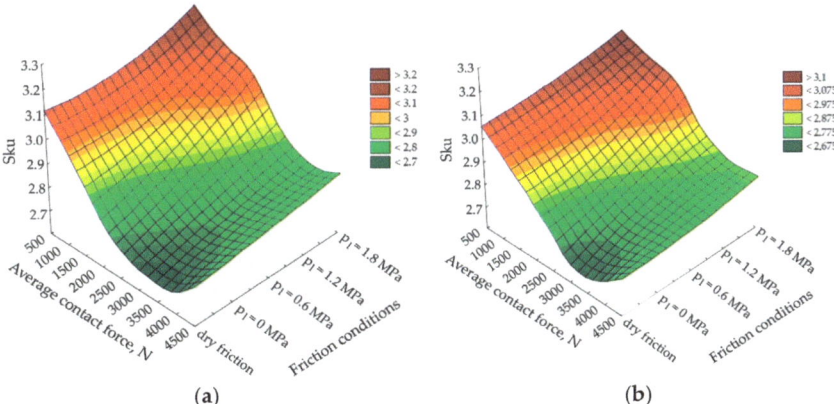

Figure 17. The effect of lubricant pressure and kurtosis of normal force on the value of the Sku parameter: (**a**) measured data, (**b**) response surface for predicted data.

The influence of lubricant viscosity on the coefficient of friction is closely related to the topography of the sheet metal surface [43]. The application of oil with high viscosity on the surface leads to the decrease in the coefficient of friction [44]. Lee et al. [45] found that high-viscosity lubricants provided lower friction coefficient values. Increasing the contact pressure results in intensified flattening of the summits of the surface asperities [39], and as a result, the value of the Sku parameter decreases with the increase in contact force. The increase in the pressure of the oil contained in closed lubricant pockets limits the mechanical interaction of the surface asperities and reduces friction by creating a pressure lubricant cushion [8]. The largest change in the Sku parameter occurs in conditions of a lack of pressurised lubrication and dry friction [38].

Detailed prediction statistics for the MLP 10-7-1 networks used for the prediction of roughness parameters Sa, Ssk and Sku are listed in Table A1 (Appendix A).

Since a neural network is nothing more than a multidimensional approximation mechanism, its use for this task is probably redundant. The surfaces obtained in the experiment (Figures 8a, 9a, 10a, 11a, 13a, 14a, 15a, 16a and 17a) and as a result of the prediction of the ANN (Figures 8b, 9b, 10b, 11b, 13b, 14b, 15b, 16b and 17b) are quite smooth, and it is most likely that the usual approximation would give a similar result.

4. Conclusions

The MLPs were used to explore the effect of friction process parameters on the value of selected parameters of the surface roughness (Sa, Ssk and Sku). The main results are as follows:

- The lubricant pressure and lubricant viscosity are variables that most significantly affect the value of the roughness parameter, Sa, of sheet metal after the friction process.
- Under lubricated conditions, if the lubricant pressure increases, the Sa value decreases. This generally applies to negative values of the kurtosis of contact force.
- The most unfavourable friction conditions due to the minimisation of the Sa roughness parameter occur at the highest average contact force and in dry friction conditions.
- Oil pressure and oil viscosity have the greatest impact on the skewness, Ssk, of the sheet metal surface after the friction process.
- Skewness, Ssk, increases with increasing lubrication pressure. After the friction process, most samples are characterised by a positive skewness, Ssk, value.
- The average value of contact force and oil pressure at almost the same level affects the kurtosis, Sku, value of the sheet surface.
- Changing the friction conditions from dry friction to oil lubrication leads to an increase in the kurtosis, Sku, value.

Author Contributions: Conceptualization, T.T., K.S. and M.S.; methodology, T.T., K.S. and M.S.; validation, T.T., K.S. and M.S.; investigation, T.T., K.S. and M.S.; data curation, T.T., K.S. and M.S.; writing—original draft preparation, T.T. and K.S.; writing—review and editing, T.T. All authors have read and agreed to the published version of the manuscript.

Funding: This research received no external funding.

Institutional Review Board Statement: Not applicable.

Informed Consent Statement: Not applicable.

Data Availability Statement: Data is contained within the article.

Conflicts of Interest: The authors declare no conflict of interest.

Appendix A

Table A1. Detailed qualitative statistics of MLP networks used to predict Sa, Ssk and Sku roughness parameters.

Parameter	Set	MLP 10-7-1 (Sa)	MLP 10-6-1 (Ssk)	MLP 10-9-1 (Sku)
	train	0.97985	−0.0136	2.64590
Minimum predicted value	test	0.98254	0.14685	2.69304
	validation	1.28784	0.10030	2.82405

Table A1. *Cont.*

Parameter	Set	MLP 10-7-1 (Sa)	MLP 10-6-1 (Ssk)	MLP 10-9-1 (Sku)
Maximum predicted value	train	1.53516	0.26168	3.14747
	test	1.46391	0.25249	3.08723
	validation	1.59470	0.20856	3.10607
Minimum value of residuals	train	−0.08946	−0.14348	−0.03739
	test	−0.08372	−0.18557	−0.08319
	validation	−0.04739	−0.14210	−0.13022
Maximum value of residuals	train	0.09547	0.16149	0.05049
	test	0.02304	0.30841	0.10875
	validation	0.05216	0.13344	0.06119
Minimum value of standardized residuals	train	−3.01634	−2.31434	−3.28748
	test	−2.34724	−1.44777	−1.51255
	validation	−1.61685	−2.03700	−2.42133
Maximum value of standardized residuals	train	3.21883	2.60478	4.43908
	test	0.064595	2.40618	1.97730
	validation	1.77976	1.91295	1.13772

References

1. Seshacharyulu, K.; Bandhavi, C.; Naik, B.B.; Rao, S.S.; Singh, S.K. Understanding friction in sheet metal forming—A review. *Mater. Today Proc.* **2018**, *5*, 18238–18244. [CrossRef]
2. Kooistra, E. Prediction and validation of galling behavior in hot sheet metal forming processes. Master's Thesis, University of Twente, Enschede, The Netherlands, March 2021.
3. Tröber, P.; Welm, M.; Weiss, H.A.; Demmel, P.; Golle, R.; Volk, W. Temperature, thermoelectric current and adhesion formation during deep drawing. *Wear* **2021**, *477*, 203839. [CrossRef]
4. Szewczyk, M.; Szwajka, K. Assessment of the tribological performance of bio-based lubricants using analysis of variance. *Adv. Mech. Mater. Eng.* **2023**, *40*, 31–38. [CrossRef]
5. Podulka, P. Detection of measurement noise in surface topography analysis. *J. Phys. Conf. Ser.* **2021**, *1736*, 012014. [CrossRef]
6. Podulka, P.; Macek, W.; Branco, R.; Nejad, R.M. Reduction in errors in roughness evaluation with an accurate definition of the SL surface. *Materials* **2023**, *16*, 1865. [CrossRef]
7. Podulka, P. Problem of selection of reference plane with deep and wide valleys analysis. *J. Phys. Conf. Ser.* **2018**, *1065*, 072017. [CrossRef]
8. Azushima, A.; Tanaka, T. Lubricant behavior trapped within pockets on workpiece surface in lubricated upsetting by means of direct fluorescence observartion technique. *CIRP Ann.* **2000**, *49*, 165–168. [CrossRef]
9. Weidel, S.; Engel, U. Surface characterisation in forming processes by functional 3D parameters. *Int. J. Adv. Manuf. Technol.* **2007**, *33*, 130–136. [CrossRef]
10. Abbott, E.J.; Firestone, F.A. Specifying surface quality: A method based on accurate measurement and comparison. *Mech. Eng.* **1933**, *55*, 569–572.
11. Kumar, A.; Gulati, V.; Kumar, P. Investigation of Surface Roughness in Incremental Sheet Forming. *Procedia Comput. Sci.* **2018**, *133*, 1014–1020. [CrossRef]
12. Shisode, M.; Hazrati, J.; Mishra, T.; de Rooij, M.; ten Horn, C.; van Beck, J.; van den Boogaard, T. Modeling boundary friction of coated sheets in sheet metal forming. *Tribol. Int.* **2021**, *153*, 106554. [CrossRef]
13. Sedlaček, M.; Podgornik, B.; Vižintin, J. Influence of surface preparation on roughness parameters, friction and wear. *Wear* **2009**, *266*, 482–487. [CrossRef]
14. Sedlaček, M.; Vilhena, L.M.S.; Podgornik, B.; Vižintin, J. Surface topography modelling for reduced friction. *Stroj. Vestn. J. Mech. Engineering* **2011**, *57*, 674–680. [CrossRef]
15. Wu, Y.; Recklin, V.; Groche, P. Strain induced surface change in sheet metal forming: Numerical prediction, influence on friction and tool wear. *J. Manuf. Mater. Process.* **2021**, *5*, 29. [CrossRef]

16. Azushima, A.; Kudo, H. Direct Observation of Contact Behaviour to Interpret the Pressure Dependence of the Coefficient of Friction in Sheet Metal Forming. *CIRP Ann.* **1995**, *44*, 209–212. [CrossRef]
17. Juanjuan, H.; Jie, Z.; Wei, Z.; Guangchun, W. Influence of metal forming parameters on surface roughness and establishment of surface roughness prediction model. *Int. J. Mech. Sci.* **2019**, *163*, 105093.
18. Klimczak, T.; Dautzenberg, J.H.; Kals, J.A.G. On the Roughening of a Free Surface During Sheet Metal Forming. *CIRP Ann.* **1988**, *37*, 267–270. [CrossRef]
19. Sigvant, M.; Pilthammar, J.; Hol, J.; Wiebenga, J.H.; Chezan, T.; Carleer, B.; van den Boogaard, T. Friction in sheet metal forming: Influence of surface roughness and strain rate on sheet metal forming simulation results. *Procedia Manuf.* **2019**, *29*, 512–519. [CrossRef]
20. Sanchez, L.R.; Hartfield-Wunsch, S. Effects on Surface Roughness and Friction on Aluminum Sheet under Plain Strain Cyclic Bending and Tension. *SAE Int. J. Mater. Manuf.* **2011**, *4*, 826–834. [CrossRef]
21. Trzepieciński, T.; Bochnowski, W.; Witek, L. Variation of surface roughness, micro-hardness and friction behaviour during sheet-metal forming. *Int. J. Surf. Sci. Eng.* **2018**, *12*, 119–136. [CrossRef]
22. Zhang, S.; Hodgson, P.D.; Duncan, J.L.; Cardew-Hall, M.J.; Kalyanasundaram, S. Effect of membrane stress on surface roughness changes in sheet forming. *Wear* **2002**, *253*, 610–617. [CrossRef]
23. Çavuşoğlu, O.; Gürün, H. Statistical evaluation of the influence of temperature and surface roughness on aluminium sheet metal forming. *Trans. Famena* **2017**, *41*, 57–64. [CrossRef]
24. EN ISO 6892-1; Metallic Materials—Tensile Testing—Part 1: Method of Test at Room Temperature. International Organization for Standardization: Geneva, Switzerland, 2019.
25. Trzepieciński, T.; Szwajka, K.; Szewczyk, M. Pressure-assisted lubrication of DC01 steel sheets to reduce friction in sheet-metal-forming processes. *Lubricants* **2023**, *11*, 169. [CrossRef]
26. Makhkamov, A. Determination of the friction coefficient in the flat strip drawing test. *Engineering* **2021**, *13*, 595–604. [CrossRef]
27. Evin, E.; Daneshjo, N.; Mareš, A.; Tomáš, M.; Petrovčiková, K. Experimental assessment of friction coefficient in deep drawing and its verification by numerical simulation. *Appl. Sci.* **2021**, *11*, 2756. [CrossRef]
28. Vollertsen, F.; Hu, Z. Tribological size effects in sheet metal forming measured by a strip drawing test. *CIRP Ann.* **2006**, *55*, 291 294. [CrossRef]
29. ISO 25178-2; Geometrical Product Specifications (GPS)—Surface Texture: Areal—Part 2: Terms, Definitions and Surface Texture Parameters. International Organization for Standardization: Geneva, Switzerland, 2012.
30. Szwajka, K.; Trzepieciński, T. On the machinability of medium density fiberboard by drilling. *Bioresources* **2018**, *13*, 8263–8278. [CrossRef]
31. Shymkovych, V.; Telenyk, S.; Kravets, P. Hardware implementation of radial-basis neural networks with Gaussian activation functions on FPGA. *Neural Comput. Appl.* **2021**, *33*, 9467–9479. [CrossRef]
32. Savitha, R.; Suresh, S.; Sundararajan, N. A fully complex-valued radial basis function network and its learning algorithm. *Int. J. Neural Syst.* **2009**, *19*, 253–267. [CrossRef]
33. Chen, J.C.; Wang, Y.M. Comparing activation functions in modeling shoreline variation using multilayer perceptron neural network. *Water* **2020**, *12*, 1281. [CrossRef]
34. StatSoft Electronic Statistics Texbook. Copyright StatSoft, Inc., 1984–2011. Available online: https://www.statsoft.pl/textbook/stathome.html (accessed on 5 August 2023).
35. Zamanlooy, B.; Mirhassani, M. Efficient VLSI implementation of neural networks with hyperbolic tangent activation function. *IEEE Trans. Very Large Scale Integr. Syst.* **2014**, *22*, 39–48. [CrossRef]
36. Mahima, R.; Maheswari, M.; Roshana, S.; Priyanka, E.; Mohanan, N.; Nandhini, N. A comparative analysis of the most commonly used activation functions in deep neural network. In Proceedings of the 4th International Conference on Electronics and Sustainable Communication Systems (ICESC), Coimbatore, India, 6–8 July 2023; pp. 1334–1339.
37. Trzepieciński, T.; Luiz, V.D.; Szwajka, K.; Szewczyk, M.; Szpunar, M. Analysis of the lubrication performance of low-carbon steel sheets in the presence of pressurised lubricant. *Adv. Mater. Sci.* **2023**, *23*, 64–76. [CrossRef]
38. Szewczyk, M.; Szwajka, K.; Trzepieciński, T. Frictional characteristics of deep-drawing quality steel sheets in the flat die strip drawing test. *Materials* **2022**, *15*, 5236. [CrossRef]
39. Gierzyńska, M. *Friction, Wear and Lubrication in Metal Forming*; Wydawnictwa Naukowo-Techniczne: Warsaw, Poland, 1983.
40. Zwicker, M.; Bay, N.; Nielsen, C.V. A discussion of model asperities as a method to study friction in metal forming. *Discov. Mech. Eng.* **2023**, *2*, 3. [CrossRef]
41. Trzepieciński, T.; Fejkiel, R. On the influence of deformation of deep drawing quality steel sheet on surface topography and friction. *Tribol. Int.* **2017**, *115*, 78–88. [CrossRef]
42. Gadelmawla, E.S.; Koura, M.M.; Maksoud, T.M.A.; Elewa, I.M.; Soliman, H.H. Roughness parameters. *J. Mater. Process. Technol.* **2002**, *123*, 133–145. [CrossRef]
43. Galdos, L.; Trinidad, J.; Otegi, N.; Garcia, C. Friction Modelling for Tube Hydroforming Processes—A Numerical and Experimental Study with Different Viscosity Lubricants. *Materials* **2022**, *15*, 5655. [CrossRef]

44. Şen, N.; Şirin, Ş.; Kıvak, T.; Civek, T.; Seçgin, Ö. A new lubrication approach in the SPIF process: Evaluation of the applicability and tribological performance of MQL. *Tribol. Int.* **2022**, *171*, 107546. [CrossRef]
45. Lee, B.H.; Keum, Y.T.; Wagoner, R.H. Modeling of the friction caused by lubrication and surface roughness in sheet metal forming. *J. Mater. Process. Technol.* **2002**, *130–131*, 60–63. [CrossRef]

Disclaimer/Publisher's Note: The statements, opinions and data contained in all publications are solely those of the individual author(s) and contributor(s) and not of MDPI and/or the editor(s). MDPI and/or the editor(s) disclaim responsibility for any injury to people or property resulting from any ideas, methods, instructions or products referred to in the content.

Article

Terahertz Non-Destructive Testing of Porosity in Multi-Layer Thermal Barrier Coatings Based on Small-Sample Data

Dongdong Ye [1], Zhou Xu [2,*], Houli Liu [3,4], Zhijun Zhang [3,5], Peiyong Wang [6], Yiwen Wu [7,*] and Changdong Yin [2,*]

[1] Huzhou Key Laboratory of Terahertz Integrated Circuits and Systems, Yangtze Delta Region Institute (Huzhou), University of Electronic Science and Technology of China, Huzhou 313001, China; ddyecust@ahpu.edu.cn
[2] School of Electrical and Automation, Wuhu Institute of Technology, Wuhu 241006, China
[3] School of Artificial Intelligence, Anhui Polytechnic University, Wuhu 241000, China; hlliu@ahpu.edu.cn (H.L.); 2230121103@stu.ahpu.edu.cn (Z.Z.)
[4] Anhui Key Laboratory of Mine Intelligent Equipment and Technology, Anhui University of Science & Technology, Huainan 232001, China
[5] Tianjin Key Laboratory of Optoelectronic Detection Technology and System, Tianjin 300387, China
[6] Xiongming Aviation Science Industry (Wuhu) Co., Ltd., No. 206, Xinwu Avenue, Wuhu 241000, China; peiyongw@xiong-ming.com
[7] Institute of Intelligent Manufacturing, Wuhu Institute of Technology, Wuhu 241006, China
* Correspondence: 101043@whit.edu.cn (Z.X.); 101138@whit.edu.cn (Y.W.); 101044@whit.edu.cn (C.Y.)

Citation: Ye, D.; Xu, Z.; Liu, H.; Zhang, Z.; Wang, P.; Wu, Y.; Yin, C. Terahertz Non-Destructive Testing of Porosity in Multi-Layer Thermal Barrier Coatings Based on Small-Sample Data. *Coatings* **2024**, *14*, 1357. https://doi.org/10.3390/coatings14111357

Academic Editor: Przemysław Podulka

Received: 29 September 2024
Revised: 22 October 2024
Accepted: 24 October 2024
Published: 25 October 2024

Copyright: © 2024 by the authors. Licensee MDPI, Basel, Switzerland. This article is an open access article distributed under the terms and conditions of the Creative Commons Attribution (CC BY) license (https://creativecommons.org/licenses/by/4.0/).

Abstract: Accurately characterizing the internal porosity rate of thermal barrier coatings (TBCs) was essential for prolonging their service life. This work concentrated on atmospheric plasma spray (APS)-prepared TBCs and proposed the utilization of terahertz non-destructive detection technology to evaluate their internal porosity rate. The internal porosity rates were ascertained through a metallographic analysis and scanning electron microscopy (SEM), followed by the reconstruction of the TBC model using a four-parameter method. Terahertz time-domain simulation data corresponding to various porosity rates were generated employing the time-domain finite difference method. In simulating actual test signals, white noise with a signal-to-noise ratio of 10 dB was introduced, and various wavelet transforms were utilized for denoising purposes. The effectiveness of different signal processing techniques in mitigating noise was compared to extract key features associated with porosity. To address dimensionality challenges and further enhance model performance, kernel principal component analysis (kPCA) was employed for data processing. To tackle issues related to limited sample sizes, this work proposed to use the Siamese neural network (SNN) and generative adversarial network (GAN) algorithms to solve this challenge in order to improve the generalization ability and detection accuracy of the model. The efficacy of the constructed model was assessed using multiple evaluation metrics; the results indicate that the novel hybrid WT-kPCA-GAN model achieves a prediction accuracy exceeding 0.9 while demonstrating lower error rates and superior predictive performance overall. Ultimately, this work presented an innovative, convenient, non-destructive online approach that was safe and highly precise for measuring the porosity rate of TBCs, particularly in scenarios involving small sample sizes facilitating assessments regarding their service life.

Keywords: thermal barrier coatings; porosity; small-sample modeling; terahertz non-destructive testing

1. Introduction

With the vigorous progress of the modern aviation industry, the optimization of the performance of aero-engines, as the core power source that drives aircraft to soar in the blue sky, was directly related to the overall efficiency and operational safety of the aircraft. Thermal barrier coatings (TBCs), as a key protective barrier for aero-engine turbine blades and other extreme high-temperature components (service temperature exceeding 1500 °C), had effectively curbed the high-temperature erosion of the substrate material by virtue of their excellent thermal insulating performance, significantly extending the service life of the engine and improving the operational efficiency [1–3]. However, under extreme service conditions (cold, hot, or heavily corrosive environments), TBCs were susceptible to defects such as pores and cracks, which not only weakened the thermal insulation performance of the coating, but also induced serious consequences such as coating spalling and functional failure [4,5]. Therefore, the accurate and efficient evaluation of the porosity of TBCs was of inestimable value for safeguarding the coating quality, predicting the service life and optimizing the preparation process.

Traditional porosity inspection methods, such as scanning electron microscopy (SEM) and X-ray diffraction (XRD), although dominant in terms of accuracy, were limited by such drawbacks as having a cumbersome inspection process, being time-consuming, having a high cost, and being destructive to samples, which made it difficult to satisfy the urgent needs of modern industry for rapid and non-destructive inspection techniques [6–8]. In this context, with the continuous innovation of non-destructive testing (NDT) technology, the terahertz (THz; the commonly used frequency range is 0.1–10 THz) NDT technology, with its unique physical properties and wide application potential, had gradually emerged, and had shown extraordinary application prospects in the field of TBC detection [9,10]. Terahertz non-destructive testing technology had unique advantages such as having a fast response, and it could non-destructively penetrate a variety of non-metallic and non-polar materials such as ceramics to achieve the fine detection of the internal structure of materials [11]. In TBC detection, THz waves could penetrate the ceramic layer and trigger the total reflection phenomenon at the interface, thus realizing the precise characterization of the internal microstructure of the coating [12]. In addition, terahertz time-domain spectroscopy (THz-TDS) further enriched the dimension of the detection information by capturing the dynamic changes of the electric field after the THz pulse interacted with the sample, which enhanced the accuracy and resolution of the detection [13].

However, facing the complexity of the structure of multilayer TBCs and the limitation of small-sample data, THz-based NDT technology still faced many challenges. The material properties and interface characteristics in multilayer structures differed significantly, which put higher requirements on the propagation and reflection mechanisms of THz waves [14]. Meanwhile, under the condition of small-sample data, how to ensure the reliability and stability of the detection results and avoid the overfitting or underfitting phenomenon of the model had become a key problem to be solved [15]. In addition, the current THz non-destructive testing technology mostly relied on large-sample data or standard samples for verification, and the research on the detection method for small-sample data was still insufficient, which resulted in it being difficult to meet the rapid detection needs in engineering. In view of this, this paper was dedicated to proposing a THz non-destructive testing method for the porosity of multilayer TBCs based on small-sample data, exploring the potential of machine-learning algorithms under small-sample conditions by integrating experimental research and numerical simulation, aiming to open up a new way for the rapid and non-destructive testing of the porosity of TBCs.

The problem with terahertz NDT is that the amount of data is too large and there is a lot of clutter. For this reason, machine learning algorithms are used in this study to solve this problem. Reference [16] summarizes the principle and application of decision tree, which makes decisions by constructing a tree structure, which is easy to understand and interpret, which can deal with both data-type and routine-type attributes, and which can make good results in a short time for large data sources, but there are drawbacks such as difficulty in dealing with missing data, easy to overfitting, and ignoring correlation between attributes. Reference [17] studied the characteristics of SVM algorithm, which is good at dealing with data nonlinear separable problems. SVM achieves linear classification by finding the maximum interval hyperplane, and finds the optimal classification hyperplane by mapping the kernel function to the high-dimensional space in nonlinear cases, which is able to solve the nonlinear problems and avoid dimensionality catastrophe. Reference [18] discusses the application of BP neural networks with self-learning characteristics in engineering fields. It is highly flexible, can handle various types of data, learns and extracts features automatically without manual feature engineering, and can accelerate training and prediction by parallel computing, but suffers from overfitting problem and complexity of model selection and parameter adjustment. References [19,20] explored the effect of SNN and GAN in small sample modeling. SNN combines the characteristics of factorial decomposition machine and neural network, which can deal with multi-domain category features, improve prediction accuracy through feature combination and pattern recognition, make full use of the limited data information, and achieve excellent performance in small sample datasets. GAN consists of a generator and discriminator, which can generate highly realistic and diversified samples through continuous confrontation and optimization, and can be used to generate a variety of samples. GAN consists of a generator and a discriminator, which generates highly realistic and diversified samples through continuous confrontation and optimization, learns the details of data distribution, generates high-quality samples, and has a wide range of prospects for application in the fields of image generation, style migration, etc. It can also be used in combination with other technologies to expand the application scenarios and enhance the performance of models.

The study of non-destructive testing techniques for the terahertz wave porosity of multilayer thermal barrier coatings based on small-sample datasets not only occupied an important position at the level of theoretical exploration, but also demonstrated a wide range of application potentials in engineering practice [21,22]. In this work, 8YSZ (8 wt% Y_2O_3-stabilized ZrO_2) was used as the raw material to prepare thermal barrier coatings using the APS method. 8YSZ has a high melting point, which ensures the stability and morphology maintenance of the coatings under a high temperature environment. Meanwhile, its low thermal conductivity can effectively control the temperature of the alloy matrix and provide good thermal insulation. In addition, 8YSZ matches the thermal expansion coefficient of the metal substrate, which can reduce the internal stress of the coating under thermal cycling conditions, thus extending the service life of the coating. Moreover, the toughening mechanism of 8YSZ brings about a longer cycle life. The core objective of this study was to reconstruct a realistic simulation model using the four-parameter stochastic growth method (QSGS) based on the microstructural characterization data of the thermal barrier coating collected after several thermal shock tests [23]. The QSGS model allows for the precise control of the pore structure and distribution by tuning parameters such as the porosity, distribution probability of growing nuclei, and directional growth probability. Numerical simulation using the QSGS model is usually more efficient than the direct processing of SEM image data, especially in terahertz non-destructive testing characterized by large data. The data generated by the QSGS model can be easily imported into various engineering software and numerical simulation tools for further analysis and processing. There are still some problems in applying the QSGS method in practice. The accuracy of the model needs to be verified by experimental data, which requires a large number of experiments and comparative analyses to ensure that the QSGS model can truly reflect the actual pore structure characteristics. When importing the QSGS model into numerical simulation software, the problem of incompatible data formats or conversion difficulties may be encountered, which requires additional data-processing work. This model aimed to simulate the real state of the thermal barrier coatings under service conditions. To achieve this goal, the finite-difference time-domain (FDTD) numerical simulation technique [24] was employed in this study to simulate the propagation process of THz waves in multilayer thermal barrier coatings, and, accordingly, the THz wave propagation model was established under different porosity conditions. FDTD was a numerical computation method that can simulate the propagation and scattering process of electromagnetic waves in complex media. FDTD had the advantages of having a high computational efficiency, wide range of application, and easy implementation, which was especially suitable for simulating and analyzing the propagation characteristics of terahertz waves in materials. To enhance the realism and reliability of the simulated signals, a Gaussian white noise of 10 dB was introduced during the simulation process to simulate the signal interference that may exist in the actual test environment. Subsequently, three noise-processing techniques were used to filter the noise-containing signals with the aim of maximizing the recovery of the signal quality. To further improve the data-processing efficiency and model-training performance, this study also applied the kernel principal component analysis (kPCA) technique [25] to perform feature extraction and data dimensionality reduction on the noise-reduced processed high-dimensional time-domain signal. Through the kPCA method, the redundancy of the input features was effectively reduced, and the key information was retained to provide optimized data inputs for the subsequent machine-learning model construction. In the model construction stage, the porosity data obtained from experimental measurements were used as the output target features of the model in this study, while the THz time-domain signals processed by noise reduction and dimensionality reduction were used as the input features. Subsequently, SNN and GAN algorithms were used to accurately characterize and predict the porosity of thermal barrier coatings under small-sample conditions. By comparing and analyzing the performance of different algorithms, the optimal model construction scheme was selected. The research results not only provided a new technical path for the fast and non-destructive detection of the porosity of thermal barrier coatings, but also provided data support and theoretical basis for the prediction and evaluation of the service life of thermal barrier coatings.

2. Materials and Methods

2.1. Thermal Barrier Coating Specimen Preparation and Testing

The specimen preparation and testing process was shown in Figure 1. In this study, the matrix material of the thermal barrier coating was selected as 8YSZ, which was prepared by the atmospheric plasma spraying (APS) technique, and the specific parameters of the preparation process were shown in Table 1 [24]. Thermal insulation coating samples with different pore structure characteristics were prepared by adjusting the technical parameters during the spraying process. To simulate and obtain the evolution of the pore characteristics of the coatings under different service stages, a unique thermal shock evaluation device was designed. Upon completion of the assessment, the samples collected at different assessment time points were systematically examined for their micro-morphological characteristics using high-resolution microanalysis. To ensure that the thermal barrier coatings were protected from mechanical damage during subsequent sample processing, the samples were carefully encapsulated using E7 epoxy adhesive (FM1000, Foshan Advanced Surface Technology Co., Ltd., Foshan, China). Subsequently, the encapsulated samples were precisely cut into small pieces of appropriate size for subsequent exhaustive analysis using wire-cutting technology (XKG200, Suzhou Hualong Dajin Electro-Machining Co., Ltd., Suzhou, China). During the cutting process, JCM1 epoxy resin (JCM1, Hangzhou Jingjing Inspection Instrument Co., Ltd., Hangzhou, China) was used to cold-set the cutting surface in order to maintain the structural integrity of the samples. A series of standardized sample preparation procedures, including step-by-step sanding of the sample surface with 150–3000 grit sandpaper and fine polishing with 0.25 μm grit polishing compound, were performed to ensure that the sample surface achieved the flatness and finish required for microanalysis. Afterwards, SEM (Zeiss EVO MA15, Carl Zeiss SMT Ltd., Cambridge, UK) was used to observe the microstructure of the treated samples. Based on the acquisition of high-resolution microstructural images, the study applied the threshold analysis method of Image J (Version 2.14.0, National Institutes of Health, Bethesda, MD, USA) image processing software to automatically calculate and extract the porosity data of the samples. To enhance the reliability of the data, the study repeated the analysis on at least 50 images of the same sample to minimize the human statistical errors and ensure the accuracy and reproducibility of the porosity assessment results.

Table 1. APS process parameters [24].

Spraying Parameter	Value
Spraying distance (mm)	100
Spraying power (KW)	30
Particle size of the alumina powder (μm)	25–45
Primary air flow (Ar) (L/min)	45
Auxiliary air flow (H_2) (L/min)	17.5
Powder feeder speed (r/min)	25
Number of preheating channels	2

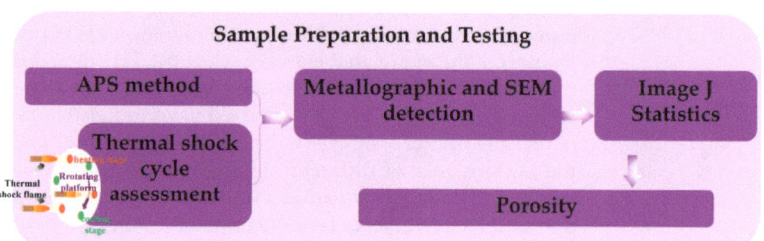

Figure 1. Schematic diagram of thermal shock test.

2.2. Terahertz Inspection Signal Obtained by FDTD Simulation

The microstructure inside the thermal barrier coating would undergo a series of complex changes in the process of thermal shock assessment, and the destructive method of detection could not meet the needs of engineering applications. Secondly, in addition to the substrate and bonding layer, there were multilayered and thin structural features inside the thermal barrier coatings, and the traditional non-destructive testing technology was difficult to meet its accuracy requirements, which brought more difficulties to the accurate characterization of its porosity. For this reason, in this study, after obtaining the porosity characteristics through experiments, the model with different porosity was reconstructed by QSGS algorithm, and the model reconstruction diagram was shown in Figure 2 [25]. The model was imported into FDTD Solutions software (Version 8.12.631 for X64, Lumerical Solutions Inc., Vancouver, BC, Canada) to model and simulate the propagation process of terahertz waves in the thermal barrier coating [26]. A white noise of 10 dB was added to the simulated signal to more closely match the real signal. A total of 64 sets of valid data were obtained upon completion.

Figure 2. Model diagram of QSGS method reconstruction.

2.3. Data Preprocessing and Small-Sample Model Construction

The original signal had more noise, which might cause misclassification and take longer time when modeling directly with it. Combining the findings of previous studies [27] and data characteristics, this study used Savitzky Golay (SG) filtering, Kalman filtering, and wavelet transform (WT) filtering for noise reduction and compared the effect of noise reduction [28–30]. SG filtering was a local polynomial regression-based filtering method, which was mainly used for smoothing the data and reducing the noise. However, it was likely to lose important information because the processed data were too smooth. Kalman filtering was an optimized, recursive digital processing algorithm that was particularly suitable for processing signals with noise and uncertainty. However, it often took a lot of time to compute, which limited the application of this algorithm in engineering. WT filtering had good localization characteristics in both time and frequency domains, and could accurately capture the instantaneous mutation and detail information of the signal, which made it possible to effectively remove the noise in the signal while retaining the important characteristics of the signal. However, it also had the problem of long processing time. In order to compare the effectiveness of each filter, it was judged by calculating the root mean square error ($RMSE$) between the filtering result of each filter and the original signal, and the smaller the error value, the better the filtering effect. The algorithm with the smallest error would be selected for filtering in the subsequent modeling process.

After the signal filtering was completed, in order to further enhance the ability of the model for engineering applications, in this study, kPCA was used to further process the data. Traditional principal component analysis (PCA) dimensionality reduction was a linear dimensionality reduction method, and, when there were nonlinear relationships between data, PCA might not be able to capture these relationships effectively. In contrast, kPCA maps the data to a high-dimensional space through a kernel function, which was able to deal with complex nonlinear relationships, and, thus, was more advantageous when dealing with nonlinear data. kPCA was essentially a nonlinear extension of PCA [31]. Its core idea was to project the data from the original space to a high-dimensional feature space through a nonlinear mapping (usually through a kernel function), and then perform PCA processing in this high-dimensional feature space. The role of the kernel function was to map the original data from a low-dimensional space to a high-dimensional space, which made the data more linearly differentiable in the high-dimensional space, thus making it easier to find the principal components of the data. In the high-dimensional feature space, kPCA found the principal components and projects the data into a new coordinate system for dimensionality reduction. In this study, the noise-reduced data were used as the input to the dimensionality reduction model, which, in turn, allowed us to obtain the eigenvalues, contributions, and scores of each principal component. The data after kPCA dimensionality reduction were used as input for subsequent modeling.

From previous studies, it could be found that machine-learning techniques were highly advantageous in constructing prediction models based on big data [16–20]. However, traditional machine-learning algorithms usually faced dilemmas such as overfitting, uneven data distribution, and poor generalization ability in small-sample modeling. In this regard, Siamese neural network (SNN) and generative adversarial network (GAN) algorithms, which were advantageous in small-sample data processing, were selected for modeling in this study [19,20]. SNN was a special type of neural network inspired by the concept of twin brother, because it consisted of two neural networks with the same structure and parameters. It consisted of two neural networks with the same structure and shared parameters and weights. These two networks worked in parallel and receive input data at the same time, and the modeling process using it was shown in Figure 3. SNN was an effective method to solve the small-sample prediction problem due to its advantages of shared weight structure, effective loss function design, powerful feature learning ability, concise training process, etc. GAN was a deep-learning model proposed by Ian Goodfellow et al. in recent years, which learned through the mutual confrontation of two neural networks—generator and discriminator. GAN had the advantages of powerful sample generation ability, flexible architecture design, data enhancement, and generalization ability, as well as potential data mining and understanding when dealing with small-sample data, and the modeling process using it was shown in Figure 4.

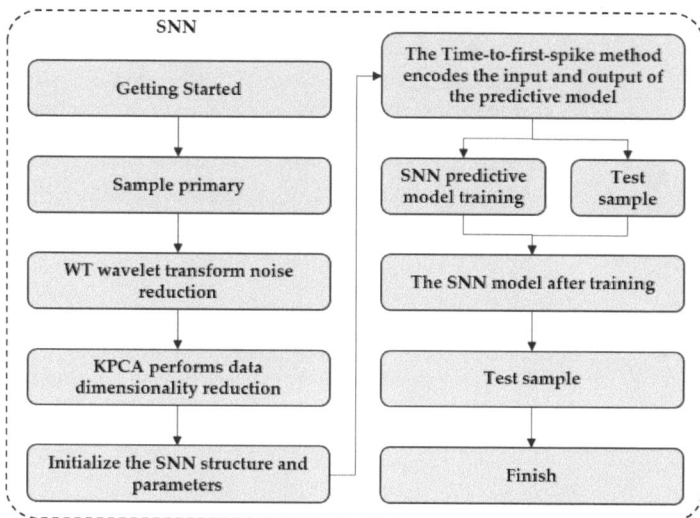

Figure 3. Schematic diagram of SNN.

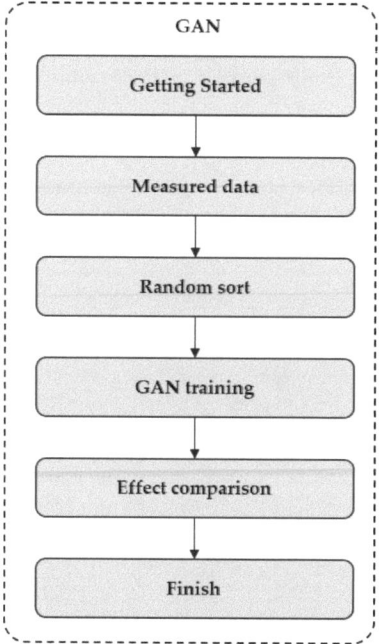

Figure 4. Schematic diagram of GAN.

A dataset containing 64 sets of samples was collected in this study, and, to ensure the comprehensiveness and robustness of the model evaluation, an 8-fold cross-validation strategy was used. Specifically, the dataset was randomly divided into eight mutually exclusive and identically sized subsets (folds), after which one of the subsets was selected as an independent test set each time by means of a loop, while the remaining seven subsets were combined as the training set. This process was repeated eight times until each subset was given a chance to be used as a test set. For the model prediction results in each iteration, four quantitative assessment metrics were used for systematic evaluation, including root mean square error (RMSE), mean absolute error (MAE), mean absolute percentage error (MAPE), and squared correlation coefficient (R^2), which measure the model prediction performance in different dimensions, respectively [31]. The evaluation metrics were defined as follows:

$$RMSE = \sqrt{\sum_{i=1}^{n}(Y_i - \hat{Y}_i)^2/n} \quad (1)$$

$$MAE = \sum_{i=1}^{n}|Y_i - \hat{Y}_i|/n \quad (2)$$

$$MAPE = \sum_{i=1}^{n}\frac{|Y_i - \hat{Y}_i|}{Y_i}/n \quad (3)$$

$$R^2 = \left[\frac{\sum_{i=1}^{n}\left(\hat{Y}_i - \overline{\hat{Y}}\right)(Y_i - \overline{Y})}{\sqrt{\sum_{i=1}^{n}\left(\hat{Y}_i - \overline{\hat{Y}}\right)^2}\sqrt{\sum_{i=1}^{n}(Y_i - \overline{Y})^2}}\right]^2 \quad (4)$$

where n was the number of samples, $\hat{Y}_i$ was the predicted value, Y_i was the actual value, $\overline{\hat{Y}}$ was the average of the predicted value, and $\overline{Y}$ was the average of the actual value. RMSE, MAE, and MAPE belong to the error—the lower the value, the better—and its ideal value is 0. R^2 is the accuracy of the prediction—the bigger the value, the better—and its optimal value is 1, at which time the accuracy of the prediction is 100%.

3. Results

3.1. Experimental Results Statistics

Figure 5 shows the microstructure of the thermal barrier coating before and after the thermal shock assessment and the resultant figure processed by Image J software. From the figure, it could be seen that the microstructure of the thermal barrier coating in the sprayed state usually contained a large number of unbonded defects (small pores) and irregular large pores. After comparing Figure 5a,c, it could be found that the coating prepared by the atmospheric plasma spraying process had healed the internal pores under the high-temperature thermal shock test, and the number of small internal pores was rapidly reduced. This was one of the main factors to improve the thermal conductivity of the thermal barrier coating in service, and it was also something to be avoided. To investigate the evolution of the pores of the thermal barrier coating after high-temperature service, this study was carried out by adjusting the time of the thermal shock assessment, and the statistical results of the porosity after different times of the thermal shock assessment are shown in Figure 6. It could be clearly seen in Figure 6 that, in the initial stage of the thermal shock assessment, the porosity decreased with the rapid sintering of the small internal pores of the coating, which was in line with the conclusion of the previous report [32]. The sintering rate of the coating was controlled by the large pores, which were difficult to be healed by sintering, which made the porosity of the coating basically remain within a certain range after the sintering of the small pores [33]. Therefore, it was necessary to regulate the process and the size of the raw materials to adjust the proportion of large pores and small pores, which required a reasonable porosity to ensure the thermal radiation performance of the coating, but also a reasonable pore structure to ensure that it was not easy to be destroyed.

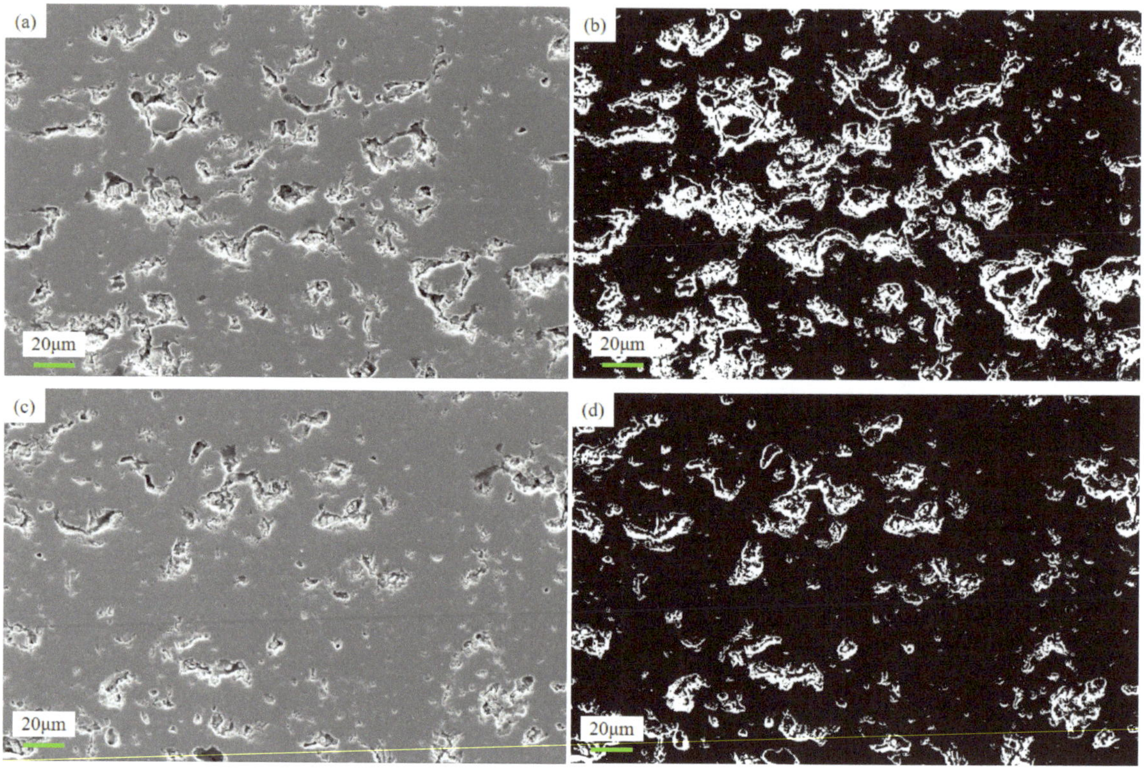

Figure 5. Micrographs of thermal barrier coatings before and after thermal shock assessment: (**a**) SEM image before assessment; (**b**) processing results of Image J software for SEM image before assessment; (**c**) SEM image after assessment; and (**d**) processing results of Image J software for SEM image after assessment.

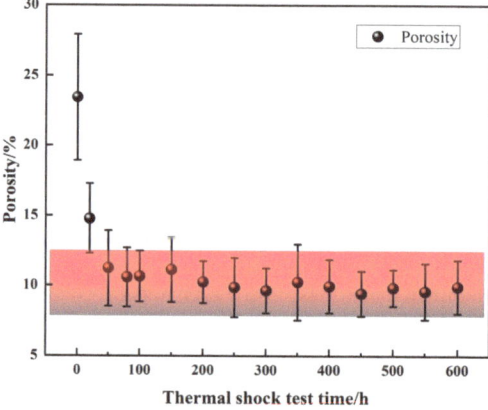

Figure 6. Diagram of the statistical results of porosity after different times of thermal shock assessment.

3.2. Signal Simulation and Noise Reduction

After obtaining the pore pattern inside the coating, this study used these data as the basis for reconstructing the model using the QSGS method, and the results of the reconstructed model with different porosities were shown in Figure 7. The constructed model was imported into the FDTD simulation software (Version 8.12.631 for X64, Lumerical Solutions Inc., Vancouver, BC, Canada) to obtain the terahertz time-domain spectroscopy detection signals under different porosities. Taking the three-layer model with a porosity of 10% as an example, noise was added to the ideal detection signal to obtain a signal close to the actual one, as shown in Figure 8. From Figure 8, it could be seen that the signal after adding noise made it difficult to extract the eigenvalues of the signal above three layers. From the processing effect in Figure 8, the effect of the three noise reduction methods was very close to each other, and it was impossible to judge which way was more superior, and further comparisons needed to be made. In this study, the *RMSE* between the noise reduction signal and the original signal was used for comparison, and the comparison results were shown in Table 2. From Table 2, it could be seen that the noise reduction using wavelet transform had the best effect, followed by SG filtering, and Kalman had the worst noise reduction effect. The best wavelet transform would be used for data noise reduction in the subsequent process.

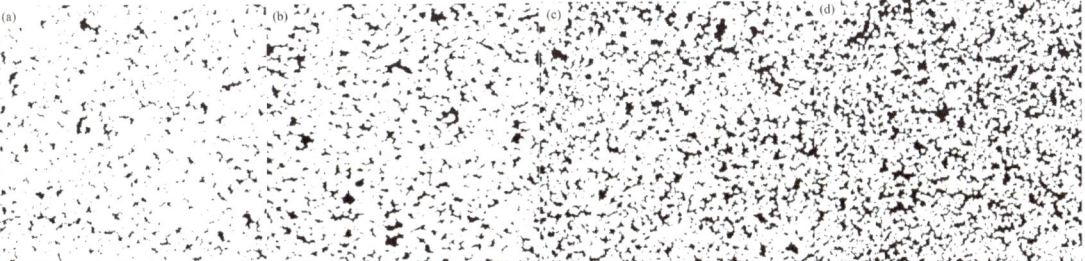

Figure 7. Diagrams of reconstructed models by QSGS method for coatings with different porosities: (**a**) 10%; (**b**) 15%; (**c**) 20%; and (**d**) 25%.

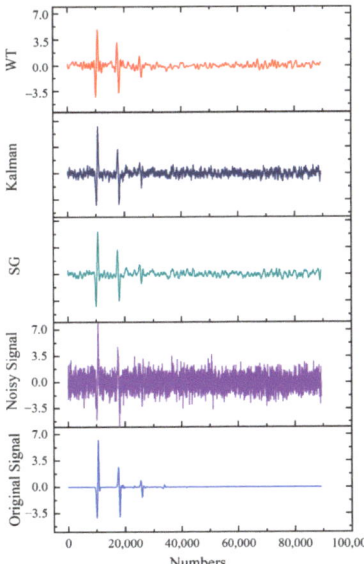

Figure 8. Comparison of noise reduction effects of different noise reduction methods.

Table 2. RMSE of different noise reduction methods after noise reduction.

Noise Reduction Method	RMSE
SG	1.4721
Kalman	2.8539
WT	2.1894

3.3. Comparison of Various Machine-Learning Approaches

The data after the noise reduction process still contained a large number of invalid points, which needed further processing before they could be used for modeling. The dimensionality of the data after FDTD simulation was 64×4096, and the kernel principal component analysis was used for the dimensionality reduction in the signal after noise reduction. Figure 9a showed the contribution of each component factor and the total contribution after the dimensionality reduction. From the figure, it could be seen that the dimensionality of the FDTD simulated signal was reduced from 64×4096 to 64×47, and the cumulative contribution of the first 16 factors had exceeded 85%. Figure 9b showed the contribution and total contribution of each component factor after the dimensionality reduction using the conventional PCA method. The dimension after the dimensionality reduction using PCA was 64×63 and the cumulative contribution of the first 16 factors was only 70%, indicating that the weight shared of the kPCA dimensionality reduction factors was more concentrated. This was beneficial to the subsequent modeling, which could further improve the modeling efficiency under the premise of guaranteeing the effect. The two methods of dimensionality reduction results showed that the relationship between the terahertz time-domain spectral data and porosity was mainly nonlinear. The dimensionality-reduced data were used as inputs to a machine-learning model for the characterization and prediction of the porosity of thermal barrier coatings under small-sample conditions.

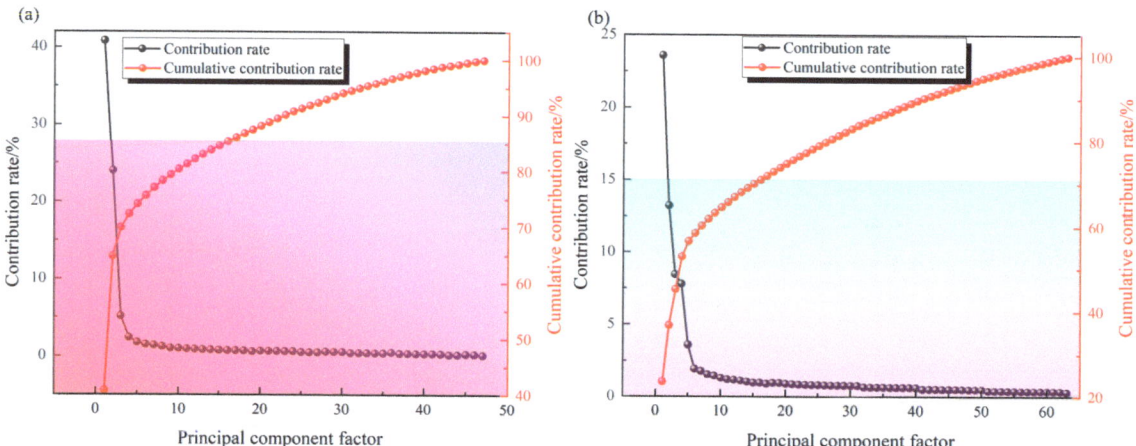

Figure 9. Contribution and total contribution curves of the component factors for different methods of dimensionality reduction: (**a**) kPCA; and (**b**) PCA.

To provide uniformity in the training of SNN and GAN models, both used the mean square error (MSE) as the error loss function for both. Although the training error could theoretically be reduced by increasing the training time, from the efficiency point of view, it may take a longer time to reduce the 1% error during subsequent training, which was not conducive to advancing the engineering application of the cover system. Therefore, from the global point of view, the number of training times was uniformly set to 500 times, and stopped immediately after reaching the number of training times. The model trained with

500 iterations is used as the final model, and the prediction of coating porosity can then be made for small samples. The final training errors of the SNN model and the GAN model under the eight-fold cross-validation were shown in Table 3. It could be seen from Table 3 that the training error of the GAN model is smaller, which indicated that the GAN model could predict the training data better during the training process. Because the data of FDTD in this study are more similar, it led to the SNN model could not make a better distinction between the two inputs during training, which, in turn led to the model predicting the training data poorly during learning.

Table 3. Final training error of SNN and GAN models under 8-fold cross-validation.

Final Training Error	SNN	GAN
1	0.66	0.35
2	0.47	0.34
3	0.55	0.18
4	0.52	0.32
5	0.63	0.26
6	0.68	0.21
7	0.71	0.30
8	0.59	0.24
Average	0.60	0.28

To further verify the prediction effect of the above models on unknown data, four quantitative assessment indicators were used to evaluate the prediction effect of the SNN model and GAN model, and the evaluation results were shown in Table 4. From Table 4, it could be seen that the error performance index of the SNN model indicated that it had a certain prediction performance for porosity under small-sample conditions, but the accuracy still needed to be improved. In contrast, the GAN model showed excellent performance, with an error value close to 0 and an average R^2 value of more than 98%. All four evaluation metrics indicated that the GAN model had a high accuracy and robustness in predicting the porosity of thermal barrier coatings under small-sample conditions. Therefore, it was reasonable to believe that the GAN model could meet the high requirements for characterizing the porosity of thermal barrier coatings under small-sample conditions.

Table 4. Evaluation index value of predictive effectiveness of SNN model and GAN model under 8-fold cross-validation.

RMSE	SNN	GAN	MAE	SNN	GAN	MAPE	SNN	GAN	R^2	SNN	GAN
1	13.199	1.678	1	11.147	1.430	1	0.133	0.017	1	0.404	0.991
2	20.620	0.645	2	16.761	0.530	2	0.184	0.005	2	0.244	0.982
3	14.638	1.493	3	13.063	1.151	3	0.143	0.012	3	0.159	0.981
4	8.688	1.592	4	7.174	1.305	4	0.085	0.015	4	0.006	0.989
5	7.052	0.531	5	6.444	0.439	5	0.063	0.004	5	0.116	0.959
6	6.443	0.699	6	5.392	0.539	6	0.057	0.006	6	0.005	0.996
7	7.719	0.761	7	6.550	0.637	7	0.066	0.006	7	0.007	0.974
8	6.275	0.535	8	5.039	0.395	8	0.048	0.004	8	0.613	0.967
Average	10.579	0.992	Average	8.946	0.803	Average	0.097	0.009	Average	0.194	0.980

In this study, the porosity of thermal barrier coatings was taken as the object of research, and the application of the SNN model and GAN model under small-sample data was explored and the effect of the two was compared. The results showed that the GAN model was more accurate in predicting the porosity of thermal barrier coatings under small-sample conditions, and it should have an R^2 value close to 1 and an error value close to 0. The reason was mainly due to the fact that the SNN model focused on simulating the behavior of biological neurons and dealing with complex perceptual tasks, while the GAN model focused more on generative tasks, which continuously optimized the quality of the generated data through the antagonistic training of the generator and the discriminator. Specifically, in training the SNN model for 500 times, the prediction results of each training had a large discrepancy with the real values, indicating that even the optimal SNN model could not achieve the desired prediction results. Compared with the dual-input twin network model, the GAN model had significant advantages in terms of the naturalness and realism of the generated data, the freedom of the model design, and the training efficiency. Aiming at the problem of noisy raw data and many invalid data points, this study proposed to use WT for the data noise reduction and kPCA for the dimensionality reduction, which not only reduced the computation time of the model, but also enhanced the robustness of the model. Superior prediction accuracy could be obtained without excessive training. The WT-kPCA-GAN model proposed in this study combines the advantages of the WT, kPCA, and GAN models, and its performance on the test dataset was comparable to that of the training dataset, which proved that the WT-kPCA-GAN model had some practical applications in the non-destructive testing of the porosity of thermal barrier coatings under the conditions of small samples.

Currently, the assessment of the porosity of thermal barrier coatings mainly relied on destructive testing techniques, and this status quo greatly restricted its wide application in engineering practice. Although non-destructive testing techniques such as ultrasonic testing had been developed, it was still difficult to completely circumvent the interference of ultrathin thermally grown oxide layers, metal bonding layers, and substrate materials during the implementation process, resulting in persistent challenges such as the insufficient accuracy of the detection results and difficulties in signal characterization. In view of this, this study introduced the terahertz non-destructive testing (NDT) technique, which exhibited significant advantages in characterizing non-metallic ceramic layers. Due to the strong impermeability of terahertz waves to metallic materials, the acquired time-domain spectral signals could purely reflect the optical properties of the ceramic layer, effectively avoiding the interference of the metallic layer. However, the application of terahertz inspection technology to thermal barrier coatings faced many technical bottlenecks, especially the difficulty of directly preparing thermal barrier coating specimens with precisely controlled microstructural features, which directly affected the reliability and accuracy of the inspection results. To overcome this problem, this study proposed to use numerical simulation to virtually construct and simulate thermal barrier coating models with different microstructural features as a complement to and verification of experimental studies. In the future, standard samples of thermal insulation coatings with clear and diversified microstructural features would be prepared through diversified pretreatment strategies and process optimization, and the sample database would be gradually expanded on this basis, so as to lay a solid foundation for the engineering application of terahertz technology in detecting the porosity of thermal insulation coatings in small samples, and to push forward the technological advancement and practical application in the field.

4. Conclusions

The aim of this study was to explore an innovative method based on terahertz non-destructive testing technology combining the WT, kPCA, and GAN algorithms in order to accurately measure the porosity of thermal barrier coatings under small-sample conditions. The microstructural characteristics of the thermal barrier coating were obtained by SEM analysis, and a simulation model was constructed on this basis, and 64 sets of valid data samples were successfully collected. Subsequently, the FDTD method was used to simulate the generation of terahertz time-domain signals with different porosity characteristics, and 10 dB white noise was artificially introduced to simulate the actual detection environment. To improve the signal quality, this study compared the effects of three filtering techniques, namely, WT, SG filtering and Kalman filtering, and the results showed that WT performs optimally in terms of noise reduction performance, followed by SG, while Kalman filtering was relatively ineffective. In view of the high-dimensional characteristics of the noise-reduced data, this study further compared the efficacy of the traditional PCA and the new kPCA in data dimensionality reduction, and found that the kPCA was capable of generating more concentrated principal component factors, thus optimizing the subsequent modeling process. In the model construction stage, this study constructed and compared two porosity prediction models based on the noise-reduced terahertz time-domain data as the input and the experimentally measured porosity as the output, using the SNN and GAN algorithms. Through a multi-fold cross-validation strategy, the GAN-based regression prediction model was finally established, which exhibited a lower training error under eight-fold validation. In the porosity prediction practice, the proposed hybrid WT-kPCA-GAN model exhibits a near-zero error value, with a prediction accuracy of more than 95.9%, and an average accuracy of 98% for the eight-fold validation, which was significantly better than that of the SNN model. This performance difference was mainly attributed to the fact that the SNN model's ability to simulate the behavior of biological neurons and complex perceptual tasks was not fully demonstrated when dealing with the specific type of terahertz spectral data in this study, compared to the GAN model's greater adaptability to the data.

In this study, an innovative hybrid machine-learning framework was proposed, which effectively integrated terahertz non-destructive testing techniques and showed excellent performance in characterizing the porosity of thermal barrier coatings under small-sample conditions, providing strong technical support and theoretical backing for the lifetime prediction and quality assessment of thermal barrier coatings. In view of the current challenges in the preparation of diverse microstructured thermal barrier coatings, subsequent research would focus on the in-depth exploration of the microstructure evolution law under experimental conditions. Meanwhile, the integration of simulation and machine-learning technology not only broadened the path of novel material development, but also significantly reduced the cost of research and development, which was expected to have a broad application prospect and economic benefits.

Author Contributions: Conceptualization, Z.X. and D.Y.; methodology, Z.X. and D.Y.; software, Z.X. and D.Y.; validation, C.Y., Y.W., H.L., Z.Z. and P.W.; formal analysis, Z.X., Y.W. and D.Y.; investigation, Z.X., C.Y., Y.W., H.L., Z.Z., P.W. and D.Y.; resources, Z.X., C.Y. and D.Y.; data curation, Z.X., Y.W., H.L., Z.Z. and P.W.; writing—original draft preparation, Z.X. and D.Y.; writing—review and editing, C.Y., Y.W., Z.Z. and P.W.; visualization, Z.X., C.Y., Y.W., H.L., Z.Z., P.W. and D.Y.; supervision, Z.X., C.Y. and D.Y.; project administration, C.Y., Y.W., H.L., Z.Z., P.W. and D.Y.; funding acquisition, Z.X., C.Y., Y.W. and D.Y. All authors have read and agreed to the published version of the manuscript.

Funding: This research was funded by the Open Research Fund of Huzhou Key Laboratory of Terahertz Integrated Circuits and Systems (No. HKLTICY23KF05, D.Y.), the Open Project of Tianjin Key Laboratory of Optoelectronic Detection Technology and System (No. 2024LODTS122, D.Y.), the Open Research Fund of Anhui Key Laboratory of Mine Intelligent Equipment and Technology (No. ZKSYS202201, D.Y.), the Xiongming Aviation Science Industry (Wuhu) Co., Ltd. Industry-University-Research Project (HX-2024-09-085, D.Y.), the Undergraduate Teaching Quality Improvement Program Project of Anhui University of Technology (No.2023jyxm76, D.Y., No.2023szyzk39, D.Y.), the Science and Technology Plan Project of Wuhu City (No. 2023jc40, Y.W., No. 2023yf131, Z.X.), the 2023 Anhui Province Scientific Research Preparation Plan Project (No. 2023AH052399, Z.X., No. 2023AH052384, C.Y., No. 2023AH052380, Y.W.), the Natural Science Research Project of Wuhu Institute of Technology (No. wzyzr202419, Z.X., No. wzyzr202332, Y.W.), and the Opening Project of Automotive New Technique of Anhui Province Engineering Technology Research Center (No. QCKJ202209B, Y.W.).

Institutional Review Board Statement: Not applicable.

Informed Consent Statement: Not applicable.

Data Availability Statement: Data is contained within the article.

Conflicts of Interest: Author Peiyong Wang was employed by the company Xiongming Aviation Science Industry (Wuhu) Co., Ltd. The remaining authors declare that the research was conducted in the absence of any commercial or financial relationships that could be construed as a potential conflict of interest. The authors declare that this study received funding from Xiongming Aviation Science Industry (Wuhu) Co., Ltd. The funder was not involved in the study design, collection, analysis, interpretation of data, the writing of this article or the decision to submit it for publication.

References

1. Padture, N.P.; Gell, M.; Jordan, E.H. Thermal barrier coatings for gas-turbine engine applications. *Science* **2002**, *296*, 280–284. [CrossRef] [PubMed]
2. Vaßen, R.; Jarligo, M.O.; Steinke, T.; Mack, D.E.; Stöver, D. Overview on advanced thermal barrier coatings. *Surf. Coat. Technol.* **2010**, *205*, 938–942. [CrossRef]
3. Darolia, R. Thermal barrier coatings technology: Critical review, progress update, remaining challenges and prospects. *Int. Mater. Rev.* **2013**, *58*, 315–348. [CrossRef]
4. Miller, R.A. Thermal barrier coatings for aircraft engines: History and directions. *J. Therm. Spray Technol.* **1997**, *6*, 35–42. [CrossRef]
5. Saini, A.K.; Das, D.; Pathak, M.K. Thermal barrier coatings-applications, stability and longevity aspects. *Procedia Eng.* **2012**, *38*, 3173–3179. [CrossRef]
6. Jiao, D.; Shi, W.; Liu, Z.; Xie, H. Laser multi-mode scanning thermography method for fast inspection of micro-cracks in TBCs surface. *J. Nondestruct. Eval.* **2018**, *37*, 1–10. [CrossRef]
7. Klement, U.; Ekberg, J.; Kelly, S.T. 3D analysis of porosity in a ceramic coating using X-ray microscopy. *J. Therm. Spray Technol.* **2017**, *26*, 456–463. [CrossRef]
8. Klement, U.; Ekberg, J.; Creci, S.; Kelly, S.T. Porosity measurements in suspension plasma sprayed YSZ coatings using NMR cryoporometry and X-ray microscopy. *J. Coat. Technol. Res.* **2018**, *15*, 753–757. [CrossRef]
9. Li, R.; Ye, D.; Xu, Z.; Yin, C.; Xu, H.; Zhou, H.; Yi, J.; Chen, Y.; Pan, J. Nondestructive Evaluation of Thermal Barrier Coatings Thickness Using Terahertz Time-Domain Spectroscopy Combined with Hybrid Machine Learning Approaches. *Coatings* **2022**, *12*, 1875. [CrossRef]
10. Fischer, B.M.; Wietzke, S.; Reuter, M.; Peters, O.; Gente, R.; Jansen, C.; Vieweg, N.; Koch, M. Investigating material characteristics and morphology of polymers using terahertz technologies. *IEEE Trans. Terahertz Sci. Technol.* **2013**, *3*, 259–268. [CrossRef]
11. Zhong, S. Progress in terahertz nondestructive testing: A review. *Front. Mech. Eng.* **2019**, *14*, 273–281. [CrossRef]
12. Niijima, S.; Shoyama, M.; Murakami, K.; Kawase, K. Evaluation of the sintering properties of pottery bodies using terahertz time-domain spectroscopy. *J. Asian Ceram. Soc.* **2018**, *6*, 37–42. [CrossRef]
13. Liebermeister, L.; Nellen, S.; Kohlhaas, R.; Breuer, S.; Schell, M.; Globisch, B. Ultra-fast, high-bandwidth coherent cw THz spectrometer for non-destructive testing. *J. Infrared Millim. Terahertz Waves* **2019**, *40*, 288–296. [CrossRef]
14. Guo, J.W.; Yang, L.; Zhou, Y.C.; He, L.M.; Zhu, W.; Cai, C.Y.; Lu, C.S. Reliability assessment on interfacial failure of thermal barrier coatings. *Acta Mech. Sin.* **2016**, *32*, 915–924. [CrossRef]
15. Zhang, Y.; Ling, C. A strategy to apply machine learning to small datasets in materials science. *npj Comput. Mater.* **2018**, *4*, 25. [CrossRef]
16. Tong, W.; Hong, H.; Fang, H.; Xie, Q.; Perkins, R. Decision Forest: Combining the predictions of multiple independent decision tree models. *J. Chem. Inf. Comput. Sci.* **2003**, *43*, 525–531. [CrossRef]
17. Sapankevych, N.I.; Sankar, R. Time series prediction using support vector machines: A survey. *IEEE Comput. Intell. Mag.* **2009**, *4*, 24–38. [CrossRef]
18. Cui, K.; Jing, X. Research on prediction model of geotechnical parameters based on BP neural network. *Neural Comput. Appl.* **2019**, *31*, 8205–8215. [CrossRef]
19. Jagtap, A.B.; Sawat, D.D.; Hegadi, R.S.; Hegadi, R.S. Verification of genuine and forged offline signatures using Siamese Neural Network (SNN). *Multimed. Tools Appl.* **2020**, *79*, 35109–35123. [CrossRef]
20. Gonog, L.; Zhou, Y. A review: Generative adversarial networks. In Proceedings of the 2019 14th IEEE Conference on Industrial Electronics and Applications (ICIEA), Xi'an, China, 19–21 June 2019; IEEE: Piscataway, NJ, USA, 2019; Volume 1, pp. 505–510.
21. Tu, W.; Zhong, S.; Shen, Y.; Incecik, A. Nondestructive testing of marine protective coatings using terahertz waves with stationary wavelet transform. *Ocean Eng.* **2016**, *111*, 582–592. [CrossRef]
22. Liu, H.; Ke, L. Measuring low-level porosity structures by using a non-destructive terahertz inspection system. *Opt. Laser Technol.* **2017**, *94*, 240–243. [CrossRef]
23. Wang, Q.; Zhang, H.; Cai, H.; Fan, Q.; Li, G. Reconstruction of co-continuous ceramic composites three-dimensional microstructure solid model by generation-based optimization method. *Comput. Mater. Sci.* **2016**, *117*, 534–543. [CrossRef]
24. Ye, D.; Xu, Z.; Pan, J.; Yin, C.; Hu, D.; Wu, Y.; Li, R.; Li, Z. Prediction and Analysis of the Grit Blasting Process on the Corrosion Resistance of Thermal Spray Coatings Using a Hybrid Artificial Neural Network. *Coatings* **2021**, *11*, 1274. [CrossRef]
25. Yang, L.; Liu, Q.X.; Zhou, Y.C.; Mao, W.G.; Lu, C. Finite element simulation on thermal fatigue of a turbine blade with thermal barrier coatings. *J. Mater. Sci. Technol.* **2014**, *30*, 371–380. [CrossRef]
26. Shiokawa, Y.; Date, Y.; Kikuchi, J. Application of kernel principal component analysis and computational machine learning to exploration of metabolites strongly associated with diet. *Sci. Rep.* **2018**, *8*, 3426. [CrossRef]
27. Xu, Z.; Yin, C.; Wu, Y.; Liu, H.; Zhou, H.; Xu, S.; Xu, J.; Ye, D. Terahertz Nondestructive Measurement of Heat Radiation Performance of Thermal Barrier Coatings Based on Hybrid Artificial Neural Network. *Coatings* **2024**, *14*, 647. [CrossRef]
28. Al-gawwam, S.; Benaissa, M. Robust Eye Blink Detection Based on Eye Landmarks and Savitzky–Golay Filtering. *Information* **2018**, *9*, 93. [CrossRef]
29. Zhang, Q. Adaptive Kalman filter for actuator fault diagnosis. *Automatica* **2018**, *93*, 333–342. [CrossRef]
30. Xia, Y.X.; Ni, Y.Q. A wavelet-based despiking algorithm for large data of structural health monitoring. *Int. J. Distrib. Sens. Netw.* **2018**, *14*, 1550147718819095. [CrossRef]

31. Bobzin, K.; Bagcivan, N.; Parkot, D.; Schäfer, M.; Petković, I. Modeling and Simulation of Microstructure Formation for Porosity Prediction in Thermal Barrier Coatings Under Air Plasma Spraying Condition. *J. Therm. Spray Technol.* **2009**, *18*, 975–980. [CrossRef]
32. Giolli, C.; Scrivani, A.; Rizzi, G.; Borgioli, F.; Bolelli, G.; Lusvarghi, L. Failure mechanism for thermal fatigue of thermal barrier coating systems. *J. Therm. Spray Technol.* **2009**, *18*, 223–230. [CrossRef]
33. Liu, T.; Luo, X.T.; Chen, X.; Yang, G.J.; Li, C.X.; Li, C.J. Morphology and size evolution of interlamellar two-dimensional pores in plasma-sprayed $La_2Zr_2O_7$ coatings during thermal exposure at 1300 °C. *J. Therm. Spray Technol.* **2015**, *24*, 739–748. [CrossRef]

Disclaimer/Publisher's Note: The statements, opinions and data contained in all publications are solely those of the individual author(s) and contributor(s) and not of MDPI and/or the editor(s). MDPI and/or the editor(s) disclaim responsibility for any injury to people or property resulting from any ideas, methods, instructions or products referred to in the content.

Article

Surface Tribological Properties Enhancement Using Multivariate Linear Regression Optimization of Surface Micro-Texture

Zhenghui Ge [1], Qifan Hu [1], Haitao Zhu [2,*] and Yongwei Zhu [1]

[1] College of Mechanical Engineering, Yangzhou University, Yangzhou 225009, China; zhge@yzu.edu.cn (Z.G.); ywzhu@yzu.edu.cn (Y.Z.)
[2] School of Mechanical and Electrical Engineering, China University of Mining and Technology, Xuzhou 221116, China
* Correspondence: haitaozhu@cumt.edu.cn

Abstract: This work aims to provide a comprehensive understanding of the structural impact of micro-texture on the properties of bearing capacity and friction coefficient through numerical simulation and theoretical calculation. Compared to the traditional optimization method of single-factor analysis (SFA) and orthogonal experiment, the multivariate linear regression (MLA) algorithm can optimize the structure parameters of the micro-texture within a wider range and analyze the coupling effect of the parameters. Therefore, in this work, micro-textures with varying texture size, area ratio, depth, and geometry were designed, and their impact on the bearing capacity and friction coefficient was investigated using SFA and MLA algorithms. Both methods obtained the optimal structures, and their properties were compared. It was found that the MLA algorithm can further improve the friction coefficient based on the SFA results. The optimal friction coefficient of 0.070409 can be obtained using the SFA method with a size of 500 μm, an area ratio of 40%, a depth of 5 μm, and a geometry of the slit, having a 10.7% reduction compared with the texture-free surface. In comparison, the friction coefficient can be further reduced to 0.067844 by the MLA algorithm under the parameters of size of 600 μm, area ratio of 50%, depth of 9 μm, and geometry of the slit. The final optimal micro-texture surface shows a 15.6% reduction in the friction coefficient compared to the texture-free surfaces and a 4.9% reduction compared to the optimal surfaces obtained by SFA.

Keywords: micro-texture; bearing capacity; friction coefficient; multivariate linear regression

Citation: Ge, Z.; Hu, Q.; Zhu, H.; Zhu, Y. Surface Tribological Properties Enhancement Using Multivariate Linear Regression Optimization of Surface Micro-Texture. *Coatings* 2024, 14, 1258. https://doi.org/10.3390/coatings14101258

Academic Editor: Przemysław Podulka

Received: 4 September 2024
Revised: 22 September 2024
Accepted: 24 September 2024
Published: 1 October 2024

Copyright: © 2024 by the authors. Licensee MDPI, Basel, Switzerland. This article is an open access article distributed under the terms and conditions of the Creative Commons Attribution (CC BY) license (https://creativecommons.org/licenses/by/4.0/).

1. Introduction

Tribology is the study of the mechanisms of friction and wear between the surfaces of objects, encompassing many subject areas. It is estimated that over 80% of industrial equipment damage is attributable to friction and wear [1–3]. Furthermore, energy consumption accounts for 30% to 50% of the total energy consumption around the world, resulting in significant economic losses estimated at hundreds of billions of USD [4]. The improvement of tribology properties is of great practical value to engineering equipment, as it can increase service life, save energy, and improve productivity [5,6].

The general tribological theory assumes that superior tribological performance yields smoother surfaces [7]. Consequently, previous investigations focused on enhancing frictional properties by reducing surface roughness [8,9]. However, recent studies have demonstrated that rough surfaces with micro-texture can also exhibit excellent tribological properties [10,11]. Zhang et al. conducted disc-to-disc friction and wear experiments to compare the friction coefficients of surfaces with and without textures, proving that surface textures can effectively reduce friction coefficients and wear rates [12]. However, on the one hand, a unified understanding of the critical parameters for micro-texture design is yet to be achieved, and the parameters that are commonly considered include the geometry, size, depth, area ratio, local micro-texture, etc. [13,14]. Chen et al. investigated the influence of the micro-texture dimensions on the surface tribological behaviors, indicating that the friction coefficient increased with the smaller size and larger distribution density of the micro-texture [15]. Furthermore, Xin et al. prepared 25 groups of surfaces with different distribution densities of micro-textures, observing that the smallest wear rate was obtained when the micro-textures' diameter and spacing were 30 and 150 μm, respectively [16]. Therefore, performance, including wear, energy consumption, service life, etc., can be regulated by designing the size, area ratio, depth, and geometry of the surface micro-textures. These advantages can become more evident in a hydrodynamic lubrication environment, since micro-textures provide sufficient space for lubricant flow [17–19]. On the other hand, numerous variations of micro-texture brought difficulties to micro-texture design. For instance, He et al. analyzed the tribological properties between surfaces with round, square, and ring micro-textures and smooth cemented carbide surfaces, finding that the surface with round micro-textures showed a smaller friction coefficient, with a reduction of 20% compared with smooth surfaces [20]. In contrast, Wang et al. compared the friction and wear properties of texture-free surfaces and square, circular, and triangle micro-texture surfaces, indicating that the square micro-texture exhibited the optimal surface properties [21]. The optimal micro-texture geometry obtained by He et al. and Wang et al. was different because the investigations were based on a few types and a small range of parameters. Additionally, many other similar studies have been reported, but they only focus on limited parameters and values to optimize the texture [22–25]. In contrast, the coupling effect of multiple parameters and efficient methods for optimizing values in an extensive range are still unknown.

In a short summary, recent investigations focused on limited types and ranges of variants, ignoring the coupling effect of structural parameters on the bearing capacity and friction coefficient. Therefore, in order to gain a more comprehensive understanding of the structural parameter's impact on the micro-texture on the properties, this article broadens the types and range of parameters optimized, including the micro-texture size, area ratio, depth, and geometry, and compares the optimization results of the single-factor analysis (SFA) method and a multivariate linear regression (MLA) algorithm. The coupling effect and the importance of the parameters to the surface tribological behavior were investigated as well. After optimization, the optimal friction coefficient obtained by the MLA method had a further 4.9% reduction compared with that obtained by the SFA method.

2. Simulation Model Formulation

2.1. Problem Definition and Governing Equation

Figure 1a–d shows the two-dimensional (2D) model of surfaces with different micro-textures, including square micro-texture surface (Sq), rectangular micro-texture surface (Re), circular micro-texture surface (Ci) and slit micro-texture surface (Sl). The square micro-texture was composed of 16 square notches with a width of b on the substrate surface. The micro-textures were uniformly distributed on the substrate, and the area ratio between the total micro-textures and the substrate was denoted as area ratio S. Therefore, the relationship between the parameters a, b, and S can be expressed as Equation (1). Similarly, the rectangular, circular, and slit micro-textures were uniformly distributed on the substrate surface, with sizes of c, r, and d, where r was the radius of the circular micro-texture, and d was the width of the slit micro-texture, whose length was denoted as l. Notably, the length of the square micro-texture was considered as the base to guarantee the independence of the size and area ratio. The size of other micro-textures can be calculated via Equations (1)–(4).

$$a = \sqrt{\frac{16b^2}{S}} \quad (1)$$

$$c = \sqrt{\frac{b^2}{1.5}} \quad (2)$$

$$r = \sqrt{\frac{b^2}{\pi}} \tag{3}$$

$$l = 4b \tag{4}$$

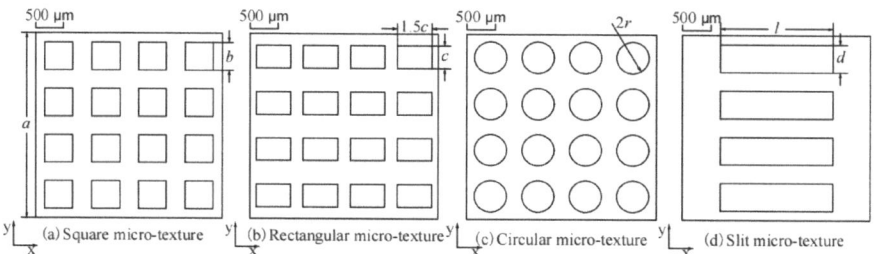

Figure 1. Two-dimensional geometric model with different surface micro-textures.

The micro-texture surfaces were coated with lubricants, and sliding friction occurred in the horizontal direction with friction pair. The lubrication model was established based on the Navier–Stokes (N–S) equation to solve the fluid domain. The following assumptions were considered:

1. The friction pair was rigid and had no deformation.
2. The fluid was incompressible with constant viscosity and density, and the volumetric forces were ignored.
3. The influence of the heat generation by friction was ignored, and the temperature was kept ambient.
4. The fluid was in laminar and constant mode.
5. Other basic assumptions of the N–S equations [26].
6. Therefore, the N–S equations and the continuity equation can be expressed as follows:

$$\text{X direction} \quad \rho\left(u\frac{\partial u}{\partial x} + v\frac{\partial u}{\partial y} + w\frac{\partial u}{\partial z}\right) = -\frac{\partial p}{\partial x} + \eta\left(\frac{\partial^2 u}{\partial x^2} + \frac{\partial^2 u}{\partial y^2} + \frac{\partial^2 u}{\partial z^2}\right) \tag{5}$$

$$\text{Y direction} \quad \rho\left(u\frac{\partial v}{\partial x} + v\frac{\partial v}{\partial y} + w\frac{\partial v}{\partial z}\right) = -\frac{\partial p}{\partial x} + \eta\left(\frac{\partial^2 v}{\partial x^2} + \frac{\partial^2 v}{\partial y^2} + \frac{\partial^2 v}{\partial z^2}\right) \tag{6}$$

$$\text{Z direction} \quad \rho\left(u\frac{\partial w}{\partial x} + v\frac{\partial w}{\partial y} + w\frac{\partial w}{\partial z}\right) = -\frac{\partial p}{\partial x} + \eta\left(\frac{\partial^2 w}{\partial x^2} + \frac{\partial^2 w}{\partial y^2} + \frac{\partial^2 w}{\partial z^2}\right) \tag{7}$$

$$\frac{\partial u}{\partial x} + \frac{\partial v}{\partial y} + \frac{\partial w}{\partial z} = 0 \qquad d \tag{8}$$

where u, v, and w represent the velocity of the fluid along the x, y, and z directions, respectively. ρ and η represent the density and dynamic viscosity of the lubricant, respectively. p represents the oil film pressure.

2.2. Boundary Conditions and Solution Method

The boundary conditions of the micro-texture fluid domain were considered based on the work performed by Liu et al. [27]. As shown in Figure 2, number 1 represents the fixed micro-texture surface, number 2 represents the smooth surface of the friction pair that moves at a velocity of 5 m/s, and numbers 3 and 4 represent the inlet and outlet boundary of pressure.

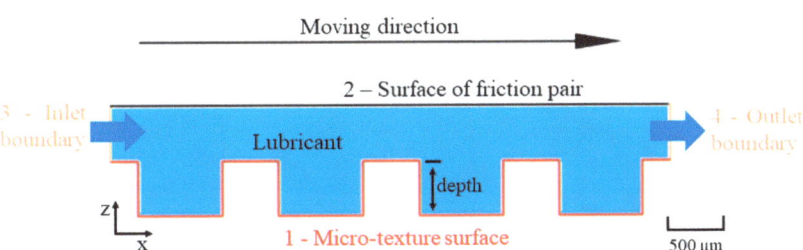

Figure 2. Boundary condition settings of micro-textured watershed.

The texture models were established in Solidworks (2024) and then imported to Ansys Fluent. The automatic meshing function was employed in Fluent (2022 R1) with a mesh tetrahedral and a mesh growth rate of 1.2. Since the Reynolds number of the fluid in the flow domain was low, the laminar flow mode was employed. The fluid density was set as 895 kg/m^3, and the dynamic viscosity was set to 0.0135145 Pa.s. The pressure with the second order and the momentum with the quick difference were coupled via the Simplec method.

2.3. Single-Factor Analysis Strategy

The parameters, including micro-texture size, area ratio, depth, and geometry, were considered. According to the literature, the most common method and parameters were orthogonal arrays of micro-texture with diameters of 0–300 μm, area ratios of 0%–64%, depths of 0–12 μm, and rectangular, circular, triangular, and rhombic geometries [19–22]. Therefore, 9, 5, and 7 values of size, area ratio, and depth, as well as 4 types of geometry, were chosen for SFA, as shown in Table 1. In the event of a change to any one of the parameters presented in the table, the remaining three parameters will remain unaltered.

Table 1. Parameter values used in the simulation with four geometries of micro-texture.

Geometry	Size (μm)	Area Ratio (%)	Depth (μm)	Fluid Density (kg/m^3)	Dynamic Viscosity (Pa.s)
Texture-free					
Square	a × a 25, 50, 100, 150, 200, 300, 400, 500, 600	10%, 20%, 30%, 40%, 50%	1, 2, 3, 5, 7, 8, 9	895	0.0135145
Rectangular	c × 1.5c				
Circular	r				
Slit	d × 1				

The hydrodynamic pressure effect is primarily observed in the vicinity of the micro-pit. As the lubricant flows into the micro-pit, a diverging wedge-shaped gap is formed due to the increased distance between the friction surfaces, resulting in a negative pressure in the diverging region. Conversely, when the lubricant flows out of the micro-pit, the distance between the surfaces decreases, leading to a converging wedge-shaped gap and a sudden increase in positive pressure in the converging region. It is noteworthy that the maximum value of the positive pressure is typically higher than the minimum value of the negative pressure, which results in an asymmetric pressure distribution at the edges of the micro-pit. This, in turn, generates an additional loading effect [18].

The bearing capacity of the lubricant film on the micro-texture surface can be calculated by an area-weighted integral of the positive pressure on the upper film surface, while an area-weighted integral of the shear stress on the moving surface along the x-direction of the lubricating film can determine the friction force. The corresponding equations can be expressed as follows:

$$F_z = \iint P \, dx \, dy \tag{9}$$

$$F_x = \iint \tau \, dx \, dy \tag{10}$$

$$\mu = \frac{F_x}{F_z} \tag{11}$$

where F_z and F_x are the bearing capacity of the lubricant film along the z- and x-directions, P is the pressure of the upper film surface, τ is the fluid shear stress, and μ is the friction coefficient.

2.4. Multivariate Linear Regression Optimization Strategy

Since the SFA method was labor-intensive, time-consuming, and less effective in the optimization of numerous parameters, the MLA algorithm was employed to investigate the influence of the micro-texture variations on the surface tribological properties. In comparison to other statistical techniques, MLA does not require the data to adhere to a normal distribution and can be employed as an initial analysis and a baseline indicator.

The data obtained in the single-factor simulation (1305 sets in total) were utilized in SPSS (SPSS26) (Statistical Package for the Social Sciences) optimization with MLA [28]. The parameter autocorrelation was initially examined by the Durbin–Watson test (D-W test) to guarantee the effectiveness of the model prediction [29]. Next, the relationship between the parameters and the properties can be expressed by a multivariate linear regression equation,

$$Y = \beta_0 + \beta_1 X_1 + \beta_2 X_2 + \beta_3 X_3 + \beta_4 X_4 + \varepsilon \tag{12}$$

where Y denotes the surface property of bearing capacity or friction coefficient, X_1, X_2, X_3, and X_4 denote the size, area ratio, depth, and geometry, β_1, β_2, β_3, and β_4 denote the regression coefficients of the variables, and ε denotes the random disturbance. Finally, the Technique for Order Preference by Similarity to Ideal Solution (TOPSIS) was employed to determine the global optimal solution out of multiple solutions [30]. In the context of the Scientific Platform Serving for Statistics Professional (SPSSPRO), the TOPSIS method represents a methodology employed for the assessment of the degree of superiority or inferiority of a solution [28]. In spatial stochastic simulations, multiple potential solutions are typically generated. The superior–inferior solution distance method can be employed to ascertain the relative merits of the solutions by calculating the distances between them. This method can assist in identifying the optimal solution or in selecting between multiple solutions [31]. It offers an objective approach to evaluating solutions, facilitating the determination of the best solution or an effective comparison between multiple solutions, enhancing the accuracy and reliability of simulation results, and circumventing subjective bias.

In order to gain a deeper understanding of regression analyses, this study employs two key statistical validation methods: the coefficient of determination (R^2) and the correlation coefficient (R). The R^2 value indicates the proportion of the variance in the dependent variable that can be explained by the independent variable, and it is an essential indicator for evaluating the explanatory power of the model. The correlation coefficient, on the other hand, quantifies the strength of the linear relationship between the parameters and the surface properties, as well as its direction. The detailed investigation procedure is shown in Figure 3.

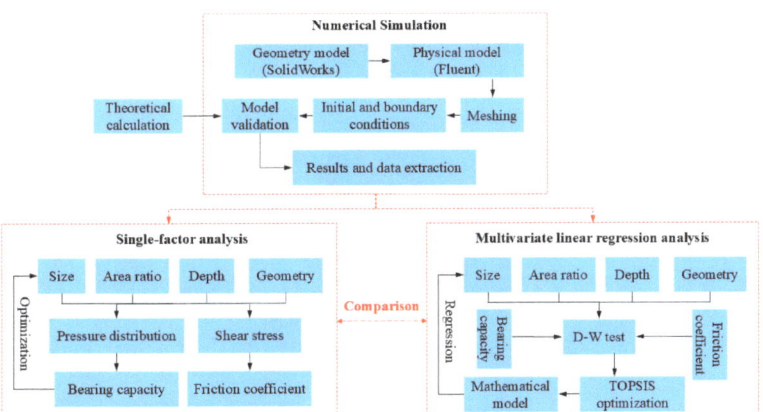

Figure 3. Flow chart of the research procedure.

3. Single-Factor Simulation Results and Discussion

3.1. Model Validation

The texture-free (Tf) surfaces with lengths of 1264.9 µm, 2529.8 µm, and 6325 µm were selected for the model validation. The length was calculated based on the square micro-texture model with an area ratio of 10% and texture lengths of 100, 200, and 500 µm. The shear stresses obtained by the simulation and theoretical calculation were compared. The equation for the shear stress of the texture-free surface is presented below [27]. The comparison results are shown in Table 2. The theoretical shear stresses for the three surfaces were all 8446.56 Pa, while the simulation results were 8545.37, 512.18, and 8475.20 Pa. The corresponding deviations were 1.17%, 0.78%, and 0.34%. The maximum deviation of 1.17% was within the allowable range, proving that the simulation model was accurate.

$$\tau = \eta \frac{du}{dz} = \eta \frac{U}{h_0} \tag{13}$$

where τ is the shear stress, η is the dynamic viscosity of the lubricant, du/dz is the velocity gradient, U is the ratio of the sliding speed, and h_0 is the film thickness.

Table 2. Comparison of theoretical and simulated values and errors of shear stress.

Length/µm	Theoretical τ_1/Pa	Simulation τ_2/Pa	Deviation δ/%
1264.9	8446.56	8545.37	1.17%
2529.8	8446.56	8512.18	0.78%
6325	8446.56	8475.20	0.34%

3.2. Influence of Micro-Texture Size on the Surface Tribological Properties

Th texture-free and square micro-texture surfaces with a depth of 5 µm and an area ratio of 50% were chosen. Figure 4 presents the oil film pressure distribution on the square micro-texture surfaces with varying sizes. It was observed that the peak oil film pressure escalates as the size enlarges. After calculation, the influence of the micro-texture size on the surface bearing capacity can be obtained, as shown in Figure 5a. The x coordinate denoted the length of the square micro-texture surface, while the size of the texture-free surface was calculated based on Equation (1). The bearing capacity variation of both surfaces exhibited a similar trend, corresponding to an increase within the length ranges from 25 µm to 100 µm, 150 µm to 400 µm, and 500 µm to 600 µm, and to a decrease within the length ranges from 100 µm to 150 µm, and 400 µm to 500 µm. The maximum and minimum values were obtained at the lengths of 600 µm and 25 µm, respectively. The surface with the square micro-texture had a higher load-bearing capacity than the texture-free surface when the size was below 500 μm, while it was lower when the size exceeded 500 μm, as the orange and blue regions show.

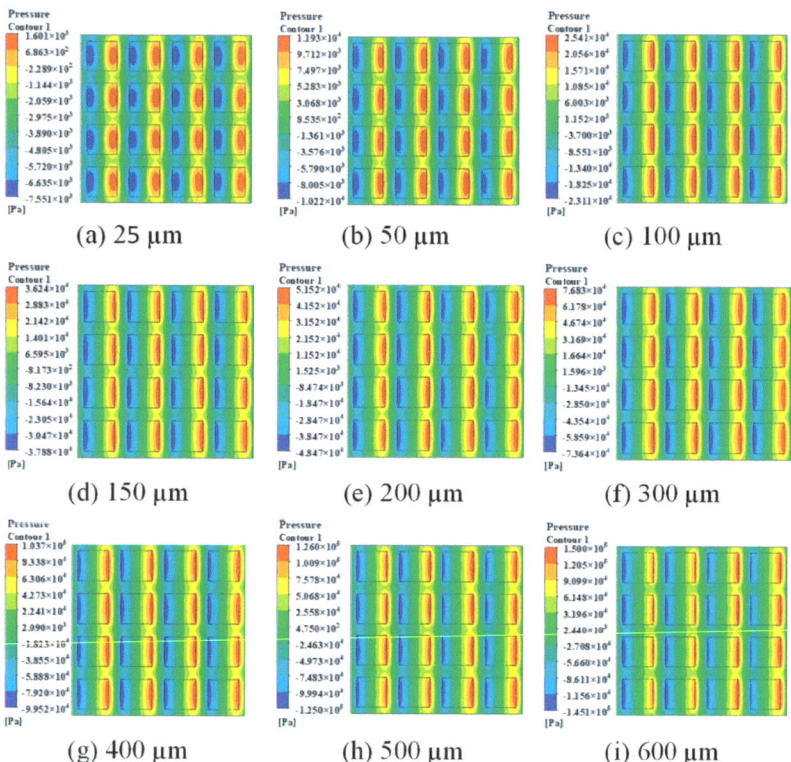

Figure 4. Pressure distribution of oil film on the square micro-texture surfaces with varying sizes from 25 μm to 600 μm under different texture sizes, at depth of 5 μm and with an area ratio of 50%.

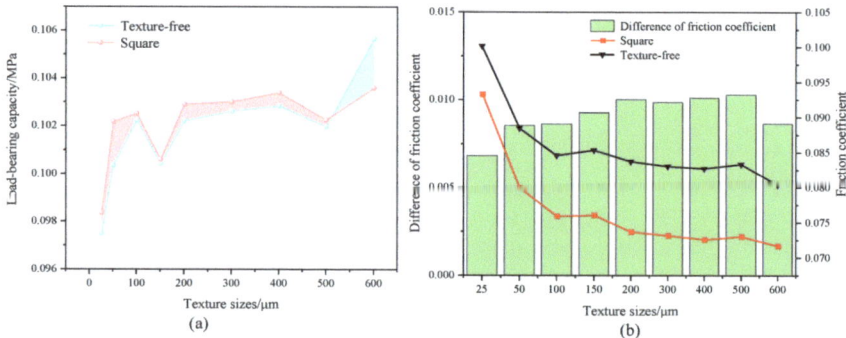

Figure 5. (**a**) Load-bearing capacity and (**b**) friction force variation of texture-free and square micro-texture surfaces with different texture sizes, at a depth of 5 μm and with an area ratio of 50%.

The influence of the micro-texture size on the friction coefficient is shown in Figure 5b. Similarly, the friction coefficient variations of both surfaces showed a similar trend within the specific length range as the bearing capacity variation. The maximum and minimum friction coefficient were also obtained at 600 μm and 25 μm, respectively. Notably, the friction coefficients of the square micro-texture surface were smaller than those of the texture-free surfaces, indicating that the square micro-texture surface had the best friction properties and bearing capacity with a size of 500 μm.

The phenomena shown above can be explained by the fact that when the surfaces of the friction pair moved relative to each other, the viscous force of the lubricant could drive the fluid between the micro-texture surface to move and thereby form a wedge-shaped gap. This flow behavior produced a hydrodynamic effect, which, in turn, increased the bearing capacity and reduced the friction coefficient. The amount of lubricant along the micro-texture surface increased with the increasing micro-texture size, enhancing the fluid hydrodynamic effect. This may be the reason for the increase in the bearing capacity and decrease in the friction coefficient with the increasing micro-texture size. It was noteworthy that sudden drops in micro-texture load-carrying capacity were observed at 150 μm and 500 μm texture sizes (see Figure 5a), which may be attributed to the change in local flow behavior caused by the size effect, breaking the oil film. When the size was small, the size effect was evident, and the local flow behavior was unstable. Thus, in some cases, the load-bearing capacity may have decreased. The size effect was weakened, and the flow behavior became stable with the increased size. Therefore, the drop at 500 μm was smaller than the drop at 150 μm.

3.3. Influence of Micro-Texture Area Ratio on the Surface Tribological Properties

Since the optimal bearing capacity and friction coefficient were obtained for the square micro-texture surface with the size of 500 μm, square micro-texture surfaces with different area ratios with a depth of 5 μm and a length of 500 μm are investigated in this section.

Figure 6 presents the oil film pressure distribution on square micro-texture surfaces with varying area ratios. It was observed that as the area ratios of the square texture escalated, the oil film pressure initially rose and, subsequently, diminished. Figure 7a shows the impact of the area ratio on the bearing capacity for the oil film on the square micro-texture surfaces and texture-free surfaces. Regarding the square micro-texture surfaces, the bearing capacity of the oil film increased as the area ratio rose from 10% to 30%, followed by a decrease from 30% to 50%, reaching its maximum and minimum at area ratios of 30% and 10%, respectively. Regarding the texture-free surfaces, the bearing capacity initially increased, followed by a decrease and then an increase. The maximum and minimum were obtained at the area ratios of 30% and 40%, respectively. However, the friction coefficient variation showed a different trend with the increased area ratio, as shown in Figure 7b. The friction coefficient initially decreased, followed by increasing and then decreasing for the texture-free surfaces, while for the square micro-texture surfaces, it continuously decreased. The maximum and minimum of both surfaces were obtained at the area ratios of 10%, 50%, 40%, and 30%, respectively. However, the maximum difference in the friction coefficient between the surfaces was obtained at the area ratio of 40%. This was because, on the one hand, the distribution of the micro-texture became denser with the increased area ratio, thus enhancing the hydrodynamic effect, resulting in the increased bearing capacity. On the other hand, the improved integrity of the oil film can reduce the direct contact between the friction pair and, thus, reduce the friction coefficient. However, the lubricant film could become unstable when the distribution of micro-texture becomes too dense, leading to a reduction in the bearing capacity.

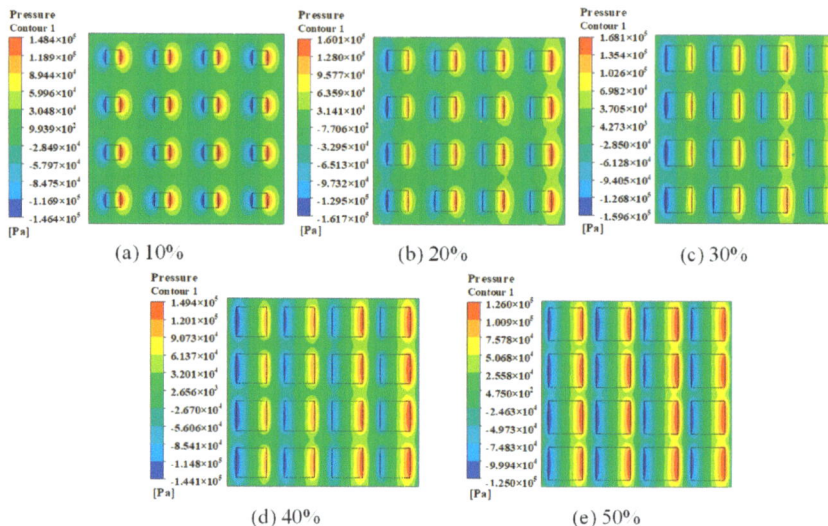

Figure 6. Pressure distributions of oil film on the square micro-texture surfaces with varying area ratios from 10% to 50% under different area ratios, at a depth of 5 μm and with a length of 500 μm.

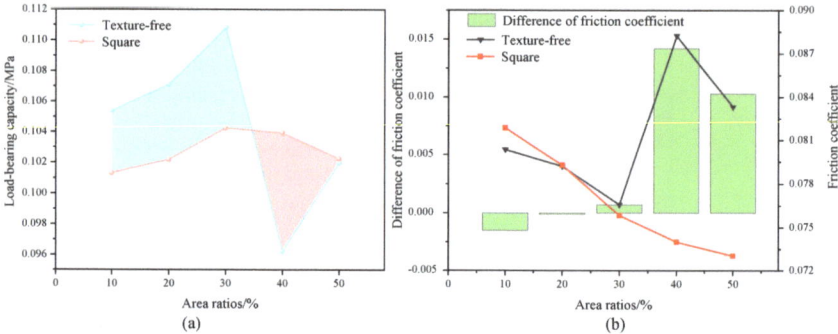

Figure 7. (a) Load-bearing capacity and (b) friction force variation of texture-free and square micro-texture surfaces under different area ratios, at a depth of 5 μm and with a length of 500 μm.

3.4. Influence of Micro-Texture Depth on the Surface Tribological Properties

Similarly, square micro-texture and texture-free surfaces with an area ratio of 40% and a length of 500 μm are discussed in this section.

Figure 8 illustrates the oil film pressure distribution on the square micro-texture surfaces with varying depths. The oil film pressure reached its maximum at the outlet boundary of the micro-texture and reached its minimum at the inlet boundary. Additionally, the maximum pressure initially increased and subsequently decreased with the increased depths. Figure 9a,b shows the variation in the bearing capacity and friction coefficients of both surfaces with increasing depth. Regarding the square micro-texture surfaces, the bearing capacity initially increased and then decreased with increasing depth. Conversely, the friction coefficient showed a reverse trend, corresponding to an initial decrease and then an increase. The maximum and minimum bearing capacities were obtained at depths of 5 μm and 9 μm, and the corresponding friction capacities were obtained at depths of 1 μm and 5 μm. Wu et al. investigated micro-texture depths between 2.86 and 7.42 μm and observed that the friction coefficient diminishes as the depth increases [31]. Conversely, in our study, we examined micro-textures ranging from 1 to 9 μm and determined that the friction coefficient would escalate after decreasing with an increase in texture depth. This was because the value range in Wu et al.'s work was limited, and the coupling effect of other parameters was ignored.

These phenomena were caused by the increase in the amount of lubricant as the depth of the micro-texture increased, offering an adequate lubrication effect to the friction pair and improving the hydrodynamic effect. However, the shear effect of the fluid gradually dominated the process when the depth exceeded the threshold, leading to the generation of a swirl phenomenon, which weakened the hydrodynamic effect. Therefore, the carrying capacity decreased, and the friction coefficient increased. Notably, the properties of the square micro-texture surfaces were better than those of the texture-free surfaces.

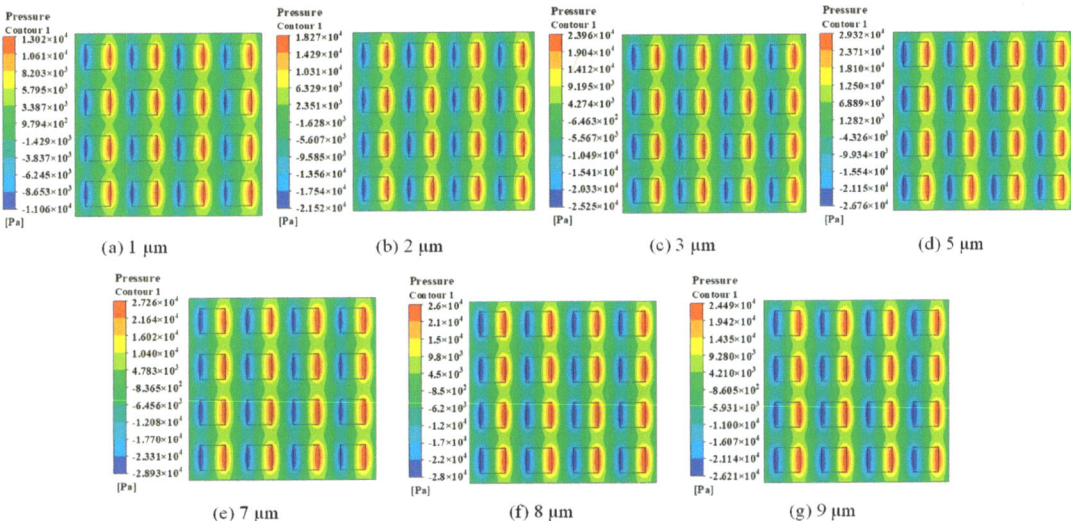

Figure 8. Pressure distributions of oil film on the square micro-texture surfaces with varying depths from 1 μm to 9 μm, with an area ratio of 40% and a length of 500 μm.

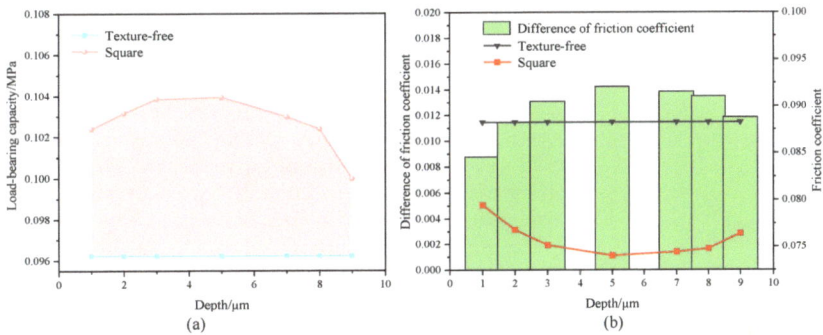

Figure 9. (**a**) Load-bearing capacity and (**b**) friction force variation of texture-free and square micro-texture surfaces at different depths, with an area ratio of 40% and a length of 500 μm.

3.5. Influence of Micro-Texture Geometry on the Surface Tribological Properties

According to the aforementioned results, five surfaces with different micro-texture geometries with a length of 500 μm, an area ratio of 40%, and a depth of 5 μm are discussed in this section.

According to the pressure distribution shown in Figure 10a–d, the pressure difference between the middle columns of the square and the circular micro-texture surfaces was diminished. This may be attributed to the large flow space, leading to a lack of lubricant in the middle of the micro-texture and, thus, impeding the hydrodynamic effect of the fluid. In contrast, the rectangular and slit micro-texture produced a conduction effect on the fluid, enhancing the inertial effect. Simultaneously, the small exit area enhanced the hydrodynamic impact, and the pressure around the region with high pressure would have been more uniform because of the transmissibility of the hydrodynamic effect.

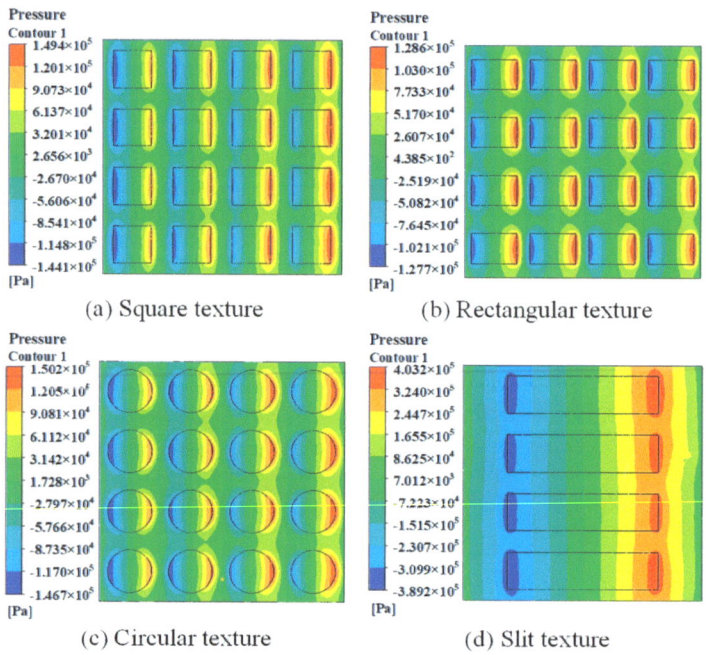

Figure 10. Pressure distribution contours of different texture shapes with a length of 500 μm, an area ratio of 40%, and a depth of 5 μm. (**a**–**d**) are pressure clouds for square, rectangle, circle, and slit, respectively.

Figure 11a shows the variation in the bearing capacity and friction coefficients for the surfaces with different micro-texture geometries. The surfaces possessing high bearing capacity can be sequenced as Sl, Sq, Ci, Re, and Tf, and the surfaces possessing low friction coefficients can be sequenced as Sl, Sq, Re, Ci, and Tf. Noticeably, all the surfaces with micro-textures exhibited higher bearing capacity and lower friction coefficient than the surfaces without micro-textures. Moreover, the properties of the rectangular and slit micro-texture surfaces were better than those of the square and circular micro-texture surfaces. He et al. conducted a comparative analysis of circular, square, and annular micro-textured surfaces, concluding that circular micro-textured surfaces exhibited superior friction reduction performance [20]. However, the findings of this paper suggest that under identical parameters, the friction reduction performance of slit micro-textured surfaces surpasses that of circular microtextured surfaces. Figure 11b shows the standard deviation of the load-bearing capacity and friction coefficient under four types of parameters, indicating the parameter sensitivity for the properties. The standard deviation of the load-bearing capacity was 0.0019, 0.0039, 0.0035, and 0.0041 for the parameters of texture size, area ratio, depth, and geometry, indicating that the geometry was the most important. Similarly, the standard deviation of the friction coefficient was 0.0077, 0.0046, 0.0066, and 0.0069 for the parameters of texture size, area ratio, depth, and geometry, indicating that the texture size was the most important.

In a short conclusion, the optimal friction coefficient of 0.070409 can be obtained under the conditions of 500 μm micro-texture width, 40% area ratio, 5 μm depth, and slit geometry, showing a reduction of 10.7% compared with texture-free surfaces under the same conditions. The bearing capacity under these conditions can reach 107,653 Pa, which is not optimal because of the difficulty of balancing the normal and shear stress distribution on the surface.

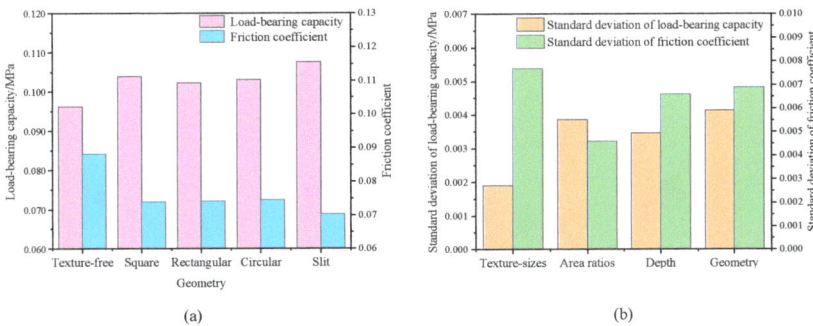

Figure 11. (a) Load-bearing capacity and friction force of different texture geometries with a length of 500 μm, an area ratio of 40%, and a depth of 5 μm. (b) Standard deviation of load-bearing capacity and friction coefficient under different variants.

4. Regression Optimization Results and Discussion

4.1. Parameter Correlation Analysis

According to the D-W test, Table 3 was obtained, where the DW values of model 1 (the dependent variable was bearing capacity) and model 2 (the dependent variable was friction coefficient) were 1.789 and 1.121, respectively, indicating that the linear model did not have autocorrelation. Furthermore, the results of the analysis suggest that the parameters had a weaker influence on the bearing capacity and a substantial impact on the friction coefficient. This can be approximated as R^2 values of 0.057 for model 1 and 0.417 for model 2, with deviations approximated as 3251.49 and 0.0038, respectively.

Table 3. Durbin–Watson test.

Model	R	R^2	Adjusted R^2	Deviation	DW
1	0.239a	0.057	0.054	3251.49086	1.789
2	0.646a	0.417	0.415	0.0038201092	1.121

4.2. Multivariate Linear Regression Optimization

The 1305 sets of data, including the parameters, bearing capacity, friction force, and friction coefficient, were imported into SPSSPRO for analysis using the TOPSIS method. The bearing capacity was designated as a positive indicator, while the friction force and coefficient were designated as negative indicators. The weights of the variables were determined using the entropy weight method. Consequently, the coefficients of the MLA equation were obtained, as shown in Table 4, and thus, the equations for models 1 and 2 can be expressed as follows:

$$Y_1 = 101692.897 - 29.1X_1 + 3.474X_2 - 2161.247X_3 + 244.656X_4 \tag{14}$$

$$Y_2 = 0.090 - 0.000319X_1 - 1.336 \times 10^{-5}X_2 - 0.011X_3 - 0.001X_4 \tag{15}$$

where Y_1 and Y_2 denote the surface properties of the load-bearing capacity and friction coefficient, and X_1, X_2, X_3, and X_4 are the size, area ratio, depth, and geometry, respectively.

According to Table 4, the significance of depth for model 1 was 0.353, which was greater than 0.05. Therefore, depth was not a significant factor. According to the importance of parameters in models 1 and 2, the parameters to which the bearing capacity is most sensitive can be sequenced as size, area ratio, geometry, and depth. Similarly, the friction coefficient's sensitivity to parameters can be sequenced as size, area ratio, depth, and geometry. The parameter sensitivity of the friction coefficient was consistent with the results obtained by Wu et al. [32], which showed that the impact of depth was more minor than that of the area ratio on the friction coefficient. Notably, the sequence was different from the results obtained by the standard deviation in the SFA, possibly because the amount of research data limited the results in the SFA.

Table 4. Coefficients of multivariate linear regression equation.

Model		USC		SC	t	S	CS	
		B	SD	Beta			T	VIF
1	(Constant)	101692.897	349.919		290.618	0.000		
	Depth	−29.100	31.326	−0.025	−0.929	0.353	0.995	1.005
	Size	3.474	0.467	0.201	7.447	0.000	1.000	1.000
	Area ratio	−2161.247	636.447	−0.091	−3.396	0.001	1.000	1.000
	Geometry	244.656	75.868	0.087	3.225	0.001	0.995	1.005
2	(Constant)	0.090	0.000		218.272	0.000		
	Depth	0.000319	0.000	0.184	−8.660	0.000	0.995	1.005
	Size	-1.336×10^{-5}	0.000	−0.516	−24.375	0.000	1.000	1.000
	Area ratio	−0.011	0.001	−0.312	−14.718	0.000	1.000	1.000
	Geometry	−0.001	0.000	−0.155	−7.318	0.000	0.995	1.005

USC—unstandardized coefficient, SC—standardized coefficient, SD—standard deviation, S—significance, CS—covariance statistic, T—tolerances.

After TOPSIS optimization, the optimal properties with a bearing capacity of 105,569.3 Pa and a friction coefficient of 0.067844 were obtained with a size of 600 μm, an area ratio of 50%, a depth of 9 μm, and a geometry of slit. Compared with the texture-free surface, the friction coefficient was reduced by 15.6%. Compared with the result obtained by the SFA, the friction coefficient was further reduced, by 4.9%.

In a short conclusion, the parameters to which the bearing capacity is most sensitive can be sequenced as size, area ratio, geometry, and depth. The parameters to which the friction coefficient is most sensitive can be sequenced as size, area ratio, depth, and geometry. The optimal friction coefficient of 0.067844 was obtained under the conditions of 600 μm micro-texture size, 50% area ratio, 9 μm depth, and slit geometry, showing a 15.6% reduction compared with the texture-free surfaces under the same conditions, and a 4.6% reduction compared with the optimal surfaces with the SFA method. The corresponding bearing capacity was 105,569.3 Pa.

5. Conclusion and Perspectives

This article numerically investigated the influence of the micro-texture size, area ratio, depth, and geometry on the surface bearing capacity and friction coefficient by using SFA. The MLA method was employed to optimize the parameters to improve the surface properties further. The following conclusions and perspectives can be drawn:

1. The optimal parameters obtained by the SFA were a slit micro-texture 500 μm in size, with a 40% area ratio, and 5 μm in depth. The corresponding bearing capacity and friction coefficient were 107,653 Pa and 0.070409, showing a reduction of 10.7% in the friction coefficient compared with those of the texture-free surfaces with the same parameters.

2. The results of the MLA algorithm indicated that the parameters to which the bearing capacity was sensitive were sequenced as size, area ratio, geometry, and depth, with size and geometry exhibiting a positive correlation, while the depth and area ratio exhibited a negative correlation. Regarding the friction coefficient, the importance of the parameters can be sequenced as size, area ratio, depth, and geometry, with all the parameters exhibiting a negative correlation.
3. The optimal parameters obtained by the MLA were a slit micro-texture 600 μm in size, with an area ratio of 50%, and 9 μm in depth. The corresponding bearing capacity and friction coefficient were 105,569.3 Pa and 0.067844, showing a 15.6% reduction in the friction coefficient compared with those of the texture-free surfaces with the same parameters and a 4.9% reduction compared with those of the optimal surfaces obtained by the SFA.
4. The MLA algorithm can analyze the micro-texture parameters within a more extensive range and broaden the understanding of the coupling effect of these parameters, which is why it is one of the promising statistical analysis methods for optimizing micro-texture parameters in the future. Similarly, other algorithms, such as random forest, deep neural networks, multimodal machine learning, etc., could also be used for comprehensive analysis and prediction in the future. More types of variants can be introduced in these algorithms for better prediction results. However, these methods require a large amount of data while being time-consuming and wasteful. Reliable mathematical models may be among the potential methods to predict properties at a low cost.

Author Contributions: Z.G., Methodology, Investigation, Writing—Review and Editing. Q.H., Investigation, Writing—Original Draft. H.Z., Methodology, Investigation, Writing—Review and Editing. Y.Z., Supervision, Writing—Review and Editing. All authors have read and agreed to the published version of the manuscript.

Funding: This work was financially supported by the Jiangsu Agriculture Science and Technology Innovation Fund (CX(21)3154), the National Natural Science Foundation of China (52175438), and the Innovation Cultivation Fund of Yangzhou University (135030421).

Data Availability Statement: Data are contained within the article.

Conflicts of Interest: The authors declare no conflicts of interest.

References

1. Holmberg, K.; Erdemir, A. Influence of Tribology on Global Energy Consumption, Costs and Emissions. *Friction* **2017**, *5*, 263–284. [CrossRef]
2. Huang, J.; Guan, Y.; Ramakrishna, S. Tribological Behavior of Femtosecond Laser-Textured Leaded Brass. *Tribol. Int.* **2021**, *162*, 107115. [CrossRef]
3. Li, S.; Yang, X.; Kang, Y.; Li, Z.; Li, H. Progress on Current-Carry Friction and Wear: An Overview from Measurements to Mechanism. *Coatings* **2022**, *12*, 1345. [CrossRef]
4. Wang, P.; Liang, H.; Jiang, L.; Qian, L. Effect of Nanoscale Surface Roughness on Sliding Friction and Wear in Mixed Lubrication. *Wear* **2023**, *530*, 204995. [CrossRef]
5. Masuko, M.; Tomizawa, K.; Aoki, S.; Suzuki, A. Experimental Study on the Effects of Sliding Velocity and Roughness Orientation of Metal Surfaces on the Function of Lubricant Additives in Controlling Friction in a Boundary Lubrication Regime. In *Tribological Research and Design for Engineering Systems*; Dowson, D., Priest, M., Dalmaz, G., Lubrecht, A.A., Eds.; Elsevier Science Bv: Amsterdam, The Netherlands, 2003; pp. 791–798.
6. Zhang, M.; Zhang, J.; Yang, Z.; Bai, J.; Guan, R. Interface of AlCrTiN/6061 laminated composite fabricated by one-pass rolling and its tribological properties. *Mater. Charact.* **2024**, *216*, 114299. [CrossRef]
7. Xie, X.; Liu, X.; He, B.; Cheng, W.; Zhang, F.; Liang, J. Fabrication and tribological behavior of laser cladding Cu/Ti$_3$SiC$_2$ reinforced CoCrW matrix composite coatings on Inconel718 surface. *Surf. Coat. Technol.* **2024**, *493*, 131289. [CrossRef]
8. Li, M.; Liu, S.; Cao, M.; Zhou, Z.; Rao, J.; Zhang, Y. Fabrication of natural mesoporous Diatom biosilica microcapsules with high oil-carrying capacity and tribological property. *Tribol. Int.* **2024**, *200*, 110170. [CrossRef]
9. Chandrasekaran, S.; Check, J.; Sundararajan, S.; Shrotriya, P. The Effect of Anisotropic Wet Etching on the Surface Roughness Parameters and Micro/Nanoscale Friction Behavior of Si(100) Surfaces. *Sens. Actuators A-Phys.* **2005**, *121*, 121–130. [CrossRef]
10. Wan, Q.; Gao, P.; Zhang, Z. Friction and Wear Performance of Lubricated Micro-Textured Surface Formed by Laser Processing. *Surf. Eng.* **2021**, *37*, 1523–1531. [CrossRef]

11. Chen, W.; Xia, M.; Song, W. Study on the Anti-Friction Mechanism of Nitriding Surface Texture 304 Steel. *Coatings* **2020**, *10*, 554. [CrossRef]
12. Na, Z.; Zhentao, L.; Muming, H.; Yancong, L.; Baojie, R.; Wang, Z. Numerical Simulation and Experimental Investigation on Tribological Performance of SiC Surface with Squamous Groove Micro Texture. *Lubr. Sci.* **2022**, *34*, 547–562. [CrossRef]
13. Hu, T.; Hu, L.; Ding, Q. Effective Solution for the Tribological Problems of Ti-6Al-4V: Combination of Laser Surface Texturing and Solid Lubricant Film. *Surf. Coat. Technol.* **2012**, *206*, 5060–5066. [CrossRef]
14. Yangyi, X.; Jing, L.; Wankai, S.; Minglin, K. Heavy Load Elastohydrodynamic Lubrication Performance of Surface Micro-textured Coating-substrate System. *Surf. Technol.* **2020**, *49*, 159–167.
15. Chen, D.; Ding, X.; Yu, S.; Zhang, W. Friction Performance of DLC Film Textured Surface of High Pressure Dry Gas Sealing Ring. *J. Braz. Soc. Mech. Sci. Eng.* **2019**, *41*, 161. [CrossRef]
16. Tong, X.; Shen, J.; Su, S. Properties of Variable Distribution Density of Micro-Textures on a Cemented Carbide Surface. *J. Mater. Res. Technol.* **2021**, *15*, 1547–1561. [CrossRef]
17. Li, D.; Yang, X.; Lu, C.; Cheng, J.; Wang, S.; Wang, Y. Tribological Characteristics of a Cemented Carbide Friction Surface with Chevron Pattern Micro-Texture Based on Different Texture Density. *Tribol. Int.* **2020**, *142*, 106016. [CrossRef]
18. Selvakumar, S.J.; Muralidharan, S.M.; Raj, D.S. Performance Analysis of Drills with Structured Surfaces When Drilling CFRP/O7075 Stack under MQL Condition. *J. Manuf. Process.* **2023**, *89*, 194–219. [CrossRef]
19. Wang, Z.; Hu, S.; Zhang, H.; Ji, H.; Yang, J.; Liang, W. Effect of Surface Texturing Parameters on the Lubrication Characteristics of an Axial Piston Pump Valve Plate. *Lubricants* **2018**, *6*, 49. [CrossRef]
20. He, C.; Yang, S.; Zheng, M. Analysis of Synergistic Friction Reduction Effect on Micro-Textured Cemented Carbide Surface by Laser Processing. *Opt. Laser Technol.* **2022**, *155*, 108343. [CrossRef]
21. Wang, Z.; Sun, L.; Han, B.; Wang, X.; Ge, Z. Study on the Thermohydrodynamic Friction Characteristics of Surface-Textured Valve Plate of Axial Piston Pumps. *Micromachines* **2022**, *13*, 1891. [CrossRef]
22. Ye, J.; Zhang, H.; Liu, X.; Liu, K. Low Wear Steel Counterface Texture Design: A Case Study Using Micro-Pits Texture and Alumina-PTFE Nanocomposite. *Tribol. Lett.* **2017**, *65*, 165. [CrossRef]
23. Chen, D. Influence of Microtextured Parameters of Dry Gas Sealing Rings on Tribological Performance. *Ind. Lubr. Tribol.* **2024**, *76*, 464–473. [CrossRef]
24. Han, J.; Jiang, Y.; Li, X.; Li, Q. Conical Grinding Wheel Ultrasonic-Assisted Grinding Micro-Texture Surface Formation Mechanism. *Machines* **2023**, *11*, 428. [CrossRef]
25. Durairaj, S.; Guo, J.; Aramcharoen, A.; Castagne, S. An Experimental Study into the Effect of Micro-Textures on the Performance of Cutting Tool. *Int. J. Adv. Manuf. Technol.* **2018**, *98*, 1011–1030. [CrossRef]
26. Jia, J.; Hao, Y.; Ma, B.; Xu, T.; Li, S.; Xu, J.; Zhong, L. Gas-Liquid Two-Phase Flow Field Analysis of Two Processing Teeth Spiral Incremental Cathode for the Deep Special-Shaped Hole in ECM. *Int. J. Adv. Manuf. Technol.* **2023**, *127*, 5831–5846. [CrossRef]
27. Liu, W.; Ni, H.; Chen, H.; Wang, P. Numerical Simulation and Experimental Investigation on Tribological Performance of Micro-Dimples Textured Surface under Hydrodynamic Lubrication. *Int. J. Mech. Sci.* **2019**, *163*, 105095. [CrossRef]
28. Dunn, P. SPSS Survival Manual: A Step by Step Guide to Data Analysis Using IBM SPSS. *Aust. N. Z. J. Public Health* **2013**, *37*, 597–598. [CrossRef]
29. Jeong, J.; Chung, S. Bootstrap Tests for Autocorrelation. *Comput. Stat. Data Anal.* **2001**, *38*, 49–69. [CrossRef]
30. Allard, D.J.-P.; Chilès, P. Delfiner: Geostatistics: Modeling Spatial Uncertainty. *Math. Geosci.* **2013**, *45*, 377–380. [CrossRef]
31. Paul, J.; Pierre, D. *Geostatistics: Modeling Spatial Uncertainty*; John Wiley & Sons: Hoboken, NJ, USA, 2012. [CrossRef]
32. Wu, F.; Liu, N.; Ma, Y.; Zhang, X.; Han, Y. Research on the influence of diamond coating microtexture on graphitization law and friction coefficient. *Diam. Relat. Mater.* **2022**, *127*, 109153. [CrossRef]

Disclaimer/Publisher's Note: The statements, opinions and data contained in all publications are solely those of the individual author(s) and contributor(s) and not of MDPI and/or the editor(s). MDPI and/or the editor(s) disclaim responsibility for any injury to people or property resulting from any ideas, methods, instructions or products referred to in the content.

Article

Improvement of Methods and Devices for Multi-Parameter High-Voltage Testing of Dielectric Coatings

Vladimir Syasko [1], Alexey Musikhin [2,*] and Igor Gnivush [3]

1. D.I. Mendeleev Institute for Metrology (VNIIM), Moskovsky Ave, 19, 190005 St. Petersburg, Russia; 9334343@gmail.com
2. LLC "KONSTANTA", Ogorodny Lane, 21, 198095 St. Petersburg, Russia
3. Faculty of Mechanical Engineering, Empress Catherine II Saint Petersburg Mining University, 21st Line, 2, 199106 St. Petersburg, Russia; klin4_g@mail.ru
* Correspondence: musikhinaleksei@gmail.com; Tel.: +7-981-129-14-73

Abstract: Currently, the high voltage testing method is widely used to detect pinholes and porosity defects in dielectric coatings. However, most modern coatings also have requirements for the minimum allowable coating thickness. Conducting tolerance tests on the thickness of dielectric coatings concurrently along with monitoring integrity within a single technological process appears promising. Additionally, mitigating the impact of various interfering parameters is crucial. This paper conducts a theoretical and experimental examination of spark formation processes in both gas and dielectrics. This analysis takes place during the identification of both through and non-through defects in dielectric coatings on conductive substrates. The principles of selecting the test voltage for the investigated dielectric coatings, considering the need to detect both through defects and inadmissible thinning, are theoretically and experimentally justified. It is suggested to utilize a probabilistic approach for evaluating the detectability of the mentioned defects. It is demonstrated that, when the dielectric strength of the coating is known, it is feasible to identify both through and non-through defects in coatings with a calculated probability under a specified test voltage. The conditions of occurrence of partial discharges in the process of testing are investigated, and measures to suppress their influence on the inspection results are proposed. The influence of the substrate surface roughness on the magnitude of the breakdown voltage during testing is considered.

Keywords: dielectric strength; holiday detection; breakdown voltage; continuity; coating; thickness

Citation: Syasko, V.; Musikhin, A.; Gnivush, I. Improvement of Methods and Devices for Multi-Parameter High-Voltage Testing of Dielectric Coatings. *Coatings* **2024**, *14*, 427. https://doi.org/10.3390/coatings14040427

Academic Editor: Binbin Zhang

Received: 9 February 2024
Revised: 21 March 2024
Accepted: 27 March 2024
Published: 1 April 2024

Copyright: © 2024 by the authors. Licensee MDPI, Basel, Switzerland. This article is an open access article distributed under the terms and conditions of the Creative Commons Attribution (CC BY) license (https://creativecommons.org/licenses/by/4.0/).

1. Introduction

Today, testing of protective dielectric coatings is carried out in most countries on pipeline transportation infrastructure facilities, production, and other structures, and is regulated by a significant amount of regulatory documentation [1–5]. One of the main requirements for such coatings is the absence of through defects, while the need to identify non-through pores, scratches, and inadmissible thinning is not regulated, although such defects also affect the protective properties of coatings and reduce the service life of the final product [6–8].

One of the most common nondestructive testing (NDT) methods [4,5] for testing dielectric coating continuity is the high-voltage method based on the occurrence of discharges at the formation of a high-intensity electric field (E) between the coating surface and a conductive substrate (Figure 1) [9]. The sensitivity of the method is achieved by differences in the dielectric strength of the flawless and defective areas of the coating.

When considering the existing methods of high-voltage testing, it becomes clear that they do not pay proper attention to the connection between the coating thickness (d_c) and its breakdown voltage (U_{dc}), which is essential for the purpose of identifying specific areas of inadmissible thinning and con-through coating defects. In addition, the influence of factors caused by the parameters of the testing items is not considered (in particular, changes in the electric field pattern at areas of roughness on the substrate surface and partial discharges caused, for example, by the undulations of the product surface), along with the conditions of inspection, in connection with which the development of a single refined methodology to detect not only con-through coating defects but also inadmissible thinning, considering the influence of interfering parameters, is a relevant task.

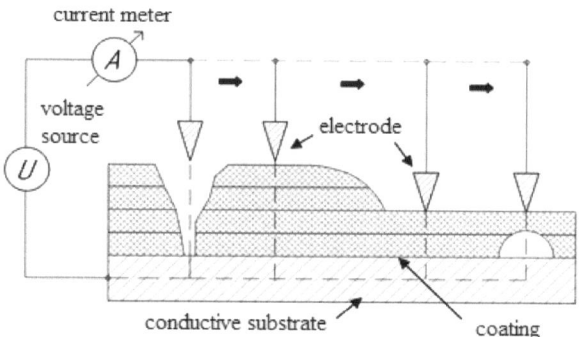

Figure 1. Schematic representation of the principle of the high-voltage testing method.

In [10], the authors proposed to increase the reliability of a high-voltage testing method by introducing a system of standard sample certification. However, the issues of identifying the thinning of the coating, through defects, and the degree of influence of interfering parameters on the occurrence of false positive and false negative test results were not covered in any way.

Works [11,12] describe a methodology for the integrated use of high-voltage testing methods and electric capacitance methods to identify the unacceptable thinning of cable products based on changes in the interelectrode electrical capacitance. It proposes to detect through defects according to standard methods of high-voltage testing. The works describe that thinning can be detected if its thickness is no more than 65% of the maximum cable thickness, and the final decision on the presence of thinning in the cable should be made by flaw detectors based on checking the readings with the thickness gauge readings. These results significantly increase the information content of monitoring the continuity of cable products; however, the electrical capacitance method is practically impossible to apply in the case of monitoring objects of arbitrary shape and coating thickness.

Work [6] describes an installation for automated monitoring of coating continuity, as well as a method for determining the geometric parameters of defects, such as crack size and shape. However, the influence of interfering parameters on the control process has not been studied in any way, and the testing voltage is set exclusively by standard dependencies.

To test paintwork, the electrolytic NDT method has become widespread in practice worldwide [13,14], based on the occurrence of electrical contact between the electrode and the substrate through a liquid electrolyte. This method has a number of disadvantages: the ability to control only through coating defects, the possibility of missing narrow pores due to the capillary effect, the low speed of inspection of most industrial objects, and requirements for the orientation of objects in space.

High-voltage testing of paintwork is limited due to the possibility of damaging a thin dielectric coating with high voltage, due to the small difference between the breakdown voltage of the air and the breakdown voltage of the coating (the electric hardening effect). Thus, if there is a need to test large areas of paintwork, it is necessary to pay closer attention to the choice of test voltage and reduce it to the minimum permissible (the air breakdown voltage). In addition, when performing testing using the high-voltage method, much attention should be paid to the sensitivity of the equipment, and to determining the criteria for which signals should be considered a signal about the presence of a defect in the coating. When monitoring paintwork, a situation may arise in which the discharge covers only part of the interelectrode gap and spreads over the surface. The signal from such a discharge can be mistakenly taken for a coating defect. Accordingly, it is necessary to develop criteria for separating full and partial discharges during monitoring.

2. Materials and Methods

First of all, it is worth separating the tasks of detecting non-through defects and inadmissible thinning due to the difference in the physical processes of spark discharge formation in the air gap of the non-through defect and in the dielectric coating material.

The mechanism for the formation of a spark discharge for air (discharge) interelectrode gaps d_c ranging from 5 μm to 50 mm is explained by the Townsend theory of electrical breakdown of gases [15,16]. If a free electron appears in a gas between two electrodes creating an electric field, then, moving towards the anode with sufficient electric field strength, it can ionize an atom or molecule of the gas upon collision. As a result, a new electron and a positive ion appear. The new electron, together with the initial one, ionizes new atoms and molecules, and the number of free electrons continuously increases until an avalanche of electrons appears. According to the above theory, a streamer is formed from electron avalanches arising in the electric field of the discharge gap, which, lengthening, covers the discharge gap and connects the electrodes, forming a spark discharge.

The intensity of electron multiplication in an avalanche is characterized by the impact ionization coefficient (the first Townsend coefficient) α, equal to the number of ionizations produced by an electron along a path of 10 mm in the direction of action of the electric field.

When analyzing the ongoing processes, it should be taken into account that during the development of an avalanche, simultaneously with electrons, positive ions are formed, the mobility of which is much less than that of electrons, and during the development of the avalanche, they practically do not have time to move in the gap to the cathode. Thus, after the passage of an avalanche of electrons, positive ions remain in the interelectrode gap, which distorts (reduces or increases) the electric field [17].

The key element of reliable high-voltage testing is to ensure conditions for independent spark discharge in places of coating defects (for example, violation of their continuity). After the first avalanche passes through the gap, the avalanche process may resume or die out. To resume the avalanche process (organize a self-discharge), at least one secondary effective electron is required, which can arise, including as a result of the passage of a primary avalanche, with an increase in the voltage applied to the electrodes.

The number of positive ions $\left(n_i^+\right)$ remaining in the interelectrode gap after the passage of the avalanche is equal to the number of electrons in the avalanche, excluding the initial electron, i.e.,

$$n_i^+ = n_0 \cdot e^{(\alpha-\eta)\cdot d_c} - 1, \qquad (1)$$

where n_0 is the number of primary electrons and η is the sticking coefficient.

It should be taken into account that not all electrons knocked out from the cathode participate in the formation of secondary avalanches. Some electrons recombine with positive ions. The overall process of formation of secondary electrons from the cathode is characterized by the secondary ionization coefficient (second Townsend coefficient) γ, which depends on the cathode material, composition, and gas pressure, while $\gamma \ll 1$. The number of secondary electrons formed after the passage of the primary avalanche with an independent discharge form must satisfy the condition:

$$\gamma \cdot \left(e^{(\alpha-\eta)\cdot d_c} - 1\right) \geq 1 \qquad (2)$$

This shows that as a result of the passage of a primary avalanche, the formation of at least one effective electron capable of igniting a secondary avalanche is necessary.

As mentioned above, during the development of an avalanche, the number of electrons and positive ions continuously increases. As the number of electrons in the avalanche head increases, the field strength at the avalanche front increases, while at the same time, the field strength decreases at the avalanche tail. This causes the electrons at the head of the avalanche to stop and possibly recombine with ions, emitting photons, which, in turn, are able to ionize neutral molecules near the tail of the primary avalanche, forming secondary avalanches. Secondary avalanches, following the lines of force and having an excess negative charge on the head, are drawn into the region of the positive space charge left by the primary avalanche. The electrons of the secondary avalanches mix with the positive ions of the primary avalanche and form a streamer, an area with the highest current density, which, when heated, begins to glow. The highest concentration of particles (current density) is formed near the cathode. For photoionization in a gas volume, the photon energy must be greater than the ionization energy. This process is successfully carried out in mixtures of gases containing components with relatively low ionization energy (including in air).

According to the above theory [18–20], the minimum breakdown voltage of a non-through coating defect can be calculated using the following formula:

$$U_{da} = \frac{B_0 \cdot P \cdot d_c}{\ln \frac{A_0 \cdot P \cdot d_c}{\ln\left(1+\frac{1}{\gamma}\right)}},\qquad(3)$$

where P is the gas pressure, E is the electric field strength, A_0 is the coefficient that depends on the composition of the gas, B_0 is the coefficient which depends on the ionization energy of the gas, and γ is the secondary electron ionization coefficient.

It can be inferred from Equation (3) that with a consistent external air temperature within a uniform field $U_{da} = f(P \cdot d_c)$, there is a quasi-constant atmospheric pressure $U_{da} = f(d_c)$ (Figure 2).

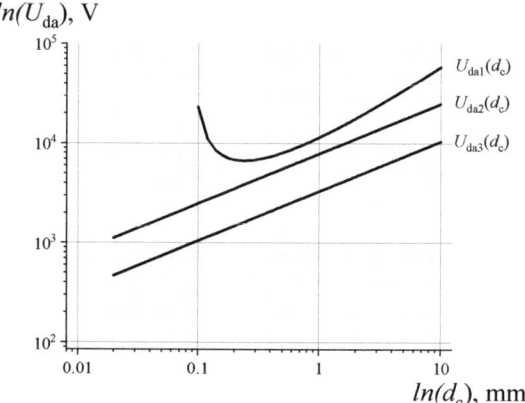

Figure 2. Calculated values of U_{da}: 1—$U_{da}(d_c)$ according to Formula (1) for atmospheric air under normal conditions, 2—$U_{da}(d_c)$ according to Formula (2) for $d_c > 1$ mm, 3—$U_{da}(d_c)$ according to Formula (2) for $d_c < 1$ mm.

Simultaneously, established techniques for determining the test voltage, accounting for the non-uniformities in the electric field during testing, provide the calculation of U_{da} based on the empirical relationship [2,3]:

$$U_{da} = M \cdot \sqrt{d_c},\qquad(4)$$

where M is a constant empirical coefficient depending on the thickness of the coating (d_c) ($M = 3294$ for coatings with $d_c < 1$ mm and $M = 7843$ for $d_c > 1$ mm).

As can be seen from Figure 2, the standardized test voltage calculation methods can be applied exclusively to detect through coating defects as they do not consider the increase in breakdown voltage of U_{dc} solid dielectrics.

In [21], an empirical dependence of the coating breakdown voltage U_{dc} is proposed for a wide range of dielectric coatings:

$$U_{dc} = \frac{K}{d_c} \cdot K_P \cdot \left(A_c^0\right)^{1.1} \cdot \exp\left(\frac{a}{b + lg(b)} + \frac{m}{n + lg(\tau)}\right), \quad (5)$$

where K represents the proportionality factor based on d_c, τ signifies the duration of applied voltage, K_P denotes the probability of breakdown, A_c^0 is the energy required for channel formation, while a, b, n, and m are constants contingent on the dielectric for the purpose of approximation.

Equation (5) is applicable for computing the dielectric strength of dielectrics within a range of thicknesses from 0.01 to 40 mm under the duration of the applied voltage pulse $\tau = 0.1$–10 µs.

Table 1 provides a comparative display of experimentally obtained and formula-calculated U_{dc} values for several solid dielectrics with a sample thickness of $d_c = 0.1$ mm [22].

Table 1. Calculated and experimental values of the dielectric strength U_{dc} for 0.1 mm thick dielectric materials.

Coating Material	U_{dc}, kV	
	Experimental Values	Calculated Values
polyethylene	6.75–7	6.2
polystyrene	5.5–7.3	4.3
fluoroplast-4	3.5	4

The larger scatter in the values U_{dc} of polystyrene can be explained by its low density and, as a consequence, the large influence of the probabilistic processes of streamer formation in the material. The calculation of U_{dc} using Equation (5) is applicable to both single-component and multilayer coatings, with their parameters available in the reference literature, such as those for anti-corrosion coatings on pipelines. However, it should be considered that the parameters required for the calculation of multi-component coatings (e.g., paint and varnish coatings) are usually not standardized. Therefore, U_{dc} multi-component coatings or coatings for which it is not possible to determine U_{dc} by calculation should be determined empirically. To experimentally determine U_{dc}, the recommendation is to employ a coating sample that is either identical to or closely resembles the one being tested in terms of composition and thickness, and apply it to a conductive substrate.

3. Results

To measure the breakdown voltages of through defects and dielectric coatings (and to calculate the electrical strengths of coatings based on the breakdown voltage), a setup was used, the structural diagram of which is shown in Figure 3. The distinctive features of each experiment are described directly in its description.

Figure 3 shows a diagram of the experimental setup. The high-voltage pulse generator creates a high voltage on the electrode (Figure 4), which is applied to the controlled sample. High-voltage pulses follow at a frequency of 50 Hz. The duration of one pulse is approximately 20 µs. Pulse amplitude can be set in the range of 0.5–40 kV. A high-voltage voltage divider is also connected to the electrode, which is needed to measure the high voltage on the electrode. The voltage divider has a division factor of 1000, and thus high voltage can be measured using a regular digital oscilloscope. When a spark discharge occurs, the discharge current flows into the discharge indication circuit. Using a current detector, the presence of a spark discharge is detected, and the device signals this using light and sound alarms.

To confirm this, an experiment was conducted to determine U_{dc}. Plates of textolite coated by aluminum were used as substrates, on which the paintwork was created—Molecules MLS 306 enamel (Figure 5). Enamel was utilized in three, six, and nine layers (the thickness of one layer was equal to 12–16 µm). The ambient air temperature was controlled by a TTZh-K thermometer and varied during the experiment from 22 to 26 °C. Atmospheric pressure was controlled by the Aneroid BAMM-1 barometer and amounted to 96.2 ± 0.3 kPa.

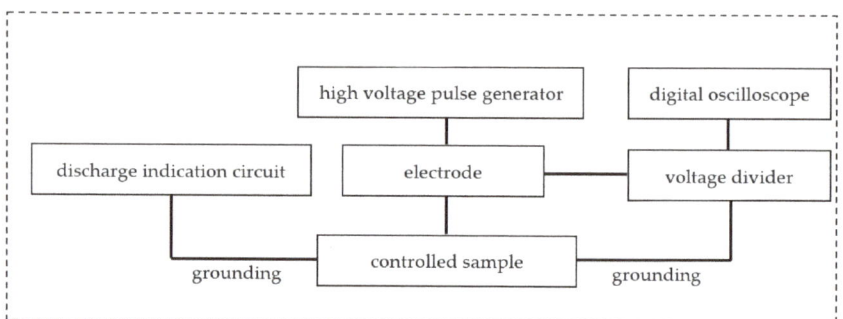

Figure 3. Experimental setup diagram.

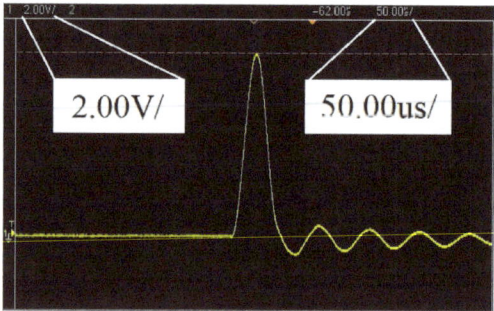

Figure 4. CH1: High-voltage pulse of testing voltage.

Figure 5. Testing objects: 1—coated aluminum sheet, 2—foil-coated laminate sheet with a fabric substrate.

After the fabrication of the specimens, the thickness was gauged at testing points where the breakdown voltage was determined. Points in this case mean the area bounded by a circle with a diameter of 5 mm. This is performed to consider the possible path of the spark discharge in the area with the lowest dielectric strength of the coating E_c in this area. A test voltage was applied to the coating and increased until it broke down, and the U_{dc} value was captured utilizing a DSO-X 2002A oscilloscope. Consequently, the correlation of $U_{dc}(d_c)$ was obtained, as presented in Figure 6.

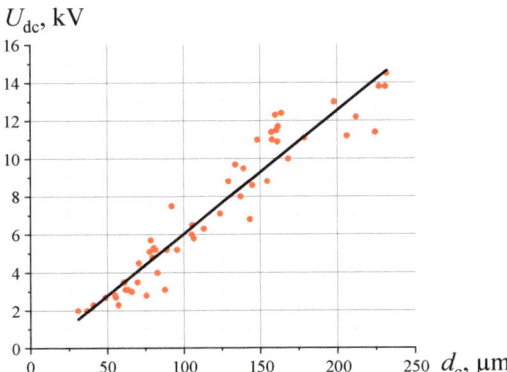

Figure 6. The relationship between the breakdown voltage of the coating U_{dc} and its thickness d_c.

As evident from Figure 6, the obtained U_{dc} values exhibit a considerable degree of variation. It is likely attributed to the development of discharge at the location with the least coating thickness and certain alterations in coating parameters [23,24]. Due to this factor, it is necessary to estimate the probabilities of detecting inadmissible thinning as a function of U_i and d_c, for which an algorithm was applied to the construction of a regression line of the $U_{dc}(d_c)$ dependence and the generation of normal distribution functions with predetermined parameters from it [25–27]. In the studied area, the function $U_{dc}(d_c)$ has a quasi-linear form. Based on this, a linear regression of the type $U = k \cdot d_c + b$ was computed using the least squares method, utilizing the acquired experimental data. Subsequently, the normal distribution function of the probability of spark discharge formation from the value of U_i was constructed:

$$P(U_i) = \frac{1}{\sqrt{2} \cdot \sigma} \int_{-\infty}^{U} e^{-\frac{(U_i - \mu)^2}{2 \cdot \sigma^2}} dU_i, \tag{6}$$

where U_i is the test voltage, k and b are parameters of the regression line, μ is the mathematical expectation, and σ is the standard deviation.

The boundaries of the confidence interval $P_{\pm}(U_i)$ concerning the regression model (Figure 7) according to [28] are as follows:

$$PP_{\pm}(U_i) = P(U_i) \pm t_P \sqrt{D} \sqrt{\frac{1}{n} + \frac{(\ln U_i - \overline{U}_i)^2}{\sum_{i=1}^{n}(\ln U_i - \overline{U}_i)^2}}, \tag{7}$$

where n is the quantity of measurements, t_P is the Student's coefficient for the 95% confidence level and $(n - 2)$ degrees of freedom, D is the variance of U_{dc}, and $\overline{U}_i$ is the mean value of U_i.

The probability dependence graph of defect detection $P(U_{dc})$ indicates the dependability of the testing process. The chart, portraying a sigmoid function, delineates the boundaries of the interval at a specified confidence level (illustrated by dashed lines). Obviously, with increasing d_c, the characteristic $P(U_{dc})$ shifts to the right. It is also advisable to construct the relationship $P(U_{dc})$ to achieve a confidence probability of 0.9 (90%) when conducting tolerance testing for coating defects (detection of areas of inadmissible thinning) [29,30]. Therefore, it is feasible to ascertain the dielectric strength of the coating (E_p) with a defect detection probability set at 90% for every examined coating sample (Figure 8).

The experimental results indicate that the estimated value of E_c within the designated thickness range is 75.4 ± 8.2 kV. Assuming a nearly constant dielectric strength across the specified thickness range, the likelihood of detecting a defect of a specific thickness (based on the dielectric strength of the coating) would be 0.8 (80%).

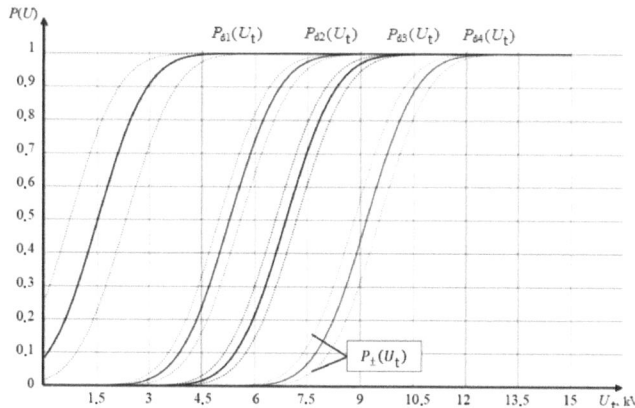

Figure 7. Breakdown probability distribution as a function of test voltage U_t for coatings with a thickness of $d_1 = 38$ μm, $d_2 = 89$ μm, $d_3 = 113$ μm, and $d_4 = 148$ μm.

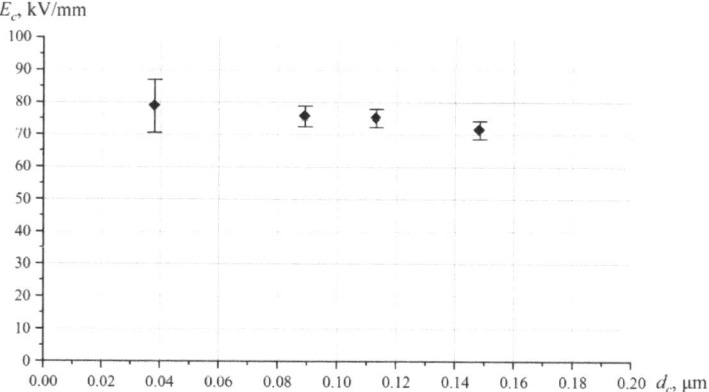

Figure 8. A graph showing the relationship between the dielectric strength of the coating ($_c$) and its thickness (d_c) for MLS 306 coating samples.

Therefore, given a recognized dielectric strength value for the coating (determined through calculation or experimentation), one can identify unacceptable thinning with a calculated probability by conducting testing with a test voltage equal to $U_i = E_c/d_c$.

At the same time, it should be considered that during testing, including for the purpose of detecting inadmissible thinning, the formation of air gaps between the electrode and the coating surface is possible (Figure 9), which can lead to the formation of surface discharges—discharges that cover part of the interelectrode gap. Such discharges do not signal a coating defect but may generate a signal mistaken for a coating defect signal (corresponding to a full discharge) at given equipment settings. Proceeding from this, it is possible to assert that partial discharges are an interfering parameter in the process of inspection, the possibility of their occurrence should be considered, and appropriate measures should be taken to eliminate their influence when building flaw detector circuits, developing electrode designs, and creating inspection techniques.

Partial discharges occur due to air gaps (d_a) between the electrode and the coating, caused, for example, by inhomogeneous coating thickness, curvature or waviness of the product surface, or inaccuracy in placing the electrode on the coating [31]. In the formed air gaps, the electric field strength can exceed the field strength in the dielectric coating because the relative dielectric constant of the coating is greater than the dielectric constant of the air

(Figure 10). In this case, favorable conditions for the formation of partial discharges are created [11,32].

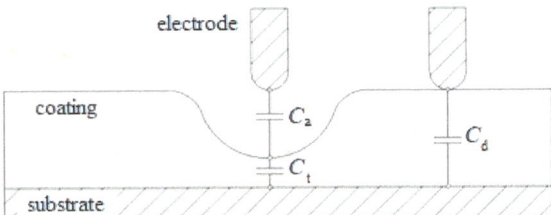

Figure 9. Equivalent scheme to explain the processes of partial discharge occurrences in the testing area; C_a is the air gap capacitance; C_d is the dielectric capacitance in the area of inadmissible thinning (coating defect); and C_a is the dielectric capacitance in the coating area of the specified thickness.

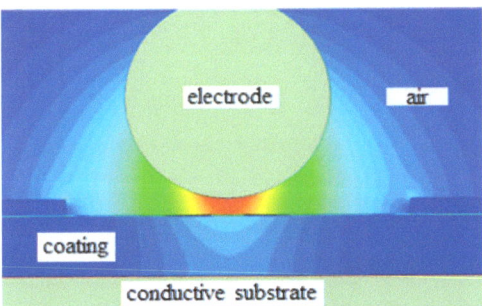

Figure 10. Finite element model of electric field strength distribution on a section with air gap 50 μm and residual coating thickness 180 μm.

The primary informative parameter carrying information about the presence or absence of defects in the coating is the amplitude of the voltage pulse on the shunt $u_R(t)$, included in series in the measurement circuit when the discharge current flows through it. It has been experimentally established [33] that in the area of small coating thickness ($d_c \approx 50$ μm) $u_R(t)$ at full discharge, the current flowing through it is smaller than the amplitude of $u_R(t)$ at partial discharge in the case of large coating thickness (Figure 11). It follows that for instrumental realizations of the nondestructive electric discharge testing, it is necessary to adjust the sensitivity of the instruments. At the same time, if the sensitivity of the device is set incorrectly, false alarms due to partial discharges may occur.

Adjustment of the required sensitivity level can be carried out on test or standard samples identical to or close in characteristics to the testing object. However, this approach does not exclude the possibility of human factor influence on the choice of the sensitivity level and, as a consequence, on the testing results.

Figure 12a,b shows the oscillograms of voltages at the formation of test voltage pulses and the occurrence of partial discharges: pulse $u_i(t)$ of the test voltage (in blue) and pulses of voltage drop on the shunt $u_R(t)$, included in series in the measurement circuit, are caused by the flow of partial discharge currents (in red). Figure 12c, shows the oscillograms of $u_i(t)$ and $u_R(t)$ pulses at full discharge through the interelectrode gap in the area of the through coating defect.

As can be seen from the oscillograms, in general, the amplitude of $u_R(t)$ at partial discharge is smaller than the amplitude of $u_R(t)$ at full discharge. This is due to the fact that in a partial discharge, the source of charges is the charged capacitance of the air gap (C_a), while in a full discharge, the current is due to the flow of charges from both the charged capacitances and the high voltage source. As a consequence, at the moment of complete discharge of the interelectrode gap, the voltage at the electrode $u_i(t)$ decreases to close to zero and, accordingly, the duration τ of the test voltage pulse decreases. Thus, in order to completely eliminate false positives due to partial discharges, it is proposed to estimate τ for $u_i(t)$ and fix the decrease in τ at full discharge using the scheme, the structure of which is presented in Figure 11.

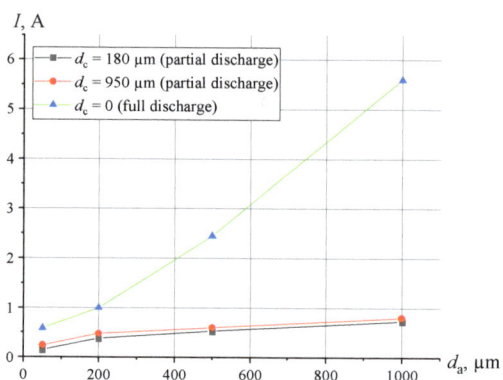

Figure 11. Dependence of the amplitude I of the discharge current pulse and calculated values of the partial discharge current on the size of the air gap d_c for three values of the coating thickness d_c.

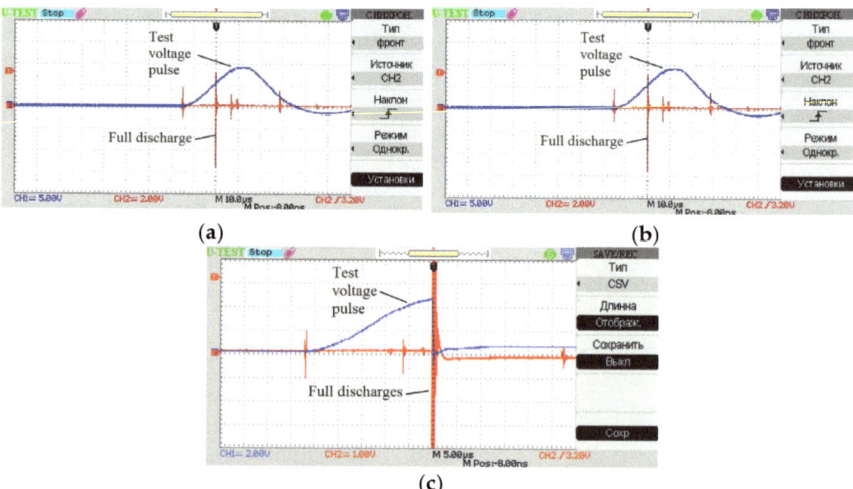

Figure 12. Oscillograms of test voltage pulses and partial discharges: (a) without full discharge (time sweep 10 μs/div., scale: CH1: 5 V/div., CH2: 2 V/div.); (b) without full discharge (time sweep 2.5 μs/div., scale: CH1: 2 V/div., CH2: 1 V/div.). (c) at full discharge (time sweep 5 μs/div., scale: CH1: 2 V/div., CH2: 1 V/div.).

The circuit proposed in Figure 13 works as follows: the voltage $u_R(t)$, the amplitude of which depends on the value of the discharge current, is fed through a low-pass filter to the comparator driver, which limits the maximum amplitude of the pulse. The comparator compares the pulse amplitude $u_R(t)$ with the set sensitivity level, and the output is a logic one or logic zero. At the same time, the circuit measures the duration τ of the pulse $u_i(t)$ of the test voltage whose amplitude is divided by a factor of 1000. The divided pulse is fed through a rectifier to cut off the negative component of the pulse. The rectified pulse is fed to a comparator, which converts the analog pulse into a digital rectangular pulse (meander). The duration τ of this pulse is measured by the timer of the microcontroller.

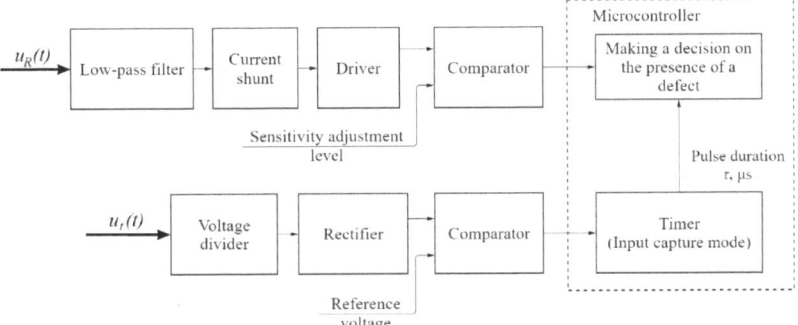

Figure 13. Structural diagram of the defect registration unit.

The scheme proposed in Figure 13 realizes the tracking of the coincidence of two conditions in the defect area: exceeding the sensitivity threshold by the voltage pulse $u_R(t)$ caused by the full discharge current and reducing the duration of the pulse $u_i(t)$, which eliminates the possibility of false positive testing results caused by partial discharges.

However, false positive testing results can also occur due to the increased roughness of the substrate. In [34,35], the influence of the inhomogeneity of the electric field formed in the testing region on the breakdown voltage of the method was shown. In these works, it is proposed to form a sharply inhomogeneous field to reduce the breakdown voltage, but the influence of the roughness of the substrate surface on the inhomogeneity of the field was not considered.

It is known that surface roughness is characterized, as a rule, by the arithmetic mean deviation of the profile along the substrate length (R_a) and the height of profile irregularities along ten points (R_z) [36]. Surface roughness parameters are set in accordance with [37]. In electric discharge testing, the substrate is one of the electrodes, the shape of which also determines the pattern of the electric field in the interelectrode gap. In this case, in a highly inhomogeneous field, regions of increased electric field strength appear, resulting in a decrease in the breakdown voltage of the gas gap. Thus, one can talk about the possible influence of surface roughness on the breakdown voltage. If the roughness is significant, the substrate should not be considered as a plane in the system of two electrodes, but as a sequence of irregularities with protrusions and depressions (Figure 14), leading to an increase in the degree of inhomogeneity of the electric field, which may entail a decrease in the value of the breakdown voltage.

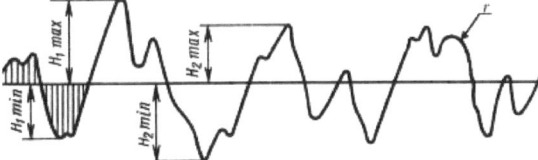

Figure 14. Surface roughness and parameters generally characterizing it.

To evaluate the effect of substrate surface roughness (e.g., after sand or shot blasting) on breakdown voltage, experiments were conducted with an Elcometer 125 Surface Comparator on shot roughness samples. In the first case, a 0.05 mm thick film with a hole simulating a defect was mounted on the sample. An electrode was placed on the surface of the film in the area of the hole. The structure was fixed with clamps (Figure 15a). In the second case, the roughness samples were painted with MLS 306 enamel (Figure 15b). In both cases, the spark breakdown occurrence voltage was recorded for different values of surface roughness.

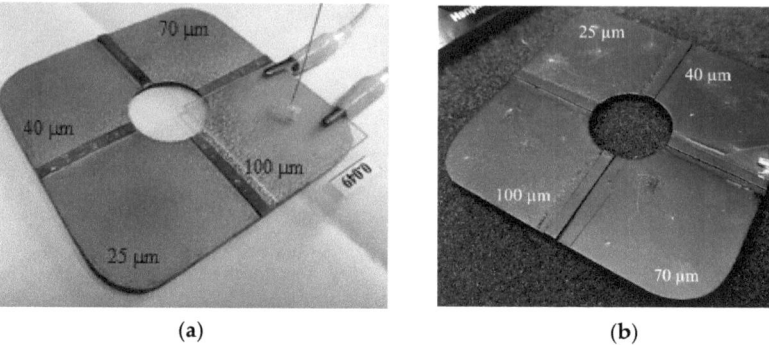

(a) (b)

Figure 15. Schematic diagram of the experiment to investigate the influence of roughness. The experimental results are summarized in Tables 2 and 3.

Table 2. Results of air gap breakdown voltage measurements for different values of R_z of the substrate in the film experiment.

R_z, μm	U_{dc}, kV			
25	1.3	1.2	1.3	
40	1.3	1.3	1.26	
70	1.36	1.3	1.28	
100	1.36	1.3	1.36	

Table 3. Results of air gap breakdown voltage measurements for different values of R_z of the substrate in the enamel experiment.

R_z, μm	U_{dc}, kV			
25	0.85	0.80	0.85	
40	0.85	0.85	0.80	
70	0.85	0.90	0.85	
100	0.85	0.90	0.90	

It can be seen from the data obtained that the breakdown voltage of the interelectrode gap simulating the through coating defect is practically unchanged when R_z is changed in the range of up to 100 μm. Presumably, this is due to the fact that the diameter D of the through defect is much larger than the distance T between neighboring protrusions, hence the spark discharge always occurs in the region between the peak and the rod. Thus, one can conclude that for defects with a diameter $D \gg T$, the roughness of the substrate has no significant influence. On the other hand, if $D < T$, the defect may form in the trough region, leading to an increase in the interelectrode gap and an increase in the breakdown voltage. In such a case, it is proposed to calculate the test stress substrated on the value $d_m + \frac{1}{2}R_z$, where d_{cm} is the maximum thickness of the testing coating.

4. Discussion

The theoretical analysis of spark formation mechanisms in gas and solid bodies, as well as the use of the regression algorithm for processing experimental results, made it possible to develop the main provisions of the methodology for identifying areas of inadmissible thinning in dielectric anti-corrosion coatings. High-voltage testing using the proposed technique increases informativeness while ensuring highly reliable results.

At the same time, the probabilistic nature of the occurrence of breakdown, the heterogeneity of the thickness of the coating samples, and the possible presence of defects in paint coating samples that locally worsen the electrical strength of coating samples contributed to an increase in the spread of experimental values of breakdown voltage, which, at this stage, makes it possible to identify, using the electric spark method, thinning only when the residual coating thickness is approximately 50% of the nominal thickness or less (this may be due to an inappropriate number of paint layers). To reduce the scatter, it is necessary to conduct further studies on interfering factors and methods for the local measurement of coating thickness under conditions of increased external electric field strength.

On the other hand, experimental and theoretical studies on the influence of partial discharges made it possible to formulate an algorithm for signal processing in electric discharge testing and a structural scheme of the device, which considers an additional informative parameter for making a decision about the presence of a defect in the coating and provides tuning from the influence of partial discharges on the inspection result. However, at this stage, tests of the instrumental implementation of the algorithm shown in Figure 11 were carried out only for samples of paintwork and organic glass. In the future, it is planned to conduct tests on a larger number of control objects (the external and internal coatings of pipes, roofing coatings, industrial paint, and varnish coatings).

Finally, theoretical and experimental analysis of the effect of the substrate roughness on the breakdown voltage of the interelectrode spacing of a given thickness has shown that there is no significant effect of the substrate surface roughness on the breakdown voltage of the interelectrode spacing in the range of R_z from 25 to 100 μm.

The obtained results make it possible to use the high-voltage method of non-destructive testing to identify through defects and an unacceptably small number of layers of paint and varnish coatings, which was previously inaccessible for testing by high-performance methods and was detected mainly visually.

Author Contributions: Conceptualization, V.S. and A.M.; methodology, V.S.; software, I.G.; validation, V.S., A.M. and I.G.; formal analysis, V.S., A.M. and I.G.; investigation, V.S., A.M. and I.G.; resources, V.S.; data curation, V.S., A.M. and I.G.; writing—original draft preparation, V.S., A.M. and I.G.; writing—review and editing, V.S., A.M. and I.G.; visualization, A.M. and I.G.; supervision, V.S., A.M. and I.G.; project administration, V.S. All authors have read and agreed to the published version of the manuscript.

Funding: This research received no external funding.

Institutional Review Board Statement: Not applicable.

Informed Consent Statement: Not applicable.

Data Availability Statement: The original contributions presented in the study are included in the article, further inquiries can be directed to the corresponding author/s.

Conflicts of Interest: Author Alexey Musikhin was employed by the company LLC "KONSTANTA", Ogorodny Lane, 21, 198095 Saint Petersburg, Russia. The remaining authors declare that the research was conducted in the absence of any commercial or financial relationships that could be construed as a potential conflict of interest.

References

1. *GOST R51164-98*; Steel Main Pipelines. General Requirements for Corrosion Protection. Izdatelstvo Standartov: Moscow, Russia, 1998.
2. *GOST 34395-2018*; Paint Materials. Spark Test Method for Continuity Inspection of Dielectric Coatings on Conductive Substrates. FSUE "STANDARTINFORM": Moscow, Russia, 2018.
3. *ASTM G62–14*; Standard Test Methods for Holiday Detection in Pipeline Coatings. ASTM International: West Conshohocken, PA, USA, 2014.

4. *NACE RP0188*; Discontinuity (Holiday) Testing of New Protective Coatings on Conductive Substrates. NACE International: Houston, TX, USA, 1999.
5. *ISO 12944-5*; Varnishes and Paints. Corrosion Protection of Steel Structures by Protective Coating Systems. ISO: Geneva, Switzerland, 2018.
6. Kharitonova, K.A.; Kisakova, D.; Salnikova, K.E.; Molchanov, V.P. Development of methods of analysis of quality and safety posts of paint coatings. In *Modern State of Economic Systems: Economics and Management*; LLC "SFK-Office" Publisher: Tver, Russia, 2020; pp. 356–363.
7. Mix, P.E. *Introduction to Nondestructive Testing: A Training Guide*; John Wiley & Sons: Hoboken, NJ, USA, 2005.
8. Singh, N.; Singh, S.P. Evaluating the performance of self-compacting concretes made with recycled coarse and fine aggregates using nondestructive testing techniques. *Constr. Build. Mater.* **2018**, *181*, 73–84. [CrossRef]
9. Klyuev, V.V. Non-Destructive Testing. In *Electrical Testing: Handbook in 7 Vol*; Mashinostroenie: Moscow, Russia, 2004.
10. Tolmachev, I.I. Metrological support for high-frequency electric spark flaw detection of protective coatings. *NDT Territ.* **2023**, *3*, 40–46.
11. Galeeva, N.S. Increase of Informativeness of Testing of Cable Products on the Basis of Complex Use of Electrospark and Electrocapacitive Methods. Ph.D. Thesis, FGAOU HE "National Research Tomsk Polytechnic University", Tomsk, Russia, 2017.
12. Redko, V.V.; Redko, L.A. Detection of air cavity type defects during electrical inspection of cable product insulation in the region of weak and strong fields. *Polzunov Bull.* **2018**, *1*, 82–87. [CrossRef]
13. Inspection and Test Methods for Coating and Lining: Low Voltage Holiday (Wet Sponge) Detection [Electronic Resource]. Available online: https://www.kuegroup.com/inspection-and-test-methods-for-coating-and-lining-low-voltage-holiday-wet-sponge-detection/ (accessed on 18 March 2024).
14. Testing for Pinholes with the Wet Sponge Method Using the Elcometer 270 Pinhole Detector [Electronic Resource]. Available online: https://www.elcometer.com/en/testing-for-pinholes-with-the-wet-sponge-method-using-the-elcometer-270-pinhole-detector (accessed on 18 March 2024).
15. Erekhinsky, B.A.; Pakhomov, A.V. Modern Technologies of Diagnostics of Gas and Gas Condensate Production Facilities. In *Applied Machinery and Equipment*; AO "Voronezhskaya oblastnaya tipografiya": Voronezh, Russia, 2017.
16. Xiao, D. Fundamental theory of townsend discharge. In *Gas Discharge and Gas Insulation*; Springer: Berlin/Heidelberg, Germany, 2016; pp. 47–88.
17. Raiser, Y.P. *Physics of Gas Discharge: Study Guide*, 2nd ed.; Nauka: Moscow, Russia, 1992.
18. Davies, M.A.S. Holiday detection: What is the correct test voltage? *Anti-Corros. Methods Mater.* **1983**, *30*, 4–12. [CrossRef]
19. Gadzhiev, Y.M.; Ibragimova, E.N. Experimental study of crack size measurement of silicate-enamel pipe coating. *Defectoscopy* **2020**, *1*, 61–65.
20. Ibragimov, N.Y.; Ibragimova, E.N. Defectoscopic installation of pipe silicate coatings crack meter. *Defectoscopy* **2017**, *11*, 55–57.
21. Trenkin, A.A.; Almazova, K.I.; Belonogov, A.N.; Borovkov, V.V.; Gorelov, E.V.; Morozov, I.V.; Kharitonov, S.Y. Investigations of the initial phase of spark discharge in air in the spear (cathode)-plane gap by laser probing. *J. Tech. Phys.* **2020**, *90*, 1–9.
22. Crichton, B.H. Gas discharge physics. In *IEE Colloquium on Advances in HV Technology*; IET: London, UK, 1996; pp. 3/1–3/5.
23. Smirnov, A.S. *Applied Physics. Physics of Gas Discharge: Studies Manual*; Publishing House of SPbSTU: St. Petersburg, Russia, 1997.
24. Mick, J.; Crags, J. *Electrical Breakdown in Gases*; Translated from English; Izdatelstvo Inostrannoi Literatury: Moscow, Russia, 1960.
25. Boychuk, A.S.; Generalov, A.S.; Dalin, M.A.; Stepanov, A.V. Probabilistic assessment of reliability of the results of ultrasonic nondestructive testing of PCM structures used in the aviation industry. *Repair. Recovery Mod.* **2013**, *9*, 36–39.
26. Vazhov, V.F.; Lavrinovich, V.A.; Lopatkin, S.A. *High Voltage Technique: A Course of Lectures*; TPU Publishing House: Tomsk, Russia, 2006.
27. Vorobyov, G.A.; Pokholkov, Y.P.; Korolev, Y.D.; Merkulov, V.I. *Dielectric Physics (The Field of Strong Fields): Textbook*; TPU Publishing House: Tomsk, Russia, 2003.
28. Grigoriev, A.N.; Pavlenko, A.V.; Ilyin, A.P.; Karnaukhov, E.I. Electric discharge on the surface of a solid dielectric. Part 1. Features of development and existence of the surface discharge. *Izv. TPU* **2006**, *1*, 66–69.
29. Chertishchev, V.Y. Estimation of probability of defect detection by acoustic methods as a function of defect size in PCM structures for binary inspection output data. *Aviat. Mater. Technol.* **2018**, *3*, 65–79. [CrossRef]
30. Lebedev, A.M. Investigation of reliability of tolerance testing. *Sci. Bull. MGTU GA* **2005**, *86*, 65–70.
31. Varivodov, V.N.; Kovalev, D.I.; Zhulikov, S.S.; Golubev, D.V.; Romanov, V.A. Prevention of partial discharges in the solid insulation of high-voltage current conductors. *Electr. Eng.* **2021**, *8*, 30–34.
32. Starikova, N.S.; Redko, V.V. Investigation of methods for testingling the integrity of insulation in the area of weak and strong electric fields. *Bull. Sib. Sci.* **2013**, *3*, 55–59.
33. Musikhin, A.S. Electric Discharge Testing of Continuity and Inadmissible Thinning of Dielectric Coatings. Ph.D. Thesis, M.N. Mikheev Institute of Metal Physics of the Ural Branch of the Russian Academy of Sciences, Yekaterinburg, Russia, 2023.
34. Syasko, V.A.; Golubev, S.S.; Musikhin, A.S. Electric discharge testing of dielectric coatings thickness. *Test. Diagn.* **2020**, *9*, 12–17.
35. Syasko, V.A.; Musikhin, A.S. High voltage testing of functional dielectric coatings with thickness from 25 µm and more. *J. Phys. Conf. Ser.* **2020**, *1636*, 1–7. [CrossRef]

36. Turaev, T.T.; Topvoldiev, A.A.; Rubidinov, S.G.; Zharaitov, Z.G. Parameters and characteristics of surface roughness. Oriental Renaissance: Innovative. *Educ. Nat. Soc. Sci.* **2021**, *1*, 124–132.
37. *ISO 25178-2:2021*; Geometrical Product Specifications (GPS). Surface Texture: Areal. Part 2: Terms, Definitions and Surface Texture Parameters. ISO: Geneva, Switzerland, 2021.

Disclaimer/Publisher's Note: The statements, opinions and data contained in all publications are solely those of the individual author(s) and contributor(s) and not of MDPI and/or the editor(s). MDPI and/or the editor(s) disclaim responsibility for any injury to people or property resulting from any ideas, methods, instructions or products referred to in the content.

Article

A New Approach for Determining Rubber Enveloping on Pavement and Its Implications for Friction Estimation

Di Yun [1,2], Cheng Tang [3], Ulf Sandberg [4], Maoping Ran [1], Xinglin Zhou [1], Jie Gao [5] and Liqun Hu [2,*]

1. School of Automobile and Traffic Engineering, Wuhan University of Science and Technology, Wuhan 430081, China; yundi@wust.edu.cn (D.Y.)
2. Key Laboratory of Pavement Structure and Material of Transportation Industry, Chang'an University, Xi'an 710064, China
3. School of Civil Engineering and Transportation, South China University of Technology, Guangzhou 510641, China; ct_tangcheng@mail.scut.edu.cn
4. Swedish National Road and Transport Research Institute (VTI), SE-581 95 Linköping, Sweden; ulf.sandberg@vti.se
5. School of Civil Engineering and Architecture, East China Jiao Tong University, Nanchang 330013, China
* Correspondence: hlq123@163.com

Abstract: The depth to which the pavement texture is enveloped by the tire tread rubber (d) is an important parameter related to contact performance. This study presents a new method (S-BAC), which relies on the ratio between the real contact area and the nominal tire-pavement contact area (S) and the bearing area curve (BAC), to measure the depth on pavements. The tire-pavement contact was simulated by contact between a non-patterned rubber block and pavement specimens. After analyzing the affecting factors, the new method was compared with previous methods by the d values and the application on the relationship between pavement texture parameters and friction. The results reveal that though there is a linear regression between the d obtained with the S-BAC and previous methods, the d values obtained with different methods differ. Applying the S-BAC method can strengthen the relationship between texture parameters and friction more than other methods.

Keywords: pavement texture; enveloping; rubber penetration depth; contact area; bearing area curve; friction coefficient

Citation: Yun, D.; Tang, C.; Sandberg, U.; Ran, M.; Zhou, X.; Gao, J.; Hu, L. A New Approach for Determining Rubber Enveloping on Pavement and Its Implications for Friction Estimation. *Coatings* **2024**, *14*, 301. https://doi.org/10.3390/coatings14030301

Academic Editor: Przemysław Podulka

Received: 26 January 2024
Revised: 25 February 2024
Accepted: 26 February 2024
Published: 29 February 2024

Copyright: © 2024 by the authors. Licensee MDPI, Basel, Switzerland. This article is an open access article distributed under the terms and conditions of the Creative Commons Attribution (CC BY) license (https://creativecommons.org/licenses/by/4.0/).

1. Introduction

The pavements have a multiscale surface roughness, it is practically impossible for the tires to establish complete contact with all parts of the pavement surface. The partial contact of the tire tread rubber with the pavement makes it difficult to find a meaningful relationship between performances on the pavement surface and simple texture indicators [1,2]. Therefore, for a better description of the pavement surface influence on the tire/pavement interaction, it is essential to find methods by which to quantify the enveloping effect of the tire rubber on the pavement surface.

1.1. Enveloping Profile and Rubber Penetration Depth

Assuming that when the tire load (F) is reduced, the rubber block only touches the pavement at the top of the highest asperity, which here is denoted as a height hmax (Figure 1). Increasing the force F, the average plane of the bottom surface of the rubber block will move downwards by a distance (d), and the rubber will envelop part of the pavement texture. In this case, the deformation on the bottom surface of the rubber block refers to the enveloping profile (red line). The distance between the highest asperity and the lowest point enveloped by the rubber is defined as the rubber penetration depth (d). Both the enveloping profile and rubber penetration depth can help explain how the tire deforms into the space between aggregate particles, how the aggregate penetrates tread rubber, and can indicate the partial contact characteristics between the tire and the pavement.

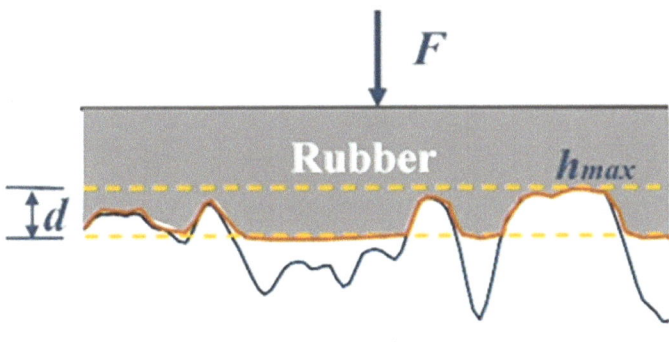

Figure 1. The enveloping profile (red line) and the rubber penetration depth.

1.2. Status of Current Research on Rubber Enveloping of Pavement Texture

The methods to obtain the penetration depth and to model the enveloping profile can be classified into three categories: physical methods that depend on the laws on which the objects move and behave through space and time, test methods that always rely on experiments, and empiric-mathematical methods, which obtain the enveloping profile by mathematical calculation, while the critical parameters in the calculation process are obtained from testing and statistics.

(1) Physical methods: Clapp and Fong, Klein, and Hamet et al. proposed the solution for contact depth based on elastic half-space deformation under linear load [3]; Persson submitted a multi-scale contact mechanism that excludes arbitrary surface roughness without any prior [4]. Dubois et al. divided the pavement texture into several scale ranges and considered the related parts in terms of contact [5]. Kane and Edmondson described the viscoelastic constitutive relationship of tread rubber using the Kelvin model. They studied the rubber enveloping on pavement texture based on the equal of the integral of interfacial stress and vertical load [6].

However, the application of theoretical models in the tire-road contact must be verified [7]. Mahboob Kanafi and Tuononen [8] and Xiao et al. [9] calculated the average penetration depth based on Persson's contact mechanism with conditional assumptions. The numerical simulation has also been used to study tire-road contact. Srirangam [10] and Zheng et al. [11] integrated the pavement texture into tire-road contact FEA simulation. However, the meshing placed high requirements on the computing ability of computers due to the pavement roughness.

(2) Test method: Matilainen and Tuononen [12] embedded sensors between the rim and the tread, but the buffering effect from the tread rubber weakened the perception of the embedded sensor on the enveloping. Wang et al. [13] tested the height changes of the rubber block surface on the specimens and calculated the penetration depth by subtracting the rubber compression from the height changes. The solution of the rubber compression depended on the elastic assumption. Du et al. [14], Hartikainen et al. [15], and Woodward et al. [16] respectively observed the peeling off of colored powder, asphalt, or paint on the aggregate surface under the rubber to obtain the enveloped area. However, the wear process required more loads, so these methods underestimated the envelope area. Ejsmont and Sommer [17] evaluated the depth of tire tread deformation by casting an imprint of tire-road contact in self-vulcanizing rubber. Chen [18], Lu et al. [19], and Yang et al. [20] tested the skid resistance and the corresponding pavement texture, took the depth corresponding to the good correlation between the texture indicators and the skid resistance as the depth of the enveloped pavement profile. Chen et al. [21], Wang et al. [22], Gao et al. [23],

and Yu et al. [24] used Fuji Prescale film to test tire-pavement contact characteristics. However, the Prescale film directly obtained is point cloud data, from which extracting indicators reflecting contact characteristics is still a problem.

(3) Empiric-mathematical method: von Meier et al. took the second derivative of the pavement profile of no more than $d*$, a parameter reflecting the tire hardness, as the tire deformation line [25]. In the ROSANNE project, the area (S) enclosed by a baseline and the pavement profile above the baseline was controlled, and the part of the pavement profile was taken as the enveloped pavement—indentor method [3]. Andersen [26] proposed a method similar to the indentor method but took the bearing area ratio as the control parameter. The values of the control parameters in these methods depend on the experimental test, which should be close to the real contact since the parameters are the calculation or iteration termination conditions for the enveloping profile. However, the method by von Meier recommends $d*$ values based on the tire deformation on the surface made by steel spheres with different radii and distribution distances. The S value in the indentor method was recommended by the deformation of plasticine in the 'pavement' formed by long wooden strips with triangular sections of different spacing under the rolling tires.

In general, determining the depth of rubber penetrated in the pavement becomes necessary due to the incomplete contact between the tire and pavement with multiscale surface roughness. However, most of the existing methods for penetration depth rely on exact theoretical analysis with several conditional assumptions, or on the tests with complicated testing processes and less time efficiency, or they cannot reflect the distribution characteristics of the pavement texture. To this end, this study intends to propose a new method that can reflect the pavement morphology and the conditions under which the contact occurred, to test the rubber penetration depth on the pavement. The content includes:

- introducing the measurement principle and discussing its feasibility;
- analyzing the affecting factors for the new methods; and
- evaluating its application by comparing it with the latest methods.

2. Materials and Methods

2.1. Measurement Principle

The new procedure in this study relies on the rubber-pavement contact area and the bearing area curve (BAC) of the pavement (Figure 2a). The BAC is also called the areal material ratio curve [27]. It characterizes the variation in sectional area as one moves from the top to the bottom of the pavement surface. Figure 2a shows the accumulative distribution of the areal ratio of the material bearing the load on the surface from the highest point down. Every rough surface has a certain BAC, determined by its morphology.

Then, suppose the projected areal ratio of the contact area between the tire and pavement $q = S2/S1$ (Figure 2b) can be measured. In that case, the rubber penetration depth (d) can be determined according to the BAC. In Figure 2b, S1 is the nominal contact area between the tire and pavement, which is in the same order of size as that of a tire-pavement interface, while S2 is the projection of the part of the particle enveloped by tread rubber onto the horizontal plane. From here onwards, the new method proposed in this study is referred to as the S-BAC method.

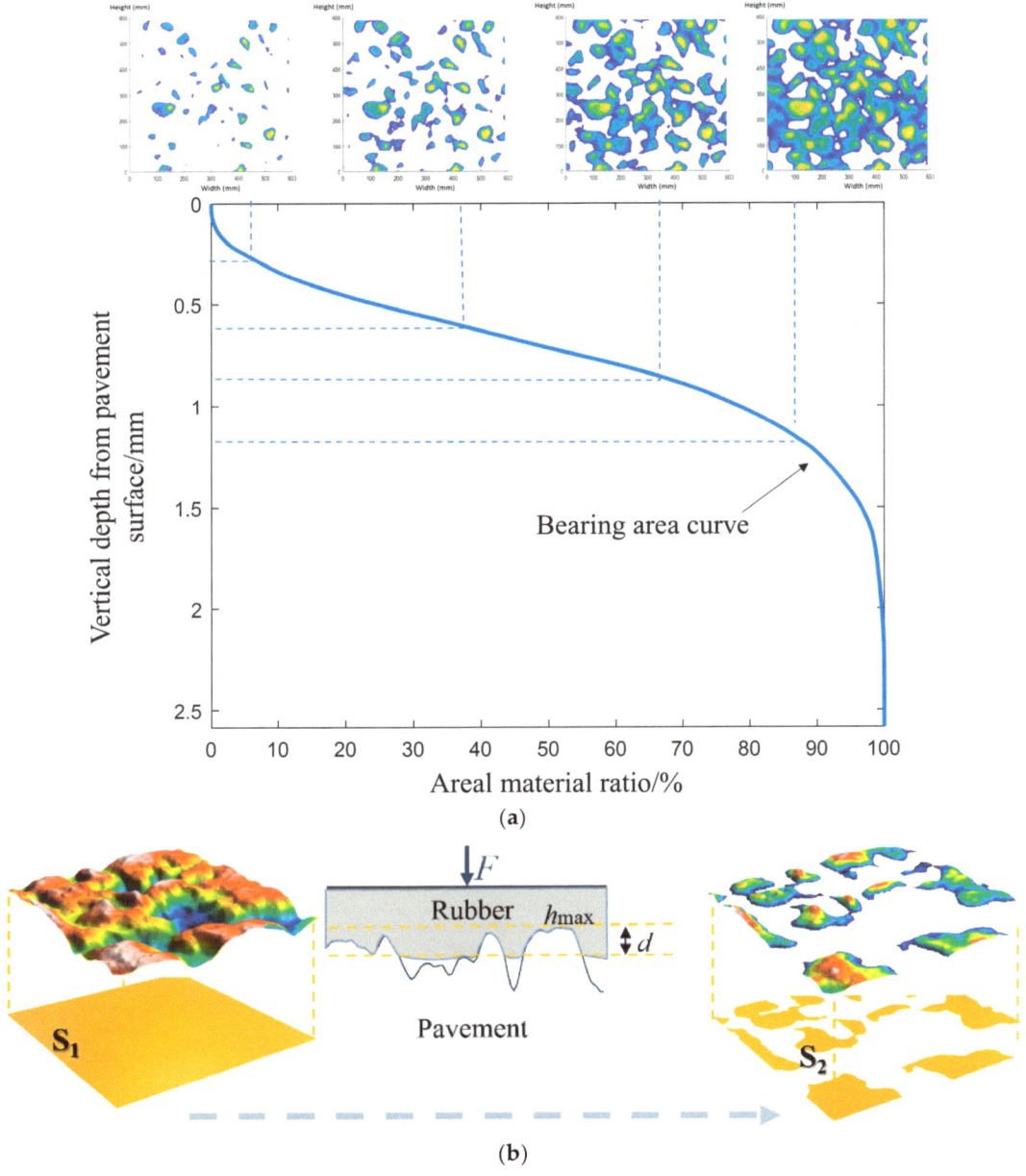

Figure 2. Schematic representation of the S-BAC method for obtaining the rubber penetration depth. (**a**) Bearing area curve of a pavement surface; (**b**) Projection of nominal and actual contact areas onto the horizontal plane.

2.2. Test Objects

2.2.1. Rubber Block Representing Tire Tread

The contact between the tire and pavement is determined by many factors, such as load, pavement texture, tire inflation, tire tread pattern, tire age, and tread rubber hardness [28]. Furthermore, the vehicle operation, such as acceleration, deceleration, turning, and road alignment (e.g., vertical and horizontal slope), could significantly affect the pressure distribution and contact. This study simplified the contact between the tire and pavement as a static contact between a rubber plate and the pavement specimens under a vertical load.

There are various tread patterns, which will cause different geometric stiffness properties of tread and change the rubber penetration depth [29]. Thus, it is virtually impossible to designate a reference tire with a reasonably representative tread pattern. The simplification can better highlight the effect of different pavement textures on the rubber penetration depth. Selecting a smooth rubber block with the same hardness as the tire tread rubber is crucial, as hardness determines tread deformation under load and is closely related to the type and distribution of tread pattern.

2.2.2. Specimen Preparation

In principle, when using the S-BAC method to obtain the penetration depth between a rubber block and the pavement, the pavement specimens can be obtained by various methods, such as directly drilling from the road, molding indoors, or copying pavement textures.

To not damage the road and fully represent the pavement texture, this study recorded the surface texture by a 3D laser scanner in the field (Step 1 in Figure 3). Then, specimens with the texture of the pavement were replicated in light-colored (white) resin materials using 3D printing technology (Step 3 in Figure 3). The reproduced specimen covers a length and a width of 60 mm, which is sufficient for enveloping.

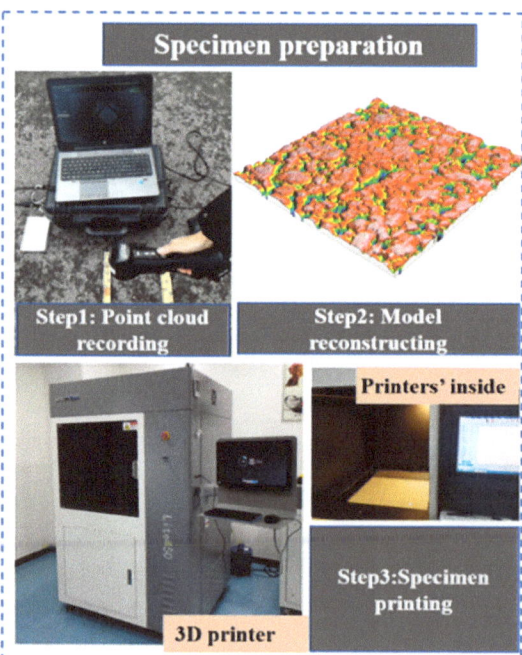

Figure 3. Production procedure for the specimens.

Previous studies and experience have suggested that texture wavelengths in the range of 5~50 mm contribute more to the rubber penetration depth than shorter texture wavelengths [30]. Mogrovejo et al. [31] and Goubert and Sandberg [3] used a sampling spacing of 0.2 and 0.5 mm, respectively, to obtain the pavement profile and the penetration depth for a tire on the pavement. Thus, the HandyScan 3D scanner (Creaform[TM], Lévis, QC, Canada), which can provide a resolution of 0.1 mm, an accuracy of 0.04 mm, and the sampling area is not limited, was used to obtain the pavement texture [32]. An industrial 3D printer lite600 (UnionTech^{TM}, Shanghai, China) based on stereolithography, with a high printing resolution of 0.1 mm and 0.05 mm in horizontal and vertical directions, was used to minimize the loss of captured texture details. Moreover, the elastic modulus of the resin materials used for 3D printing was comparable to that of typical asphalt pavements (about 3000 MPa at 25 °C) after curing.

In this way, both the texture and hardness of the printed specimens could reasonably well realistically represent the pavement surface, at least with relevance to the rubber penetration depth. Additionally, the elastic modulus of the printed specimens was much higher than that of rubber (about 8 MPa at 25 °C). The specimen's compressive deformation and its effect on the contact patches could be ignored.

2.3. Contact Area Measurement

2.3.1. Rubber/Pavement Contact Area Marked Using the Staining Method

Instruments that can precisely provide specific loads are essential for this purpose. This study used the universal test machine (UTM) (IPC^{TM}, Liscate, Italy), which can precisely control the temperature (resolution 0.1 °C) and load (± 1 N). A self-designed fixture was used with the UTM (Figure 4). A rubber block chosen to represent the tire tread, the bottom of which was stained, was glued to the underside of the steel plate of the fixture.

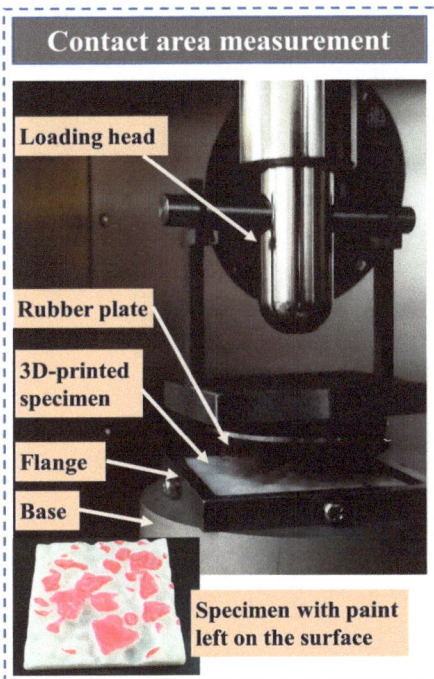

Figure 4. Loading device, self-designed fixture, and a specimen with paint on the surface.

During the test, the specimen was fixed on the base using a flange, and the base was accurately aligned with the loading head (Figure 4). The UTM works according to set processes, at which the load, load duration, and loading waveshape can be designed in advance. In this study, the contact time was 10 s, during which the load was constant. After squeezing the rubber against the specimen, the stained area on the specimen surface marked the contact area (Figure 4). From now onwards, the contact area measurement is referred to as 'the staining method'.

A potential issue with staining is that some excessive paint may be pressed somewhat outside of each local contact area, thus blurring the edges and overestimating the contact area. The paint, in this case, had an estimated average thickness of only 9~20 μm, leaving little paint available to be squeezed outside the contact.

2.3.2. Contact Area Ratio

According to Section 2.1, the proportion of the real contact area to the nominal contact area needs to be considered to determine the tire penetration depth using the S-BAC method. This study used scanning equipment, such as a printer with a scan function, to record the image of the stained specimens' surfaces (Figure 5), and the region of interest can be obtained using a clipping tool. Then, image processing was used to obtain the corresponding binary image and accurately calculate the contact area ratio.

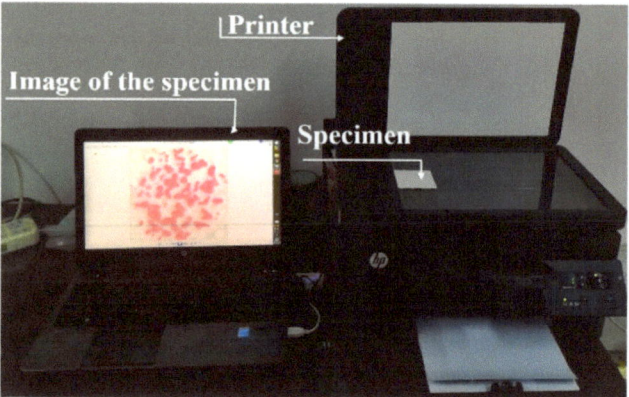

Figure 5. Obtaining the image of the specimens with paint left on the surfaces.

This study developed an algorithm (Method 1) in MATLAB based on the Lab color space and compared it with the conventional gray threshold segmentation method (Method 2) [33]. Interpretations using the two methods are shown in Figure 6.

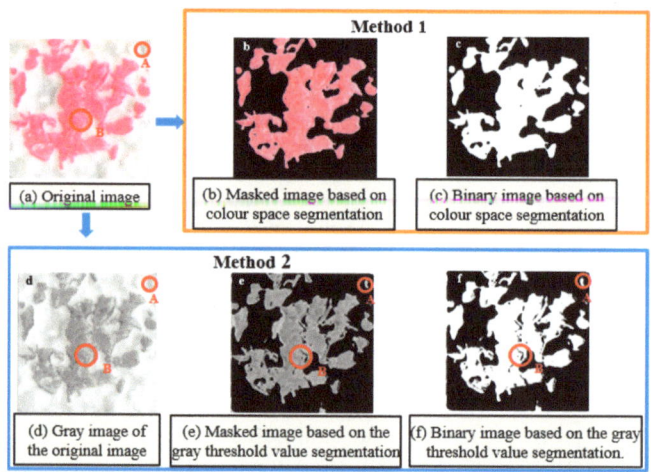

Figure 6. Comparison between the two area computation methods.

The contact area recognition of Method 1 compares the chromatic components of each pixel with pre-set thresholds to classify pixels into target and background classes (Figure 6b). In contrast, Method 2 first converts the original image (Figure 6a) into a gray image (Figure 6d). The pre-set thresholds for both Methods 1 and 2 are determined using the K-means clustering method. Method 2 could incorrectly interpret the non-contact area as a contact area (e.g., region A) or misinterpret the contact area as a non-contact area (e.g., small 'cracks' in region B). These errors could be attributed to the limitation of gray threshold segmentation in effective segment regions with similar gray levels but different colors (Figure 6e). Notably, in images with distinct color differences, the contact area in red can be identified more accurately. The ratio of real contact area to nominal contact area can be calculated using the following equation after producing the binary images.

$$\text{Contact area ratio (\%)} = \frac{Num_{pixel=1}}{Num_{total}} \times 100 \quad (1)$$

Here, $Num_{pixel=1}$ is the number of pixels with a pixel value 1, Num_{total} is the total number of pixels in the nominal contact area.

2.3.3. Resolution of the Staining Method

Previous studies have indicated that the observation scale significantly affected the observed contact area between two rough surfaces [34]. When observed at a larger scale (i.e., poorer resolution, ignoring the finer parts of the macrotexture), the observed area becomes larger than the true area. The pavement texture consists of roughness that extends across several scales, which makes the contact area dependent on the observed scale.

The staining method introduced in Section 2.3.1 regards the area with paint as the contact area. When the rubber block is squeezed onto surfaces with a finer texture by a load above a certain threshold, the stain may be squeezed to fill the space between the contact regions, and the detected contact area would no longer decrease. In other words, the staining method cannot help identify the contact area if it is pictured with a too high resolution.

Several artificial surfaces were produced with different 'smoothness' to investigate the resolution at which the contact can be detected using the staining method. The different smoothness was obtained by applying low-pass filters to the original artificial surface. The artificial surface was generated by designing a Gaussian height probability distribution based on the power spectral density (PSD) curve, as shown in Figure 7 [35]. The shape of the PSD curve was determined by the fractal dimension (H), the root-mean-square height of the surface (Sq), and the nominal maximum aggregate size ($NMAS$) of the asphalt pavement [36]. Based on measurements of these factors from the real pavement surface, this study set the Sq as 0.5 mm, the H as 2.2, and the $NMAS$ as 12.6 mm. The low-pass filtering cut-off, and also the minimum texture wavelength (λs) on a surface, ranged from 12.6 mm down to 0.2 mm. Nine digital surface variants with different λs were then produced by a 3D printer as physical samples representing surfaces at different observation resolutions (Figure 8). These physical samples were used as specimens in the setup shown in Figure 4 to determine the rubber penetration depth.

Applying the staining method to the artificial surfaces, the images in Figure 8 were obtained, where the number in each image represents the λs of the surface. The reddish region shows the contact area between the specimen and the flat rubber block for a load of 50 N. The change in the red area on the surface from picture to picture shows the effect of the observation resolution on the contact area. For the larger λs printed surface, the observed contact area is relatively high. When λs decreases, the contact area gradually shrinks, and the contact spots get separated. At the lower λs, the contact area becomes relatively independent of the λs.

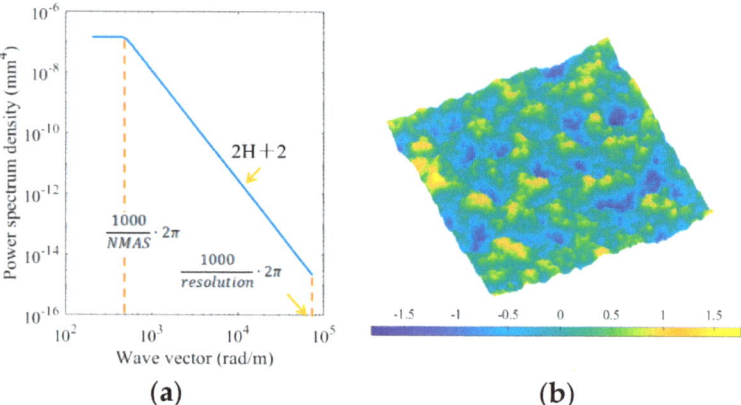

Figure 7. Generating a surface with Gaussian height probability distribution based on the power spectral density. (**a**) power spectrum density curve; (**b**) artificially designed surface.

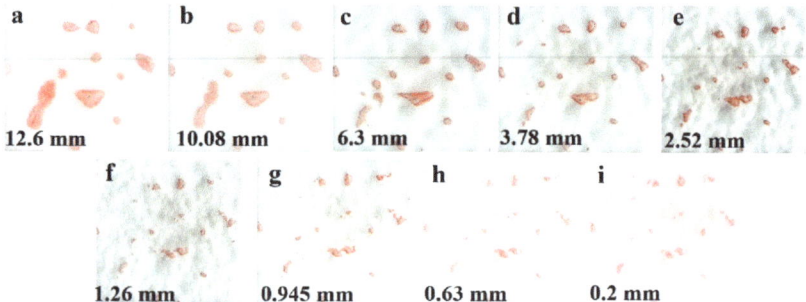

Figure 8. Contact area on artificially designed surfaces with different minimum texture wavelengths (50 N).

Figure 9 shows the contact areas obtained on surfaces with different λs under different loads and rubber hardness conditions. The rubber hardness of the tested rubber blocks included 55 and 68 Shore A, measured according to GB/T 531.1 [37]. The applied forces were 50, 500, 1000, 1500, and 2000 N. The x-axis refers to the λs applied to each surface. The y-axis shows the measured contact area ratio, which is the proportion of the contact area to the area of the rubber bottom. Figure 9 shows that when λs begins to decrease, the contact area ratio rapidly decreases. However, the paint, when squeezed by the rubber, can fill the emerging fine texture on the surface. The detected contact-area ratio does not decrease when the roughness provided is fine enough. For the artificially designed surfaces with Gaussian height probability distributions, the texture wavelength below which the ratio starts to become constant appears at 0.6 mm.

Compared to the artificially designed surfaces, the profile of the top part of the real pavement surface is flatter, which means that the pressure distribution at the interface between the rubber and pavement would be more uniform. However, using test surfaces with such fine-textured peaks as were generated in this study makes it possible to determine the proper acquired resolution.

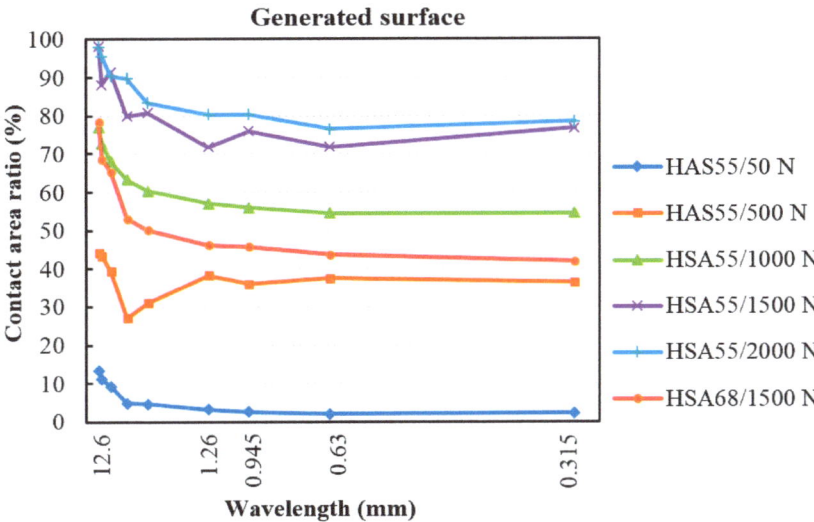

Figure 9. Contact areas determined for surfaces with different minimum texture wavelengths under different loads and rubber hardness conditions.

Nevertheless, the microtexture of the real pavement, with dimensions comparable to the thickness of the paint, would be rougher than on the generated and 3D-printed surfaces. Therefore, the paint is less likely to be squeezed out of the contact area on the pavement and thus avoids showing larger contact areas than the actual ones.

This study conservatively estimates that the determination of the rubber-pavement contact area, which can be estimated using the staining method, requires a resolution in texture wavelength of approximately 0.6 mm.

2.4. Bearing Area Curve and Analysis Scale

The bearing area curves (BACs) can be obtained using 3D surface texture data. A smaller sampling interval can capture more surface details than long ones, making the measured data more realistic. However, the analysis in Section 2.3.3 shows that the staining method could not detect the contact area for texture wavelengths shorter than 0.6 mm. According to Section 2.1, the BAC should be determined with the same resolution as the contact area. Generally, most scanners (or other texture-recording devices) operate with 0.1 mm or higher resolutions. Hence, ideally, the captured texture wavelengths are longer than 0.2 mm. Figure 9 shows that it may not be necessary to filter the surface data to remove the roughness with shorter texture wavelengths than around 0.6 mm as the contact areas do not depend on these wavelengths.

Figure 10 shows the BACs calculated from artificial surfaces with different 'smoothness' by applying low-pass filtering with cut-offs at texture wavelengths of 0.2, 0.5, 0.63, 0.945, 1.26, and 2.5 mm (using a Gaussian filter response). After filtering (removing) the short waves, it appears in Figure 10 that the top part (higher height) accounts for a higher ratio of the surfaces' areal material; in other words, the texture decreases.

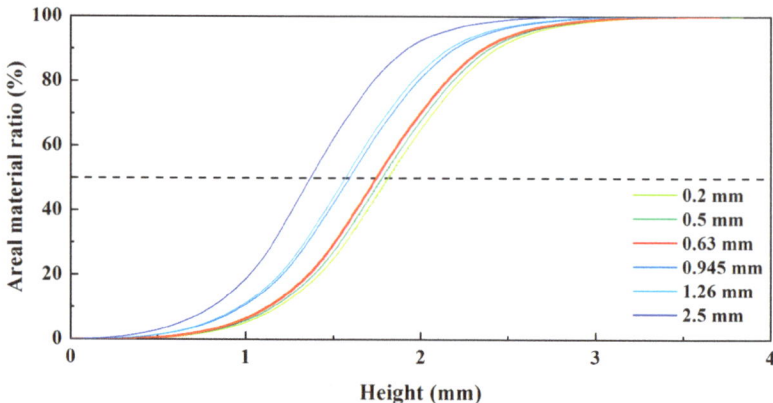

Figure 10. Bearing area curves calculated for artificially designed surfaces with texture wavelengths cut-off at 2.5 to 0.2 mm.

This study uses the contact area ratio measured with the staining method and the BAC of the specimen's surface data to obtain the penetration depth of rubber on pavements. The analysis in Section 2.3.3 suggested that the measured contact area ratio was most appropriate for a resolution of about 0.6 mm, expressed as the texture wavelength. Therefore, the resulting rubber penetration depth should be larger than the true value if the surface data used to calculate the BAC contains texture wavelengths below 0.6 mm. Specifically, when the areal material ratio is 50%, the penetration depths obtained for surfaces with λs of 0.2, 0.5, and 0.63 mm are 1.816, 1.781, and 1.751 mm, respectively. Consequently, for an artificial surface with Gaussian height probability distribution, the error caused by the analysis resolution of the BAC was less than approximately + 5%, based on a reference with a minimum texture wavelength of 0.2 mm.

For a realistic asphalt pavement, the surface generally has a negative skewness, and the top part of the profile is relatively flat due to the rolling process during construction and the accumulated vehicle load, as illustrated in Figure 11. The numbers in brackets indicate the λs of the surfaces. The upper profile curve represents the case of the pavement, and the lower one represents the case of the artificial surface with a Gaussian height probability distribution.

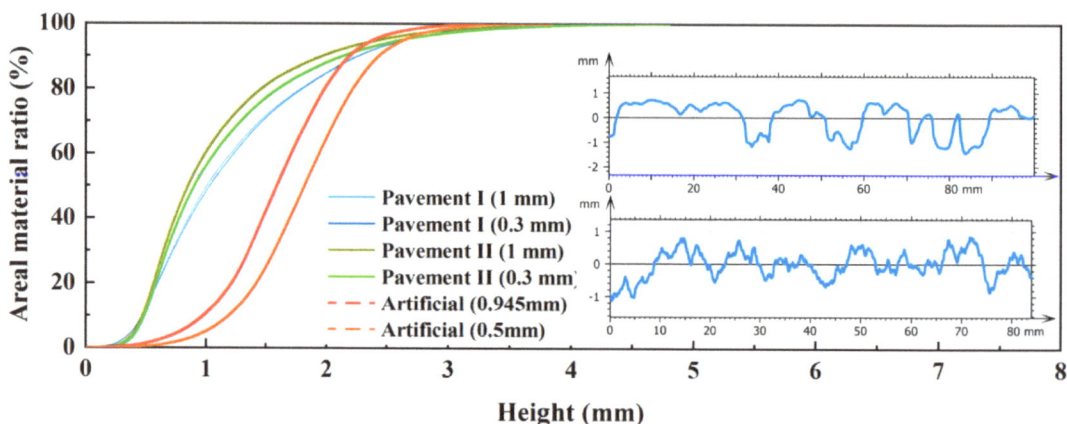

Figure 11. Bearing area curves and profiles obtained from the artificial surface and two pavements, respectively.

According to the BACs, the analysis scale of texture wavelength makes a smaller difference for the pavement surfaces than for the artificial surfaces. The profiles in blue (Figure 11) show that the top part of the pavement is flatter than on the artificial surface. In addition, the texture at the top part of the pavement contributes less to the surface height than the middle and bottom parts do. Therefore, when the staining method from Section 2.3 is used to determine the contact area, the error in the penetration depth obtained using the BAC is below 5%, even without filtering the texture with a texture wavelength below 0.6 mm.

3. Comparison of Different Methods to Determine Rubber Penetration Depth

3.1. Test Objects for the Comparison

This section compares the penetration depth obtained using the S-BAC, indentor method, and the enveloping method proposed by von Meier et al. [25]. This study compares the results by applying the three methods on the same surfaces. The surface texture data were captured on: (1) 29 asphalt pavement sections, the mean texture depth (MTD) of which ranges from 0.31 to 1.41 mm, and (2) 15 artificial surfaces designed by a similar method as described in Section 2.3. The texture on these surfaces was captured and calculated by the HandyScan 3D scanner (Creaform^{TM}, Lévis, QC, Canada).

3.2. Application of the S-BAC Method

The procedure to determine the rubber penetration depth for each sample using the S-BAC method is shown in Figure 12. The detailed information is as follows:

1. Prepare the specimens by the method described in Section 2.2.2. The size of specimens should be sufficient for enveloping. This study tests the 60×60 mm^2 specimens produced by a 3D printer.
2. The fixture described in Section 2.3.1 is attached to the loading device (e.g., the UTM). Then, the specimen and the fixture are centered. The loading head is raised to a suitable height (e.g., 15 mm above the specimen), and the rubber block's bottom is painted.
3. The loading process is set up and started using the user interface. After loading, the head is raised, and the specimen with the paint on the surface is dried.
4. The contact area ratio is determined using the method described in Section 2.3.2.
5. The surface morphology of the specimen is scanned, and the accumulative height probability distribution (BAC) of the specimen is determined.
6. According to the test principles in Section 2.1, the rubber penetration depth is obtained using the contact area ratio obtained in Step 4 and the BAC obtained in Step 5.
7. The same procedure is repeated for each of the 44 samples introduced in Section 3.1.

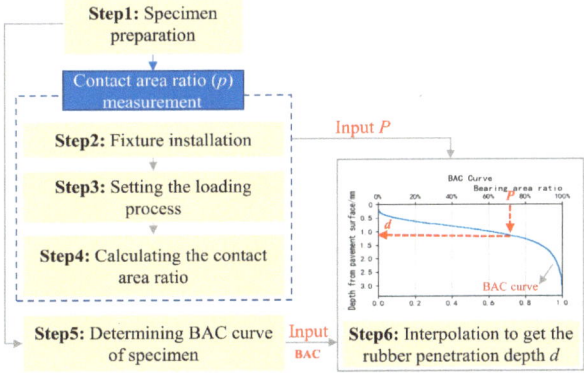

Figure 12. Procedure to determine the rubber penetration depth.

The vertical loads for the (S-BAC method) measurements were 50, 500, 1000, 1500, and 2000 N, and the corresponding pressures were 0.018, 0.18, 0.35, 0.53, and 0.71 MPa. Two rubber blocks were used in this study, with a hardness of 55 and 68 Shore A, respectively, 6 mm thick and 60 mm in diameter.

3.3. S-BAC Method and Enveloping Method by von Meier et al. [25]

Figure 13 shows the enveloping profiles obtained using the method by von Meier et al. [25], where the parameter d^* depended on the mechanical characteristics of the tire. The value $d^* = 0.054$ mm^{-1}, derived from a measurement of the deformation of a tire pressed onto various idealized profiles, was suggested [25].

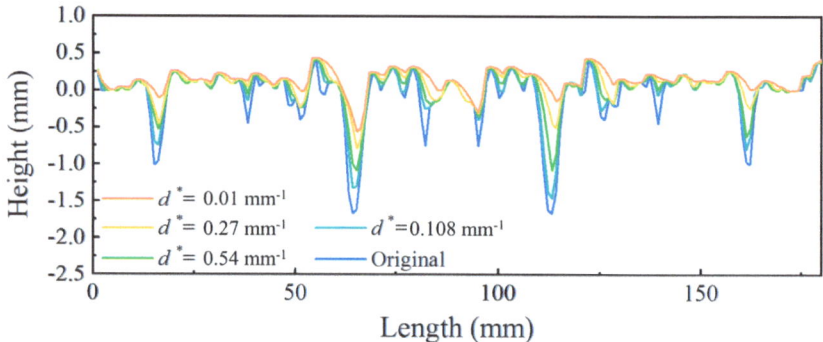

Figure 13. Effect of the d^* on enveloping profiles obtained using the method by von Meier et al. [25].

Figure 13 shows that the penetration depth of the tire into the pavement texture was greatly affected by the width between two adjacent asperities in the pavement profile. This study used a baseline with a length of 100 mm to calculate the enveloping profiles. Each profile was divided into two sections. The difference between the average of the highest and lowest points for these two sections was used as the rubber penetration depth (Figure 14). This could reduce the effect of space between adjacent asperities. The results for 10 profiles for each surface were taken as the representative value.

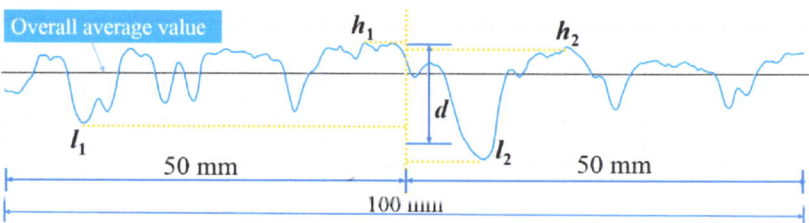

Figure 14. Schematic of the terms baseline, and the penetration depth obtained using the method by von Meier et al. [25].

Figure 15 shows the penetration depth obtained using the S-BAC method and the method by von Meier et al. [25]. The penetration depth obtained using the two methods is identical for the solid gray line ($y = x$). The solid red line refers to the regression line between the two variables. For 44 samples, when Pearson's r (correlation coefficient) is higher than 0.29 ($\alpha = 0.05$), there is a significant correlation between the two variables. There is a strong correlation between the penetration depth obtained by the S-BAC method and the von Meier method, except for the HSA68/1500 N condition.

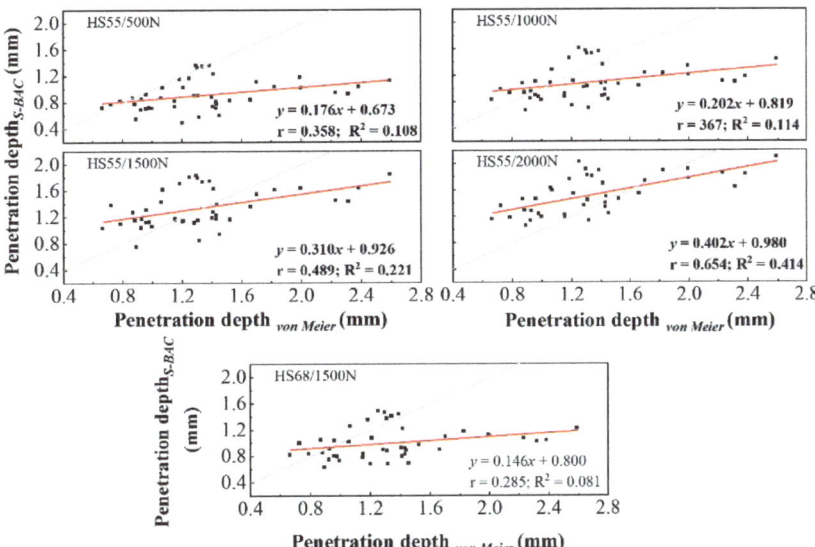

Figure 15. Comparison of penetration depth between the S-BAC and von Meier method.

Most penetration depths for the 44 surfaces obtained by the S-BAC method are below that obtained by the von Meier method for low-pressure conditions. The two penetration depths are equal on surfaces with small penetration depths obtained using the von Meier method. As the pressure increases or the rubber hardness lowers, the tail of the red line gradually rises.

3.4. S-BAC Method and the Indentor Method

Figure 16 shows the enveloping profiles obtained by the indentor method with different S. The value of S was recommended as 6 or 10 mm² by the ROSANNE project [3]. In this study, the height difference between the horizontal line's height at which a specific S can be satisfied and the pavement profile's highest point refers to the penetration depth obtained by the indentor method. The penetration depth can be calculated by iteration against S, and the error is 0.01 mm². The larger S results in a higher value of penetration depth (Figure 16).

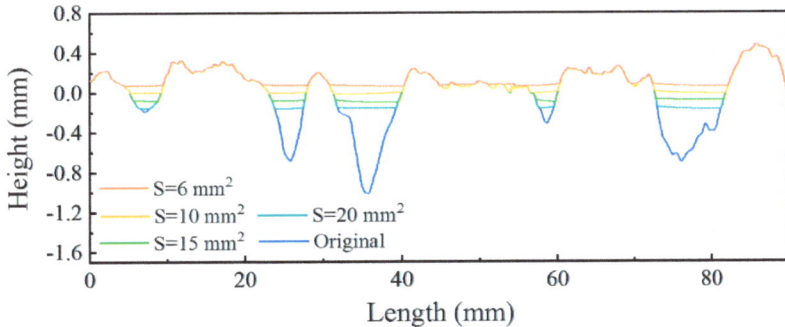

Figure 16. Tire-road enveloping profile obtained by the indentor method with different S.

Figure 17 compares the penetration depth obtained by the S-BAC and indentor methods. The depth obtained by the indentor method is the average of the values calculated from 10 profiles extracted from the specimens' surfaces.

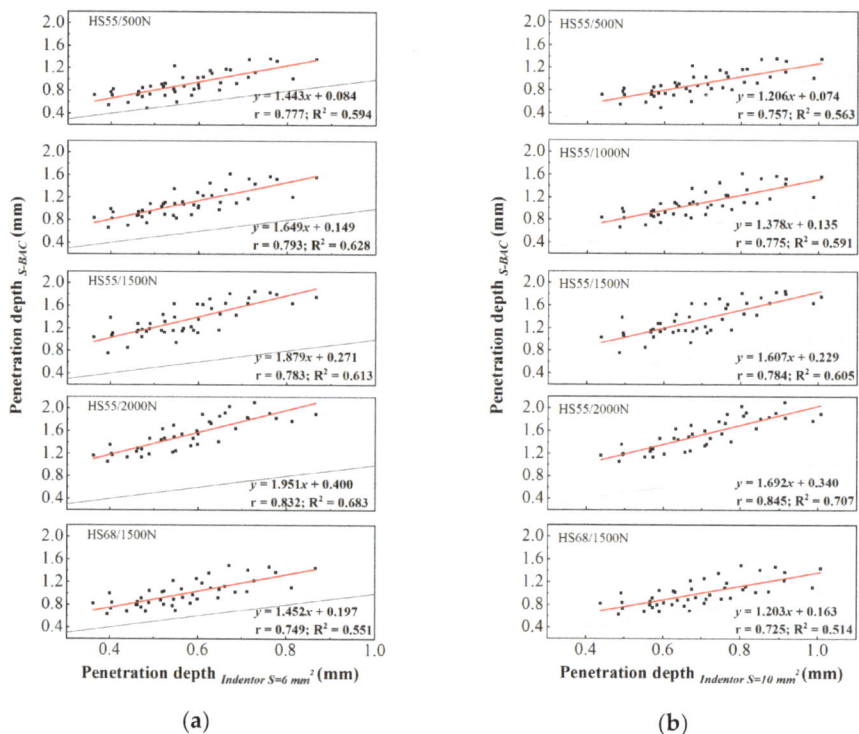

Figure 17. Relationship between rubber penetration depth obtained using the S-BAC and indentor method. (**a**) indentor $S = 6$ mm^2; (**b**) indentor $S = 10$ mm^2. The gray lines are the isoclinic line, the red lines are fitted lines.

Figure 17 reveals a close linear relationship between the penetration depth obtained by the S-BAC and indentor methods. The penetration depths obtained by the former are always higher than the latter under the test conditions in this study. More specifically, the value of depth obtained using the S-BAC method is about 1.2~3 times larger than for the indentor method. The multiple gradually increased with increasing pressure or decreasing rubber hardness.

4. Application of Rubber Penetration Depth Obtained from Different Methods

4.1. Friction Coefficient and Pavement Texture Measurement

This study chose 23 asphalt concrete test sections in Xi'an, China. The pavement texture was recorded using the HandyScan 3D scanner introduced in Section 2.2.2. The friction coefficient on these sections was measured using the T2GO shown in Figure 18 (SARSYS-ASFT™, Köpingebro, Sweden). The device creates a fixed slip of 20%, which effectively simulates the braking action.

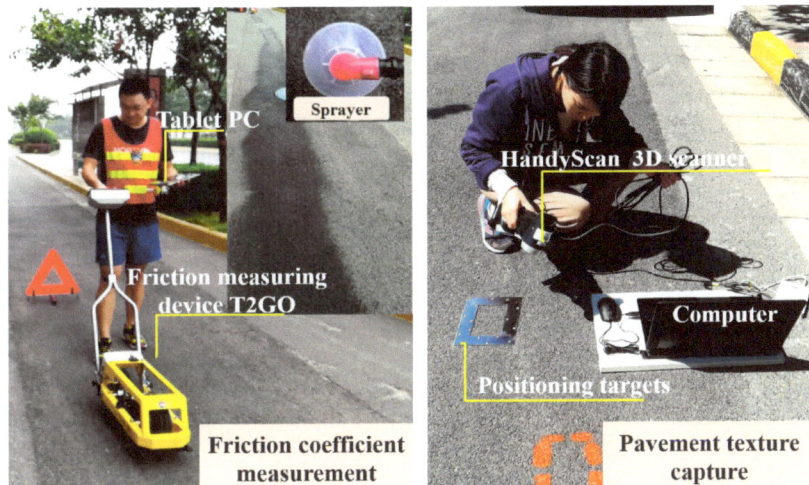

Figure 18. Measuring the low-speed friction coefficient and capturing the pavement texture.

The friction recorded speed during the test was 2~3 km/h, and the measurement length on each test section was about 15 m. Each test was repeated twice. The measurement interval for friction was set to 30 mm (0.1 feet) with the interface controls located at the top of the handlebar. Based on the walking speed during sprinkling (1.5~2 km/h), the sprinkling width (0.3 m), the water outflow speed (15~25 mL/s), and ignoring evaporation, the water film thickness was estimated at approximately 0.1~0.2 mm. Following the standard correction factor of 0.003 per 1 °C, the friction measurements obtained using the T2GO device (SARSYS-ASFT^{TM}, Köpingebro, Sweden) were adjusted to a reference temperature of 20 °C [38]. The sampling interval for the pavement texture was 0.2 mm, and the scanning area was 400 × 280 mm^2. Three positions, evenly distributed along the test path, were scanned in each section.

Due to the limitation of scanning instruments, only pavement macrotexture was captured in the field. However, microtexture significantly affects the friction coefficient between the rubber and the pavement [39]. This study took two measurements to equate the effect of the microtexture on the measured friction coefficient. First, we wet the pavement before measuring the friction coefficient; second, we chose test sites with a slight difference in microtexture. In this case, the water occupies part of the microtexture on the pavement. Furthermore, most of the test sections we chose were on the right side of the right-most lane on the road, where the aggregates suffered little abrasion, and the microtexture was still almost covered by asphalt.

Figure 19 shows the measured friction coefficient. The water film thickness during friction measurement was 0.1~0.2 mm, so there was sufficient time for a low-speed test to eliminate the water between the tire and the pavement surface. The nominal pressure between the friction-measuring device and the pavement was 0.52 MPa, and the rubber hardness was about 68 HSA [40]. Therefore, the depth of the T2GO's tire penetration in the pavement was tested using the S-BAC method at 0.52 MPa, 68HSA, under dry conditions. Table 1 shows the tire penetration depth on the pavement.

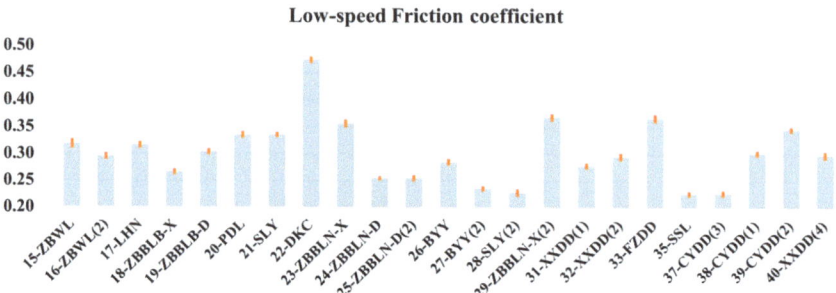

Figure 19. Measured friction coefficient by T2GO device.

Table 1. Rubber penetration depth obtained by different methods in mm.

Test Sections	S-BAC Method		Indentor Method				von Meier $d^* = 0.054$ mm^{-1}	
			$S = 6$ mm^2		$S = 10$ mm^2			
	Depth	Stdev	Depth	Stdev	Depth	Stdev	Depth	Stdev
15-ZBWL	0.84	0.05	0.49	0.04	0.60	0.05	1.12	0.32
16-ZBWL(2)	1.12	0.02	0.68	0.01	0.84	0.01	2.08	0.16
17-LHN	1.04	0.12	0.56	0.01	0.69	0.01	1.60	0.10
18-ZBBLB-X	1.11	0.13	0.62	0.05	0.78	0.08	1.89	0.32
19-ZBBLB-D	1.17	0.06	0.64	0.02	0.79	0.03	1.88	0.08
20-PDL	0.86	0.03	0.48	0.01	0.58	0.02	1.03	0.07
21-SLY	1.06	0.06	0.48	0.01	0.58	0.02	1.03	0.07
22-DKC	0.78	0.03	0.46	0.02	0.56	0.02	1.06	0.05
23-ZBBLN-X	1.12	0.12	0.62	0.04	0.76	0.06	1.48	0.14
24-ZBBLN-D	1.24	0.16	0.68	0.08	0.84	0.09	1.79	0.07
25-ZBBLN-D(2)	0.74	0.02	0.44	0.01	0.53	0.01	0.99	0.11
26-BYY	1.08	0.20	0.62	0.05	0.75	0.06	1.65	0.37
27-BYY(2)	1.15	0.01	0.63	0.04	0.76	0.05	1.60	0.15
28-SLY(2)	1.40	0.16	0.77	0.08	0.96	0.09	2.75	0.60
29-ZBBLN-X(2)	0.76	0.04	0.48	0.04	0.59	0.05	1.03	0.35
31-XXDD(1)	1.02	0.03	0.62	0.03	0.77	0.03	1.87	0.13
32-XXDD(2)	0.89	0.03	0.53	0.04	0.66	0.05	1.41	0.39
33-FZDD	0.84	0.04	0.50	0.03	0.62	0.05	1.12	0.17
35-SSL	0.99	0.05	0.60	0.02	0.73	0.04	1.50	0.06
37-CYDD(3)	0.81	0.14	0.44	0.08	0.53	0.09	0.72	0.13
38-CYDD(1)	0.86	0.16	0.48	0.07	0.58	0.08	1.05	0.27
39-CYDD(2)	0.78	0.07	0.45	0.03	0.55	0.04	0.92	0.22
40-XXDD(4)	0.89	0.17	0.55	0.09	0.67	0.10	1.27	0.12

4.2. Effect of Rubber Penetration Depth on Pavement Texture-Friction Relationship

After obtaining the penetration depth, the texture parameters were calculated based on cloud data above the depth. Two commonly used parameters (root-mean-square surface height Sq and root-mean-square slope of the surface Sdq), considered to significantly affect the actual contact area and pressure distribution, which plays a pivotal role in determining the friction coefficient formation, were selected [41]. The calculation of these two parameters required Equations (2) and (3). The average of the parameters at three positions on the test path was used to represent each test section.

$$Sq = \sqrt{\frac{1}{A} \iint_A z^2(x,y)dxdy} \tag{2}$$

$$Sdq = \sqrt{\frac{1}{A}\iint_A \left[\left(\frac{\partial z(x,y)}{\partial x}\right)^2 + \left(\frac{\partial z(x,y)}{\partial y}\right)^2\right] dx dy} \tag{3}$$

Here, $z(x, y)$ is the surface height measured from the average plane with $z = 0$, A is the evaluation area.

Figure 20 shows the effect of the penetration depth obtained by different methods on the relationship between the pavement texture parameters and friction coefficient. The parameters Sq and Sdq calculated based on different penetration depths are shown at the horizontal axes of Figure 20a,b, respectively. The low-speed friction coefficient is at the vertical axes. The red zones are shown at the 95% confidence band to compare these equations' prediction accuracy, and the smaller confidence interval indicates the more accurate prediction. Table 2 shows the details of the equations in Figure 20.

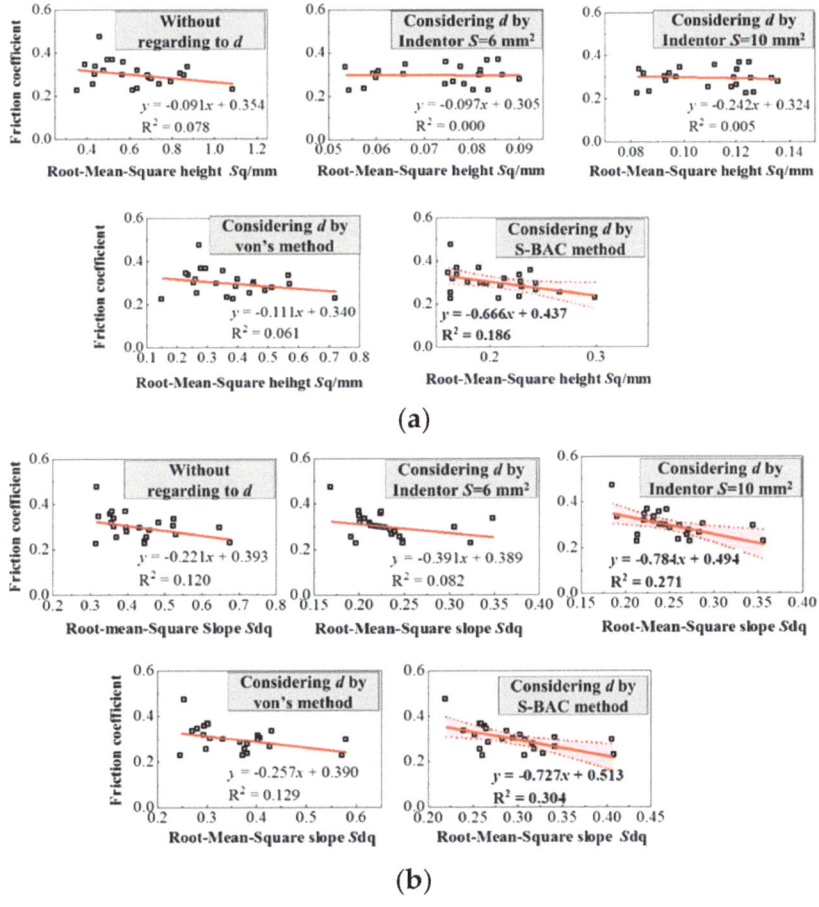

Figure 20. Effect of penetration depth on the relationship between texture parameters and the friction coefficient. (**a**) Texture parameter Sq and the friction coefficient; (**b**) Texture parameter Sdq and the friction coefficient.

Table 2. Parameters of the regression lines and the hypothesis testing.

Parameter	Method	Equation	a	b	Pearson's r	R^2	F Test -p
RMS height of the surface Sq	Full depth	$y = a \times x + b$	−0.091	0.354	−0.279	0.078	0.197
	S-BAC		−0.666	0.437	−0.432	0.186	0.040 *
	Indentor $S = 6$ mm^2		−0.097	0.305	−0.018	0.000	0.935
	Indentor $S = 10$ mm^2		−0.242	0.324	−0.068	0.005	0.758
	Von Meier		−0.111	0.340	−0.246	0.061	0.258
RMS slope of the surface Sq	Full depth		−0.221	0.393	−0.346	0.120	0.105
	S-BAC		−0.727	0.513	−0.551	0.304	0.006 *
	Indentor $S = 6$ mm^2		−0.391	0.389	−0.287	0.082	0.183
	Indentor $S = 10$ mm^2		−0.784	0.494	−0.521	0.271	0.011 *
	Von Meier		−0.257	0.390	−0.360	0.129	0.092

*: being significant with a 95% confidence.

The R2 and confidence intervals indicate that the relationship between the texture parameters (Sq, Sdq) and the friction coefficient is statistically significant when calculating the penetration depth based on the S-BAC method. Though Sq and Sdq have a weak relationship with the low-speed friction coefficient, the S-BAC method can enhance the prediction of the friction coefficient from texture parameters, compared with the von Meier and indentor methods.

The weak relationship may be because the pavement texture capture equipment can only obtain macrotexture. This study took measures expected to separate the contribution of macrotexture to friction during the friction measurement. However, microtexture dominates the low friction coefficient, which makes the friction coefficient unsensitive to the macrotexture difference.

5. Discussion

The von Meier, indentor, and S-BAC methods in this study emphasize different factors affecting tire-pavement contact. This may make the difference to the rubber penetration depth (Figures 15 and 17) and its effect on the relationship between the pavement texture and friction coefficient (Figure 20). The results obtained by the S-BAC and von Meier method are only close to each other under specific conditions. This study believes that the inherent parameter of the von Meier method causes the differences in penetration depth as the pavement morphologies in this study cover most of the pavement texture on asphalt concrete pavement (MTD ranges from 0.31~1.41 mm). Similarly, 'S' in the indentor method was derived from the contact between the tire and the surfaces with wooden triangular elements (1 cm, 0.5 cm, and 2 cm separated from each other) [3], which causes the differences in rubber penetration depth [42].

For future studies, the S-BAC method proposed in this study could be improved in the following aspects:

Firstly, a thick rubber block should be chosen to simulate the contact between the tire tread and the pavement. The 6-mm rubber block may resemble non-mobile friction testers, such as the British pendulum tester and dynamic friction tester. However, tires would have a rubber tread of significantly more than 8 mm to accommodate a tread depth of 8 mm, and the tread sits on a rather flexible belt, so the 6-mm thick rubber block on a steel plate would not resemble an actual tire.

Secondly, pavement specimens bigger than 60 × 60 mm^2 should be used. In practice, the rubber blocks used in this study were plain, while the vehicle tires in the previous two methods had various patterns. These patterns divide the tire tread into smaller rubber blocks (smaller than the rubber block used in the S-BAC method). The von Meier and indentor methods used pavement profiles of 100 and 90 mm, respectively. Hence, a 60 × 60 mm^2 specimen may be sufficient for enveloping for car tires. However, the size is too small to represent macrotexture.

The specimen size should be about 0.3×0.3 m^2 to represent macrotexture according to the smallest acceptable size of specimens in ISO 13473-1.

In addition, the S-BAC method can also be extended to field tests with complex conditions if there are reliable methods to extract the real tire-pavement contact area from the nominal contact area. For example, Du et al. [14] and Woodward et al. [43] obtained the tire-pavement contact area by marking the abrasion area via colored powder or painting. After investigating the scale of the recognized contact area, the S-BAC method can be applied to obtain the penetration depth for dynamic contact, which can provide a helpful reference for road performance evaluation and prediction.

6. Conclusions

This study describes a new method to measure the rubber penetration depth on pavement based on the contact area measured using a staining method and the bearing ratio curve of the pavement surface. After discussing the factors that affect the test results of this method and comparing it with previous statistical methods, the following conclusions can be drawn:

- The contact area recognized by the staining method is at a scale of about 0.6 mm. Removing roughness with texture wavelengths shorter than 0.6 mm via filtering is better before measuring the bearing area curve. Otherwise, the penetration depth may be overestimated (but not by more than 5% for surface data with a sampling interval below 0.1 mm).
- The penetration depth obtained using the S-BAC method is close to that obtained using the von Meier method for some conditions. For surfaces with larger space between consecutive asperities, the penetration depth obtained using the S-BAC method is close to that using the von Meier method when the pressure increases, or the rubber hardness decreases.
- A strong linear correlation exists between the penetration depth obtained with the S-BAC method and the indentor method. The former is about 1.2~3 times higher than for the indentor method, and the multiple gradually increases with higher pressure or softer rubber.
- When calculating the root-mean-square height of the surface (Sq) and the root-mean-square slope of the surface (Sdq) based on texture data above the penetration depth, the relationship between these two pavement texture parameters and the low-speed friction coefficient is clearly enhanced after considering the penetration depth obtained via the S-BAC method.

Compared with the empirical methods, the S-BAC method did not depend on any controlling parameters from experience. It considers factors such as pressure and the properties of the rubber, which have a significant effect on the penetration depth. The experiment indicates that the S-BAC method strengthens the relationship between texture parameters and the low-speed friction coefficient. In future, the rubber and specimen size should be studied to represent the actual condition more reasonably, and exploring the application of the extended S-BAC method in field tests could be helpful. Furthermore, more reliable results may be obtained if the microtexture can be captured and considered in the evaluation process in future studies.

Author Contributions: Conceptualization, D.Y. and L.H.; methodology, D.Y. and C.T.; software, C.T.; validation, D.Y. and L.H.; data curation, J.G.; writing—original draft preparation, D.Y.; writing—review and editing, U.S.; visualization, D.Y. and M.R.; funding acquisition, X.Z. and L.H. All authors have read and agreed to the published version of the manuscript.

Funding: This research was funded by the Fundamental Research Funds for the Central Universities, CHD, grant number 300102213512; National Natural Science Foundation of China, China, grant number 52172392, 52268068; Key research and development Project of Hubei Province, China, grant number 2021BAA180.

Institutional Review Board Statement: Not applicable.

Informed Consent Statement: Not applicable.

Data Availability Statement: Data are contained within the article.

Conflicts of Interest: The authors declare no conflicts of interest.

References

1. Guo, W.; Chu, L.; Yang, L.; Fwa, T. Determination of tire rubber-pavement directional coefficient of friction based on contact mechanism considerations. *Tribol. Int.* **2023**, *179*, 108178. [CrossRef]
2. Kogbara, R.B.; Masad, E.A.; Woodward, D.; Millar, P. Relating Surface Texture Parameters From Close Range Photogrammetry to Grip-Tester Pavement Friction Measurements. *Constr. Build. Mater.* **2018**, *166*, 227–240. [CrossRef]
3. Goubert, L.; Sandberg, U. Enveloping texture profiles for better modelling of the rolling resistance and acoustic qualities of road pavements. In Proceedings of the 8th Symposium on Pavement Surface Characteristics: SURF 2018, Brisbane, Australia, 2–4 May 2018.
4. Persson, B.N.J. Theory of rubber friction and contact mechanics. *J. Chem. Phys.* **2001**, *115*, 3840–3861. [CrossRef]
5. Dubois, G.; Cesbron, J.; Yin, H.; Anfosso-Lédée, F. Numerical evaluation of tyre/road contact pressures using a multi-asperity approach. *Int. J. Mech. Sci.* **2012**, *54*, 84–94. [CrossRef]
6. Kane, M.; Edmondson, V. Tire/road friction prediction: Introduction a simplified numerical tool based on contact modelling. *Veh. Syst. Dyn.* **2020**, *60*, 770–789. [CrossRef]
7. Johnson, K.J.; Xu, B.; Luo, X.; Liu, X.; Song, G.; Sun, X. *Contact Mechanics*, 1st ed.; Higher Education Press: Beijing, China, 1992.
8. Kanafi, M.M.; Tuononen, A.J. Application of three-dimensional printing to pavement texture effects on rubber friction. *Road Mater. Pavement Des.* **2017**, *18*, 865–881. [CrossRef]
9. Xiao, S.; Sun, Z.; Tan, Y.; Li, J.; Lv, H. Hysteresis friction modelling of BPT considering rubber penetration depth into road surface. *Int. J. Pavement Eng.* **2022**, *24*, 2027415. [CrossRef]
10. Srriangam, S.K. Numerical Simulation of Tire-Pavement Interaction. Ph.D. Thesis, Roskilde University Theses and Dissertations Archive, Delft University of Technology, Delft, The Netherlands, 2015.
11. Zheng, B.; Chen, J.; Zhao, R.; Tang, J.; Tian, R.; Zhu, S.; Huang, X. Analysis of contact behaviour on patterned tire-asphalt pavement with 3-D FEM contact model. *Int. J. Pavement Eng.* **2022**, *23*, 171–186. [CrossRef]
12. Matilainen, M.J.; Tuononen, A.J. Tire friction potential estimation from measured tie rod forces. In *Proceedings of the 2011 IEEE Intelligent Vehicles Symposium (IV)*; Baden-Baden, Germany, 5–9 June 2011, pp. 320–325.
13. Wang, D.; Ueckermann, A.; Schacht, A.; Oeser, M.; Steinauer, B.; Persson, B. Tire Road Contact Stiffness. *Tribol. Lett.* **2014**, *56*, 397–402. [CrossRef]
14. Du, Y.; Qin, B.; Weng, Z.; Wu, D.; Liu, C. Promoting the pavement skid resistance estimation by extracting tire-contacted texture based on 3D surface data. *Constr. Build. Mater.* **2021**, *307*, 124729. [CrossRef]
15. Hartikainen, L.; Petry, F.; Westermann, S. Frequency-wise correlation of the power spectral density of asphalt surface roughness and tire wet friction. *Wear* **2014**, *317*, 111–119. [CrossRef]
16. Woodward, D.; Millar, P.; McQuaid, G.; McCall, R.; Boyle, O. PSV tyre/test specimen contact. In *Proceedings of the 8th RILEM International Symposium on Testing and Characterization of Sustainable and Innovative Bituminous Materials*; Springer: Dordrecht, The Netherlands, 2016.
17. Ejsmont, J.; Sommer, S. Selected aspects of pavement texture influence on tire rolling resistance. *Coatings* **2021**, *11*, 776. [CrossRef]
18. Chen, D. Study on Image-based Texture Analysis Method and Prediction of Skid Resistance and Tire/Pavement Noise Reduction of HMA. Ph.D. Thesis, Chang'an University, Xi'an, China, 2015.
19. Lu, J.; Pan, B.; Liu, Q.; Sun, M.; Liu, P.; Oeser, M. A novel noncontact method for the pavement skid resistance evaluation based on surface texture. *Tribol. Int.* **2022**, *165*, 107311. [CrossRef]
20. Yang, G.; Wang, K.C.; Li, J.Q. Multiresolution analysis of three-dimensional (3D) surface texture for asphalt pavement friction estimation. *Int. J. Pavement Eng.* **2021**, *22*, 1882–1891. [CrossRef]
21. Chen, B.; Zhang, X.; Yu, J.; Wang, Y. Impact of contact stress distribution on skid resistance of asphalt pavements. *Constr. Build. Mater.* **2017**, *133*, 330–339. [CrossRef]
22. Wang, D.; Wang, G.; Li, Z.; Shao, S.; Sun, Y. Evaluation of Anti-sliding Durability of Asphalt Mixture Based on Pressure Film Technology. *China J. Highw. Transp.* **2017**, *30*, 1–9.
23. Gao, J.; Sha, A.; Huang, Y.; Liu, Z.; Hu, L.; Jiang, W.; Yun, D.; Tong, Z.; Wang, Z. Cycling comfort on asphalt pavement: Influence of the pavement-tyre interface on vibration. *J. Clean. Prod.* **2019**, *223*, 323–341. [CrossRef]
24. Yu, M.; Xu, X.; Wu, C.; Li, S.; Li, M.; Chen, H. Research on the prediction model of the friction coefficient of asphalt pavement based on tire-pavement coupling. *Adv. Mater. Sci. Eng.* **2021**, *2021*, 6650525. [CrossRef]
25. von Meier, A.; Van Blokland, G.J.; Descornet, G. The influence of texture and sound absorption on the noise of porous road surfaces. In Proceedings of the PIARC 2nd International Symposium on Road Surface Characteristics, Beilin, Germany, 23–26 June 1992.
26. Andersen, L.G. Rolling Resistance Modelling: From Functional Data Analysis to Asset Management System. Ph.D. Thesis, Roskilde University Theses and Dissertations Archive, Roskilde University: Roskilde, Denmark, 2015.
27. Zhu, S.; Huang, P. Influence mechanism of morphological parameters on tribological behaviors based on bearing ratio curve. *Tribol. Int.* **2017**, *109*, 10–18. [CrossRef]

28. Li, T.; Burdisso, R.; Sandu, C. Effect of Rubber Hardness and Tire Size on Tire-Pavement Interaction Noise. *Tire Sci. Technol.* **2019**, *47*, 258–279. [CrossRef]
29. Okonieski, R.E.; Moseley, D.J.; Cai, K.Y. Simplified Approach to Calculating Geometric Stiffness Properties of Tread Pattern Elements. *Tire Sci. Technol.* **2003**, *31*, 132–158. [CrossRef]
30. Persson, B.N.J.; Albohr, O.; Tartaglino, U.; Volokitin, A.I.; Tosatti, E. On the nature of surface roughness with application to contact mechanics, sealing, rubber friction and adhesion. *J. Physics Condens. Matter* **2004**, *17*, R1–R62. [CrossRef]
31. Mogrovejo, D.E.; Flintsch, G.W.; Katicha, S.W.; Izeppi, E.D.d.L.; McGhee, K.K. Enhancing Pavement Surface Macrotexture Characterization by Using the Effective Area for Water Evacuation. *Transp. Res. Rec.* **2016**, *2591*, 80–93. [CrossRef]
32. Sha, A.; Yun, D.; Hu, L.; Tang, C. Influence of sampling interval and evaluation area on the three-dimensional pavement parameters. *Road Mater. Pavement Des.* **2020**, *22*, 1964–1985. [CrossRef]
33. Zelelew, H.M.; Papagiannakis, A.T.; Masad, E. Application of digital image processing techniques for asphalt concrete mixture images. In Proceedings of the 12th International Conference of International Association for Computer Methods and Advances in Geomechanics (IACMAG), Goa, India, 1–6 October 2008; pp. 119–124.
34. Brown, C.A.; Hansen, H.N.; Jiang, X.J.; Blateyron, F.; Berglund, J.; Senin, N.; Bartkowiak, T.; Dixon, B.; Le Goïc, G.; Quinsat, Y.; et al. Multiscale analyses and characterizations of surface topographies. *CIRP Ann.* **2018**, *67*, 839–862. [CrossRef]
35. Pérez-Ràfols, F.; Almqvist, A. Generating randomly rough surfaces with given height probability distribution and power spectrum. *Tribol. Int.* **2019**, *131*, 591–604. [CrossRef]
36. Jacobs, T.D.B.; Junge, T.; Pastewka, L. Quantitative characterization of surface topography using spectral analysis. *Surf. Topogr. Metrol. Prop.* **2017**, *5*, 013001. [CrossRef]
37. GB/T 531.1; Rubber, Vulcanized or Thermoplastic-Determination in Indentation Hardness-Part 1: Durometer Method (Shore Hardness). Chinese Standard: Beijing, China, 2008.
38. Mahboob Kanafi, M.; Kuosmanen, A.; Pellinen, T.K.; Juhani Tuononen, A. Macro-and micro-texture evolution of road pavements and correlation with friction. *Int. J. Pavement Eng.* **2015**, *16*, 168–179. [CrossRef]
39. Roy, N.; Kuna, K.K. Image texture analysis to evaluate the microtexture of coarse aggregates for pavement surface courses. *Int. J. Pavement Eng.* **2022**, *24*, 2099854. [CrossRef]
40. Yun, D.; Hu, L.; Tang, C. Tire–Road Contact Area on Asphalt Concrete Pavement and Its Relationship with the Skid Resistance. *Materials* **2020**, *13*, 615. [CrossRef]
41. Li, Q.; Luo, J. *Contact Mechanics and Friction: Physical Principles and Applications*; Tsinghua University Press: Beijing, China, 2011.
42. Yun, D.; Tang, C.; Ran, M.; Zhou, X.; Hu, L. Enveloping profile calculation method for enhancing the efficiency of pavement skid resistance prediction. *J. Southeast Univ. (Nat. Sci. Ed.)* **2023**, *53*, 130–136.
43. Woodward, D.; Millar, P.; Tierney, C.; Ardill, O.; Perera, R. The friction measuring tire/road surface space interface. In Proceedings of the 5th International SaferRoads Conference, Auckland, New Zealand, 21–24 May 2017.

Disclaimer/Publisher's Note: The statements, opinions and data contained in all publications are solely those of the individual author(s) and contributor(s) and not of MDPI and/or the editor(s). MDPI and/or the editor(s) disclaim responsibility for any injury to people or property resulting from any ideas, methods, instructions or products referred to in the content.

Review

Functionalized and Biomimicked Carbon-Based Materials and Their Impact for Improving Surface Coatings for Protection and Functionality: Insights and Technological Trends

Aniket Kumar [1], Bapun Barik [1], Piotr G. Jablonski [2,3,*], Sanjiv Sonkaria [4,*] and Varsha Khare [2,4,*]

1. School of Materials Science and Engineering, Chonnam National University, Gwangju 61186, Korea
2. Laboratory of Behavioral Ecology and Evolution, School of Biological Sciences, Seoul National University, Seoul 08826, Korea
3. Museum and Institute of Zoology, Polish Academy of Sciences, Wilcza 64, 00-679 Warsaw, Poland
4. Soft Foundry Institute, Seoul National University, Kwanak-gu, Seoul 08826, Korea
* Correspondence: piotrjab@behecolpiotrsangim.org (P.G.J.); ssonkaria64@snu.ac.kr (S.S.); khare@snu.ac.kr (V.K.)

Citation: Kumar, A.; Barik, B.; Jablonski, P.G.; Sonkaria, S.; Khare, V. Functionalized and Biomimicked Carbon-Based Materials and Their Impact for Improving Surface Coatings for Protection and Functionality: Insights and Technological Trends. *Coatings* **2022**, *12*, 1674. https://doi.org/10.3390/coatings12111674

Academic Editor: Joaquim Carneiro

Received: 10 September 2022
Accepted: 26 October 2022
Published: 4 November 2022

Publisher's Note: MDPI stays neutral with regard to jurisdictional claims in published maps and institutional affiliations.

Copyright: © 2022 by the authors. Licensee MDPI, Basel, Switzerland. This article is an open access article distributed under the terms and conditions of the Creative Commons Attribution (CC BY) license (https://creativecommons.org/licenses/by/4.0/).

Abstract: Interest in carbon materials has soared immensely, not only as a fundamental building block of life, but because its importance has been critical to the advancement of many diverse fields, from medicine to electrochemistry, which has provided much deeper appreciation of carbon functionality in forming unprecedented structures. Since functional group chemistry is intrinsic to the molecular properties, understanding the underlying chemistry of carbon is crucial to broadening its applicability. An area of economic importance associated with carbon materials has been directed towards engineering protective surface coatings that have utility as anticorrosive materials that insulate and provide defense against chemical attack and microbial colonization of surfaces. The chemical organization of nanoscale properties can be tuned to provide reliance of materials in carbon-based coating formulations with tunable features to enhance structural and physical properties. The transition of carbon orbitals across different levels of hybridization characterized by sp^1, sp^2, and sp^3 orientations lead to key properties embodied by high chemical resistance to microbes, gas impermeability, enhanced mechanical properties, and hydrophobicity, among other chemical and physical attributes. The surface chemistry of epoxy, hydroxyl, and carboxyl group functionalities can form networks that aid the dispersibility of coatings, which serves as an important factor to its protective nature. A review of the current state of carbon-based materials as protective coating materials are presented in the face of the main challenges affecting its potential as a future protective coating material. The review aims to explore and discuss the developmental importance to numerous areas that connects their chemical functionality to the broader range of applications

Keywords: Functional carbon materials; coating technology; flame retardants; antifouling; biomimicry

1. Introduction

The elemental importance of carbon is signified by its position as the seventeenth most resourceful element that exists in the earth's outer layer, the fifth most abundant in the solar system, and the sixth most plentiful element in space [1]. According to the assessment, carbon's relative abundance ranges between 180 and 270 parts per million (ppm) [2–4]. Carbon has intriguing and unique properties, and the properties of some carbon allotropes take the form of graphite [5], graphene oxide (GO) [6,7], graphene(G) [8], reduced graphene oxide(rGO) [9], fullerenes (C_{60}) [10], carbon nanotubes (CNTs) [11], noncrystalline carbon [12], diamond morphology [13], and lonsdaleite [14]. As a result of its peculiar electronic structure, carbon can transition between different mixed orbital hybridizations, including sp^1, sp^2, and sp^3 forms. Variation in orbital hybridization has led to different structures with several other distinctive properties [15]. The carbon allotropes possess distinctive characteristics, including impenetrability to gases [16], chemical (acid, base, and salt) endurance [17,18], antibacterial potential [19,20], thermal stability [21,22], environmental friendliness, and, more critically, high specific surface area [23]. It has been found that carbon allotropes can also provide a fertile research ground for enhancing protective surface coatings because of their flexible surface chemistry and useful property diversity, making them excellent research targets for advanced material development.

In general, coatings are materials or multilayers that are applied to the surface of bulk materials to preserve the primary or undercoat material. Coatings that are synthetic or engineered can be used either individually or in combination as composites that result in altered functional properties. Individual or property combinations are represented by hierarchical patterns that are morphologically and functionally different and are tunable to deliver material properties with desirable outcomes [24]. Biological damage, corrosion, fouling, and environmental damage are some of the main reasons why functional coatings are used [25]. Besides medical instruments, electronics, and everyday household appliances, a wide range of military and marine applications are available [26]. It should be noted that, in addition to the applications previously mentioned, functional coatings have also been developed to help remove contaminants such as heavy metals and dyes from different natural and fresh waters [27,28]. Due to this, protective coatings go beyond their usual functions of adsorbing toxic substances, killing bacteria, and protecting against frost, fire, and irradiation to also address corrosion, fouling, and mechanical wear as well.

It is well known that coatings are widely used for protecting materials. Nevertheless, some metal coating technologies can negatively impact the environment they are connected with. Since they possess both carcinogenic and biocidal properties, hexavalent chromium (Cr (VI)) and tributyltin (TBT) have been completely banned from use in protective coatings [29,30]. For example, Cr (VI) has been limited to a maximum of 0.1 wt% in corrosion-protective coatings [31]. TBT has also been banned from use in antifouling coatings since 2003 to safeguard against damage to marine environments [32]. Metal coating ingredients used in the form of cadmium (Cd), cobalt (Co), and copper (Cu) have also been determined to be detrimental to the environment and have shown to be carcinogenic to humans [33–35]. Cadmium has been identified to cause hazardous effects on the environment and is associated with toxicity and its use as a material has been prohibited in electronic devices and equipment since 2006 [36]. The search for ecofriendly alternatives to replace these materials and fulfill the key industrial and regulatory requirements applied to the coating industry has been challenging. However, diamond-like carbon (DLC) is an interesting material for protective coatings, with useful applications for material toughness and wear resistance, bearing low friction coefficients and thermal conductivity. Further, DLCs offer considerable chemical stability, gas-barrier capability, and high infrared-permeable properties. What is most important is that DLC coatings are biocompatible in nature [37]. Therefore, their applications are diverse, including electronics (e.g., hard disks, video tapes, and integrated circuits), cutting tools (e.g., drills, end mills, and razors), molds (e.g., optical parts and injection molding), automotive parts (e.g., piston rings, cams, clutch plates, pumps, and injectors), optical components (e.g., lenses), plastic bottle oxygen-barrier films, sanitary equipment, windows, bathtub mirrors, and decorative uses [38]. The material properties associated with DLCs have gained much recognition in recent years, increasing their potential use as protective coatings. An extensive account of DLC-based materials as protective coatings is out of the scope of this review due to its rather broad nature; a separate review dedicated to DLCs is anticipated in greater detail.

A carbon-based technology that relies on polymer-based composites is currently the forefront for driving functionalized nanomaterials with carbon as one of its core materials.

The economic impact of combing costly materials with polymers is considerable, as polymers can reduce the cost burden by lowering the requirement of expensive materials in forming composites while maintaining material performance. This strategy has paved the way for fabricating novel protective coating products, particularly graphene-based composites, comprising paints and other useful formulations. With the availability of property-based materials, a broad range of applications is possible by tuning the compositional nature at the microstructure level of the final coating to enhance the level of protection provided by functionalized carbon. Such composites provide the freedom to modify and regulate the characteristics of the final product by altering the components in isolation. Hence, a functionalized carbon coating can deliver superior protective attributes with few shortcomings, depending on the application using (1) pure carbon coating, (2) a carbon-like coating, or (3) a functionalized carbon-enhanced composite-type coating. In this regard, remarkable progress in the advancement of innovative protective coatings has been achieved, aided by such technologies through the customization of functionalized carbon.

This review aims to consider recent developments related to the application of multipurpose carbon-based coatings. With reference to preventing corrosion, fouling, thermal or radiation damage, and contamination, the distinctive physical and chemical assets of carbon allotropes exhibited by different types of coatings will be discussed. The intrinsic protective mechanisms utilized by graphene, graphene oxide (GO), and/or reduced graphene oxide (rGO) are discussed as a prelude to their applicability, which further warrants an in-depth discussion of the merits of their technological importance and potential as engineered materials in design and function. The review also brings into view an analytical perspective on key developmental challenges and highlights future prospects of carbon-based materials in the evolution of protective-coating technologies.

2. Functionalized Carbon-Based Systems

All nanometric forms of carbon discovered to date, including graphite, diamonds, graphene, monohorns of carbon, fullerene, and carbon nanotubes (CNTs), are known as carbon nanoallotropes. A pictorial representation of carbon-based nanoallotropes is shown in Figure 1. Solubility limits the use of carbon allotropes in industrial and biological applications [39]. The solubility of carbon allotrope molecules may be enhanced through functionalization, achieved by introducing polar substituents into the molecular framework as an effective method of improving the solubility of carbon allotropes. The aromatic nature of carbon allotropes in their oxidized form (rGO) is more suppressed while GO exhibits nonaromatic composition, with carbon atoms hybridized with a sp^3 configuration. Several deviations to the planarity of the structure are attributed to the presence of sp^3 carbon atoms that are bonded to oxygen moieties. A variety of covalent and noncovalent techniques can be used for functionalizing carbon allotropes [15,40]. In the case of covalent functionalization, carbon allotropes and polar substituents are covalently coupled, while weak intermolecular forces (van der Waals forces) occur during noncovalent functionalization.

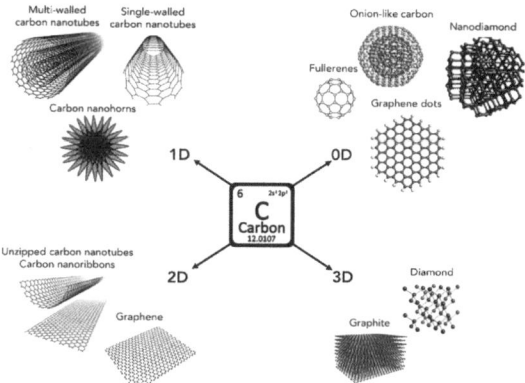

Figure 1. Carbon allotropes that are commonly found in nature. Reprinted with permission from [2] 2020 Elsevier.

2.1. Non-covalent Approaches for Functionalization

There are several non-covalent functionalization methods using several approaches. Hydrophobic and electrostatic interactions, van der Waals forces of attraction between molecules, and a bundled assemblage conforming to a condensed state feature strongly in the commonly used non-covalent functionalization approaches that are applied to carbon allotropes [41,42]. Non-covalent functionalization provides useful routes to alter surfaces, thus permitting surface functionalization without chemical disruption of the surface structural framework of the carbon allotrope and resulting in changes to the physical and electrical properties following functionalization. Graphene analogues, such as single-walled carbon nanotubes (SWCNTs) and multiwalled carbon nanotubes (MWCNTs) have, in recent years, undergone non-covalent functionalization. It is generally observed that in this mode of interaction, the πe^- density of the benzene ring(s) (e.g., pyrene and perylene) to some degree overlaps with the πe^- region of the carbon allotrope. Several non-covalent functionalization routes of GO can be seen in Figure 2. The aromatic regions of carbon allotropes are often directed to undergo π-π stacking associations. Interfaces between carbon allotropes (with supermolecules, surface-active agents, ionic liquids, etc.) can be categorized as hydrophobic. The π–π cooperativity has been reported to occur in polymers such as Kevlar [43], polyvinyl chloride, sulfonated polyaniline [44], and polyethylene glycol [45]. To investigate the electron–donor–acceptor interactions in aromatic compounds with graphene nanosheets, pyrene and perylene diimide derivatives were used. Carbon allotropes (G, GO, CNTs, etc.) work together hydrophobically with supermolecules, ionic liquids, and surfactant molecules to form stacked configurations on carbon allotropes and differ from the stacked orientation used by aromatic compounds on carbon allotropes. Both aqueous and organic solvents disperse carbon allotropes by virtue of their dispersion strength arising from chemical interactions between the carbon allotrope and the solvent environment. Electrostatic interactions play a large role in the interaction between ionic liquids and GO (conceivably with different allotrope types), in part because GO contains a variety of ionic groups, such as hydroxylates and carboxylates, which can engage in electrostatic interactions. Hydrogen bonding is not only capable of stacking and hydrophobic and electrostatic interactions but can also occur between allotropes of carbon in addition to stacking.

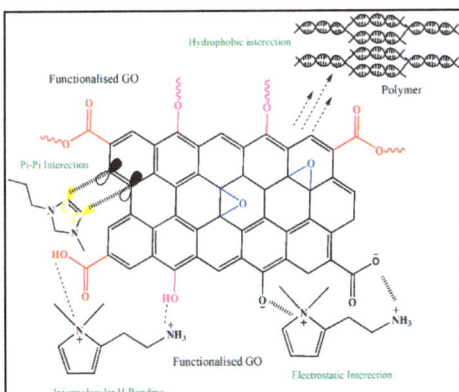

Figure 2. Non-covalent functionalization of GO.

2.2. Covalent Approaches for Functionalization

Two kinds of target sites are available on the surface of carbon-based materials such as graphene oxide (GO) and its analogues, which are amenable to undergoing covalent functionalization and can be described by the following events: (i) oxygen-containing groups such as −OH as well as −C-O-C on the basal plane, as well as −COOH at the edges, where new species might be introduced via condensation and esterification; and (ii) —C=C functionality, which requires the attacks by free radicals or dienophile species [46,47]. Commonly used strategies for functionalization of GO include: (i) acylation; (ii) amidation; (iii) silylation; (iv) esterification; (v) cycloaddition; (vi) etherification; and (vii) azidation. The —COOH groups at the edges are possibly modified by amidation or esterification. There have been several amines used in the past in order to covalently functionalize graphene oxide (GO) and, more specifically, C_{60} fullerenes, porphyrins, polythiophenes, and phthalocyanines. It is possible to activate hydroxyl groups in several ways, namely through salinization, silylation, and etherification.

There are numerous chemical processes that can be used to modify the GO, including ring-opening actions with epoxides, amine (R-NH$_2$), and alcohol (R-OH). A wide range of poly(allyl amine) and poly(vinyl alcohol) moieties can be attached to ringed structures using these reactions. Functionalization of > C=C < may be performed by free-radical-catalyzed cycloaddition reaction (Diels–Alder type reaction). Carbons undergoing sp^2 hybridization can transition to sp^3 during cycloaddition, which reduces GO's ability to conduct electrons. This is shown in Figure 3a.

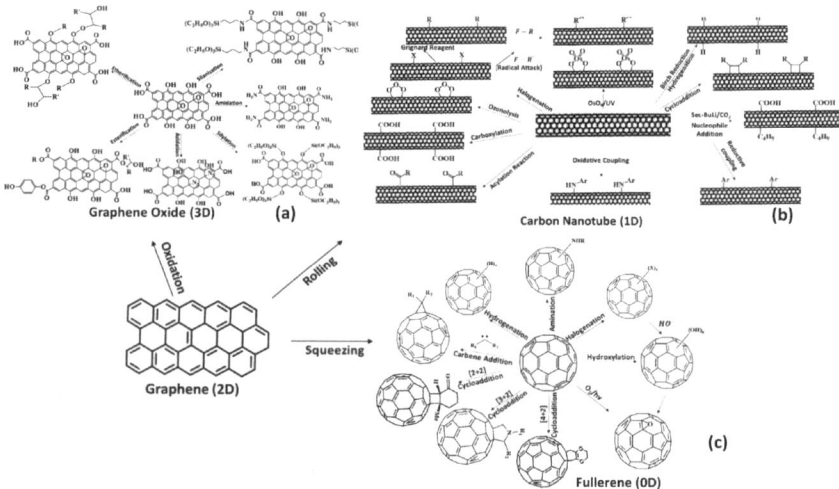

Figure 3. (**a**) A scheme showing the modification of graphene and GO using commonly applied functionalization chemical/physical strategies. (**b**) Surface functionalization of carbon nanotubes. (**c**) Surface functionalization of fullerene.

A carbon nanotube (CNT) is a cylinder of carbon with a diameter measured in nanometers (nm) and a length also measured in nanometers (nm). Carbon nanotubes are divided into single-walled carbon nanotubes (SWCNs) and multiwall carbon nanotubes (MWCNs) [41]. Van der Waals interactions can provide an avenue for multiple CNT ring connections to form MWCNTs. In material science, CNTs and their derivatives (functionalized CNTs) are used for a variety of applications due to their exceptional mechanical properties and physical properties due to their diminished size. It is important to note that CNTs exhibit exceptional electrical conductivity and are prone to corrosion in aqueous media, as they enhance electron transfer. Despite their various advantages, CNTs are thus limited in their application as corrosion inhibitors, especially in aqueous phases. The desirable properties of anticorrosion can, however, be achieved by covalently functionalizing them. Surface modification of SWCNTs and MWCNTs covalent means can be accomplished using chemical strategies (Figure 3b). Such changes allow the adaptation of CNTs (SWCNTs and MWCNTs), allowing them to undergo a number of physiochemical changes as a result of covalent functionalization, which can be described by: (i) An enhancement in overall dispersion properties; (ii) interfacial compatibility assessed by adhesion and diminished hydrophobicity; (iii) a reduction in aggregation; (iv) a reduction in mechanical properties; and (vi) a reduction in electrical conductance. The functionalized state of GO and CNTs (SWCNTs and MWCNTs) retains the capacity to undergo further chemical refinement by targeting peripheral moieties, which remain exposed for further functionalization [43,44]. The biological and industrial applications of other allotropes of carbon can also be enhanced through covalent functionalization using different chemical species, as can GO and CNTs.

The chemistry of fullerene has been thoroughly studied and investigated to date. The functionalization chemistry of fullerene is more or less similar to the approaches used to functionalize carbon nanotubes and graphene oxide, as shown in Figure 3c.

3. Adaptation of Carbon as a Protective Coating via Surface Functionalization

3.1. Functionalized Coated Carbon for Anticorrosion Applications

Recent investigations have established that corrosion barrier coatings that have been fabricated from functionalized carbon, graphene, and graphene oxide and their composites, are effective anticorrosion layers, which also take the form of alloys used in the range of electrolytes. The reported use of chemically modified GO, however, has been rather limited, with relatively fewer examples of anticorrosive properties compared to its metallized counterparts. This may partly be explained by lower levels of polar electrolytes than in nonpolar electrolytes. Only a few reports have been published about the use of carbon support that has either been functionalized or modified with suitable organic compounds as effective corrosion inhibitors in an aqueous solution. Anticorrosion behavior of different organic functional groups, which can be used to functionalize carbon supports for designing an effective corrosion inhibitor system in an aqueous solution, are summarized in Table 1. Due to its inherent binding tendency and high surface area, chemically modified GO exhibits a relatively high protection capacity.

Recently, covalent modifications and self-polymerization have been used to functionalize graphene oxide (GO) with polydopamine (PDA), which was then tested as a corrosion inhibitor in HCl solutions against carbon steel. At 100 mg/L, GO-PDA achieved almost 90% inhibition efficiency, which is a high level of corrosion protection for carbon steel. An investigation of the interaction between immersion time and concentration was conducted. A hydrophobic film was formed in micro–nano structures by lamellar functionalized GO adsorbed onto the surface of carbon steel [48]. The composite carbon material of this dense protective film was extremely hydrophobic, inhibiting the penetration of corrosive agents, as shown in Figure 4a. In this study, a transition from a GO-surface that had a negatively charged surface to a positively charged GO-PDA that significantly enhanced the adsorption at the material intersection. The contact angle measurement study, as shown in Figure 4b–e, evidently showed that the GO-PDA cover layer was hydrophobic in nature, which was corroborated by contact angle measurements at the interface.

Table 1. Different chemically functionalized carbon-based anticorrosion coatings in aqueous phases.

S.N.	Functional Group	Medium	Metal	Concentration of Inhibitor	Inhibition Efficiency (%)
1 [49]	1-tetradecylpyridazin-1-ium iodide	1.0 HCl	Carbon Steel	10^{-3} M	97.6
2 [50]	Schiff base	1mM HCl	Steel	1 mM	92.0
3 [51]	Benzotriazole	0.5 M Na_2SO_4	Galvanized steel	2.9×10^{-3} M	90
4 [52]	L-cysteine	1 wt% NaCl	Alloy Steel	400 ppm	65.7
5 [53]	Polylactic acid/2-mercaptobenzothiazole	1 M HCl	Alloy aluminum	0.5 mass ration	33.83
6 [54]	Imidazoline	1.65 wt% NaCl con. 1 g/L acetic acid	Carbon steel	0.1 g/L	99.3

Table 1. Cont.

S.N.	Functional Group	Medium	Metal	Concentration of Inhibitor	Inhibition Efficiency (%)
7 [55]	N-(quinolin-8-yl) quinoline-2-carboxamide	1 M HCl	Mild steel	0.3 mmol L^{-1}	94
8 [56]	(6-methyl-3-oxopyridazine-2-yl) acetate	1 mM HCl	Mild steel	10^{-1} mM	66.7
9 [57]	Benzotriazole	3.5 wt% NaCl	Copper	0.45 mM	92.31
10 [58]	2-(4 -nitrophenyl) benzimidazole	1 mM HCl	Carbon steel	1 mM	93.7
11 [59]	Adenosine	0.5 M HCl	Tin	10^{-3} M	73.0
12 [60]	1,8-Naphthyridine	1 M HCl	Mild steel	0.00004 M	98.7
13 [61]	Octadecyl amine	10^{-3} M Na$_2$SO$_4$	Carbon Steel	100 mgs kg^{-1}	90
14 [62]	Potassium 2,4-di(2H-benzo[d][1,2,3]triazol-2-yl)benzene1,3-bis(olate)	3.5 wt% NaCl	Copper	0.30 mM	96.7
15 [63]	4-Chloro-1H-indazole	3.0 wt% wt% NaCl	Copper	0.4 mM	98.9
16 [49]	Triazole–theophylline	1 M HCl	Steel	50 ppm	87
17 [64]	4-(3-Mercapto-5,6,7,8-tetrahydro-[1,2,4]triazolo[4,3-b][1,2,4,5]tetrazin-6-yl)phenol	1 M HCl	Mild Steel	0.5mM	67
18 [65]	1,2,4-Triazole	3 wt% NaCl	Copper	10 mM	98
19 [66]	N2,N4,N6-tris(3-(dimethylamino)propyl)-1,3,5-triazine-2,4,6-triamine	0.5 M	Mild Steel	1.0 mM	95.8
20 [67]	N-ethyl-N,N,Ntrioctylammonium ethyl sulfate	1 M H$_2$SO$_4$	steel	100 ppm	83
21 [68]	1-benzyl-4-phenyl-1H-1,2,3-triazole	1.0 M HCl	Mild Steel	2.13 mM	81.7
22 [69]	2-amino-4,6-bis-(3-hydroxy-4-methoxy-phenyl)-5-nitrocyclohex-2-ene-1,3,3-tricarbonitrile	1 M HCl	Mild Steel	25 ppm	98.96
23 [70]	6-(Dodecyloxy)-1H-benzo[d]imidazole	1 M HCl	Mild Steel	10^{-4} M	95.0
24 [71]	(1-(pyridin-4-ylmethyl)-1H-1,2,3-triazole-4-yl) methanol	1.0 M HCl	Mild Steel	1.0 mM	90.2

It has been recently demonstrated that 1,3-diphenylprop-2-en-1-one oxime–graphene oxide (CO-GO) can be produced from chalcone, hydroxylamine, and a Cl-modified graphene. For 12 h, the mixture was continuously stirred at 80 °C at a constant speed [72]. The process was completed by subjecting CO-GO to a spinning cycle, followed by a washing process in water and ethanol, and finally drying at 60 °C, as shown in Figure 5. Further, potentiodynamic polarization (PDP) and electrochemical impedance spectroscopy (EIS) were utilized to investigate corrosion inhibition effectiveness. In addition to its outstanding corrosion inhibition performance, CO-GO exhibits a primarily anodic mechanism with a mixed-type inhibitor function of up to 94%.

A variety of techniques have been employed to confirm inhibitor adsorption onto the carbon steel surface, including scanning electron microscopy and ultraviolet-visible spectroscopy.

The presence of polar substituents on functionalized carbon not only enhance the solubility of carbon-based materials in aqueous electrolytes, but also act as an adsorption center during metal–inhibitor interaction. Surface corrosion protection is obtained by the transfer of nonbonding electrons from the functionalized polar group and/or from the electrons on the aromatic or heteroaromatic rings to the metal center. In general, functionalized carbon-based systems adsorb onto metallic surfaces in two steps. The functionalized carbon-based system is transferred from the bulk solution onto the metal surface in the initial step. Adsorption occurs through coordination bonding in the second step. Electrons transferred from a functionalized carbon-based system to a metal surface without forming bonds are known as donations. Due to the interelectronic repulsion, even metals that are already electron-rich may become thermodynamically unstable as reactions result in electron transfer. Consequently, electrons may be transferred back into heteroatoms' empty p-orbitals during the interaction of carbon-based materials. The process is known as retro (back)-donation. As a result of a phenomenon known as synergism, donations and retro (back)-donations strengthen one another.

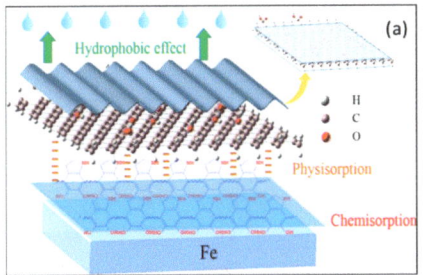
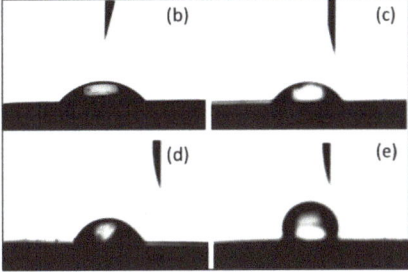

Figure 4. (**a**) The figure shows a mechanism showing how GO-PDA is inhibited by metal surfaces when it is in the solution. Contact angle measurements on the surface of carbon steel were measured for four different scenarios, including (**b**) blank, (**c**) 60 mg/L GO, (**d**) 60 mg/L PDA, and (**e**) 60 mg/L GO-PDA (**b–e**). Reprinted with permission from [48] 2022 J Mater Sci.

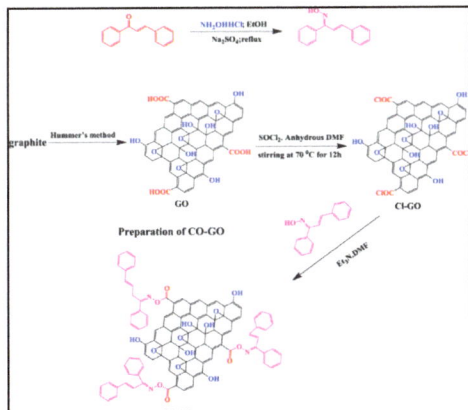

Figure 5. Scheme for the synthesis of chalcone oxime (CO)-functionalized graphene oxide. Reprinted with permission from [72] 2022 ACS Publ.

Electrostatic force of attraction is another means of interfacing functionalized nanomaterials with metallic surfaces besides chemical (coordination) bonding by means of physisorption. The adsorption of counter ions on positively charged metallic surfaces in aqueous electrolytes incite a negative charge to the metallic surface. In a separate event, protonation of a functionalized nanomaterial inhibitor group induces a positive charge and drives an electrostatic force of attraction to the negatively charged metallic site, resulting in the induction of physisorption. Figure 6 depicts chemisorption and physisorption in a pictorial manner. The most common mechanism for adsorption of organic corrosion inhibitors is the mixed mode, also known as physiochemisorption. GO and its derivatives are described in this article for their corrosion-inhibition properties.

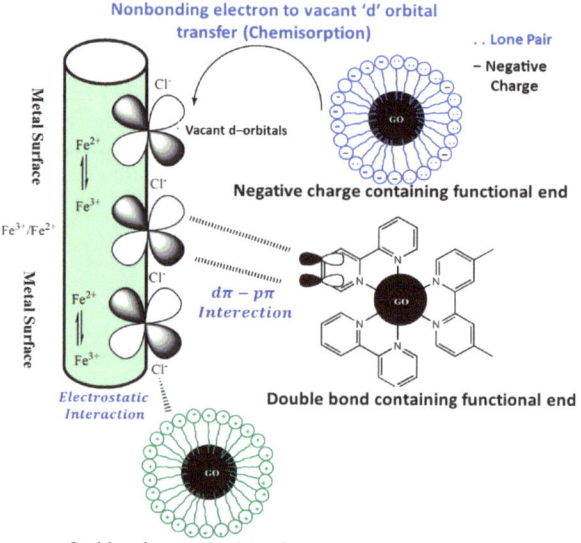

Figure 6. The interaction of metal surfaces with functionalized graphene oxide is illustrated in a series of pictures.

3.2. Functionalized Carbon for Flame-Retardant Coatings

The boundary region and basal planes of GO forming the graphene precursor constitute an abundance of epoxy, hydroxyl, carboxyl, and -C=C- groups which are freely available for functionalization [73]. Surface conversion of hydrophilic GO to its congruent hydrophobic state facilitates the dispersion of graphene sheets in the polymer matrix, as shown in Figure 7 [74]. The functionalization-based approach has been successfully applied in order to prepare organic flame retardants with functionalized graphene. With the help of this strategy, it is generally possible to covalently link organic flame retardants to graphene or GO in two steps: (i) Providing organic flame retardants with functionalities to enable their reaction with graphene or GO and (ii) reducing functionalized graphene to functionalized graphene using reducing agents. Modifiers that also serve as reducing agents can occasionally functionalize and reduce GO at the same time. A few examples of organic flame retardants functionalized with graphene, as well as the polymer composites that contain these materials, are summarized in Table 2.

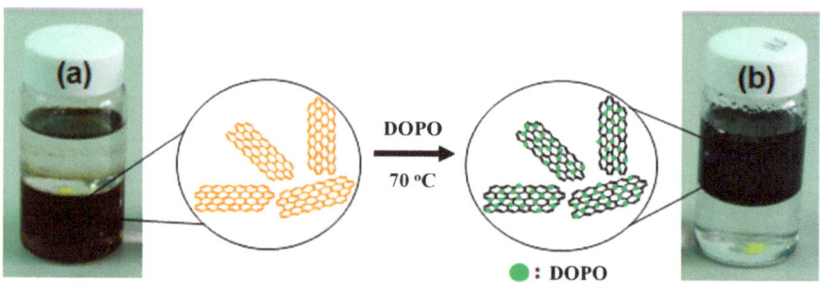

Figure 7. Photographs of (**a**) GO and (**b**) DOPO-reduced GO dispersions in water (bottom) and toluene (top) for 1 month. Reprinted with permission from [74]. 2012, Ind. Eng. Chem. Res.

Table 2. Summary of flammability of polymer nanocomposites based on functionalized graphene oxide.

S.N.	Polymer	Loading	Main Results
1 [74]	Epoxy	10 (Functionalized graphene)	Compared with pure epoxy, the char yield was increased to 30.2% from 16.7% and the GO/epoxy had a char yield of 21.1%; the limiting oxygen index (LOI) increased to 26% from 20% and the GO/epoxy had a char yield of 23%, respectively, when compared with pure epoxy.
2 [75]	Epoxy	10 (Functionalized graphene)	According to the UL-94 V-0 test (3.2 mm), the pristine graphene/epoxy composite showed a limiting oxygen index (LOI) value of 36%; however, the pristine graphene/epoxy composite had a LOI value of 26%, which was not high enough to pass the test.
3 [76]	XLPE	3 (Functionalized graphene)	Cone calorimetry observed a 29% reduction in PHRR, whereas GO/XLPE observed a 9% reduction.
4 [77]	PP	20 (Functionalized graphene)	PPA/PP composites showed a 67% and 24% reduction in PHRR and THR, respectively, whereas the PPA/PP composites showed 48% and 20% reductions, respectively.
5 [78]	Epoxy	8 (Functionalized graphene)	PPA/epoxy composites reduced both the PHRRs and THRs by 41% and 50%, respectively, whereas composites based on PPA-g-GNS/EP composites reduced them by 35% and 46%, respectively.
6 [79]	EVA	1 (Functionalized graphene)	Pure GO/EVA exhibited a reduction of 46% in PHRR and TTI of 75 s, whereas EVA/CRG–PPSPB achieved reductions of 31% and 61 s.
7 [80]	Epoxy	1 (Functionalized graphene)	A 45% reduction in PHRR was observed from pyrolysis–combustion-flow calorimetry, while that of GO/epoxy was only 18%.
8 [81]	EVA	2 (Functionalized CNTs)	24.4% reduction in PHRR observed
9 [82]	UPR	2- (Functionalized CNTs)	72% reduction in PHRR observed
10 [83]	PVDF	3 (Functionalized CNTs)	57.7% reduction in PHRR observed

burn-out limits of GO have been overcome in the past few decades by grafting several synthetic compounds onto it, including 9-oxa-10-phosphaphenanthrene-10-oxide (DOPO) [84,85], 2-(diphenylphosphino) ethyltriethoxy silane (DPPES) [78], hyper-branched flame retardants [86], polyphosphamide (PPA) [80], and poly(piperazine spirocyclic pentaerythritol bisphosphate) (PPSPB) [82]. Data on the limiting oxygen index (LOI), the UL-94 flame retardancy rating, and the peak heat-release rate (PHRR) for functionalized graphene sheets with organic flame retardants were compared; pristine graphene sheets and functionalized graphene sheets were found to exhibit better flame-retardant properties. The test results indicated that epoxy composites containing a mixture of DPPES-graphene at a loading of 10 wt% were capable of passing the UL-94 V-0 rating—in contrast to their counterparts containing equal amounts of either graphene or DPPES.

It has also been found that graphene can be functionalized with flame retardants in addition to sol–gel chemistry. Flame retardants modified with siloxane can facilitate reactions with silane-modified GO or silane-modified graphene, permitting a chemically modified interface between the retardant and graphene/GO surface formed via the silane-assisted hydrolysis of siloxane, resulting in a condensed structural composite. Figure 8 describes the step-by-step process for the functionalization of GO/PMMA composites. Several polymers, such as epoxy resins, polyuria (PU), polyvinyl alcohol (PVA), and polymethyl methacrylate (PMMA), can be made more flame-resistant by adding graphene or graphene oxide.

Flame retardants made from functionalized carbon can be effective for polymer materials because of its excellent thermal conductivity, thermal stability, and physical barrier properties. As a result of intensive research on graphene and its derivatives in epoxy resins [78], investigations have concluded that functionalized carbon is flame-retardant in three important areas. Layered structures of sp^2-hybridized carbon backbones that prevail in graphene and its derivatives can serve as physical barriers, slowing down heat transfer and preventing the initial stages of matrix decomposition. An investigation was conducted by Qu et al. to understand the mechanisms of flame retardancy of epoxy resin with the addition of POSS (octa glycidyl ether) groups to GO, as shown in Figure 9a [87]. The decomposition of the matrix is inhibited by a tumultuous path effect during the dispersion of the FGO in the epoxy resin matrix as a result of a tortuous path effect that inhibits the diffusion of heat and the degradation of combustible gases. A further benefit is that, when the OGPOSS groups contain Si, the char layer can be strengthened in such a way that heat transfer and oxygen mixing with pyrolysis gas can be prevented, thereby enhancing the flame-retardant properties of epoxy resin as well as its capability to suppress smoke emission. There is also evidence that graphene can form harmless gases (e.g., NH_3) when functionalized with organic elements (e.g., compounds containing nitrogen). This gas dilution hinders the combustion of the epoxy resin matrix material. Phytic acid (PA) and piperazine (PiP) were added to graphene oxide by Fang et al. in order to produce PiP-functionalized graphene oxide (PPGO), as shown in Figure 9b [88]. Finally, the benefit of graphene that has been functionalized by inorganic functional groups is that it leads to the formation of a dense, stable, and heat-resistant char layer and reduces the heat loss and smoke emissions from combustion processes. The combination of graphene compound (Cu^{2+}-GO) and ammonium polyphosphate (APP) can enhance flame retardation in epoxy resins, as described by Ye et al. (Figure 9c) [89].

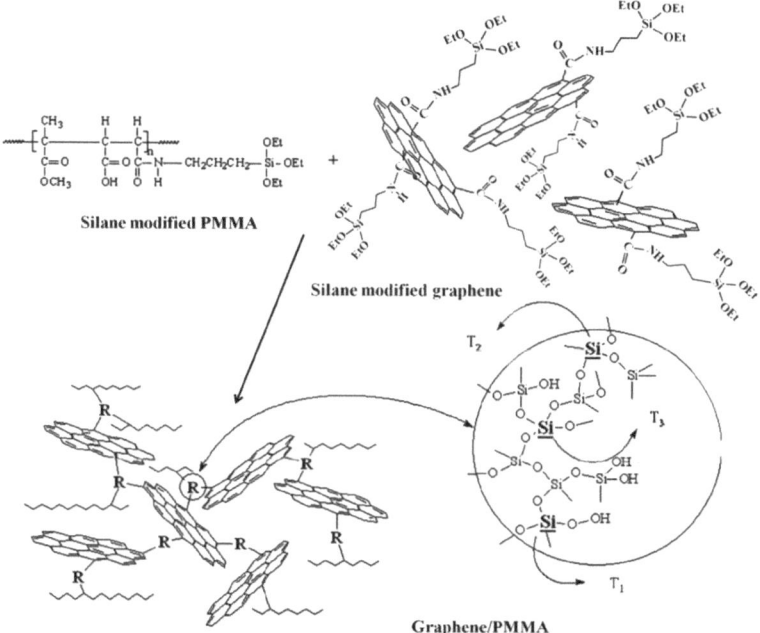

Figure 8. Schematic showing the use of the sol–gel method to prepare functionalized graphene/PMMA composites. Reprinted with permission from [86]. 2012, Mater. Chem. Phys.

3.3. Functionalized Carbon for Enhanced Antifouling Coatings

There are many problems associated with fouling on marine vessels and structures. Fouling is caused by the deposition of organic and inorganic materials, which can cause a wide variety of problems, including fouling of cargo ships, naval vessels, recreational vessels, heat exchangers, and sensors for oceanographic investigations, as well as offshore platforms such as oil rigs, desalination processes, and aquaculture systems. For the U.S. navy fleet alone, there are estimated to be maintenance costs from USD 180 million$/year to 260 million$/year due to fouling and surface setting and hardening of hard-to-remove deposits that attach to ships. This includes fuel costs, powering penalties, and maintenance expenses due to particulate fouling. TBT is currently banned in antifouling coatings, which introduces new challenges for developing antifouling technologies. Cost reduction is a major objective in this sector and the development of cost-effective materials with antifouling properties is predicted to reduce the economic burden substantially.

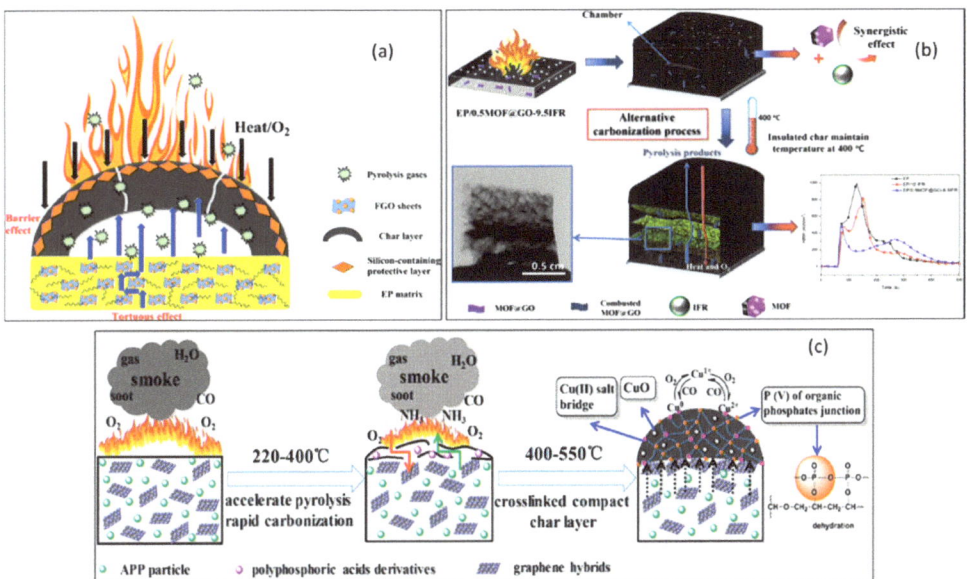

Figure 9. (a) Flame-retardant mechanisms in FGO/EP. Reprinted with permission from [87] 2020, Eur. Polym. J. (b) an EP-composite flame-retardant-mechanism model. Reprinted with permission from [88]. 2019 Carbon and (c) graphene-based EP/APP composites with flame-retardant mechanisms. Reprinted with permission from [89]. 2019, Compos. Part B Eng.

A variety of strategies have been developed as a means of preventing successive fouling of surfaces as well as fighting both inorganic as well as organic foulants. There are several strategies to maintain cleanliness on surfaces, which include surfaces with self-cleaning properties, biomimetic structural design, and the use of biocidal chemical mediators to deter fouling. Amphiphilic modified surfaces are also a recent development [90,91]. An antistick coating agent based on carbon is a promising option. Graphene surfaces can be tailored to achieve the surface polarity and wettability desired for antifouling coatings. In the presence of superhydrophilic and superhydrophobic graphene coatings, particle fouling can be prevented. Coated membranes composed of graphene can eliminate salt more effectively because of its impermeable structure. The use of carbon materials in antifouling coatings is not a novel finding since their properties are known to be associated with antifouling behavior discouraging the settlement of marine organisms and have been extensively studied in the field. A recent study of allotropic carbons as ecofriendly coatings has inspired work to harmonize the use of graphene into antifouling applications as a result of its foul-releasing properties, surface flexibility, selectivity towards protein adhesion, and covalent polymer attachment which is tunable by the degree and nature of surface functionalization.

Common organic matter that act as foulants in membrane-fouling-resistance studies are bovine serum albumin (BSA) and humic acid (HA), as shown in Figure 10a. The hydrophobic properties of these organic materials, in addition to their negatively charged surfaces, are their distinguishing characteristics. Cysteine-functionalized graphene oxide (CGO) composites can impose surface densities that are negatively charged in PES nanocomposite membranes. Increase in the CGO concentration from 0 to 0.4 wt% in the PES membrane led to improvements from −17.7 mV and −28.7 mV for surface zeta potential. Negative-charge density is attributed cysteine's high ionic charge density that resides in CGO-modified membranes, irrespective of the change in pH [1].

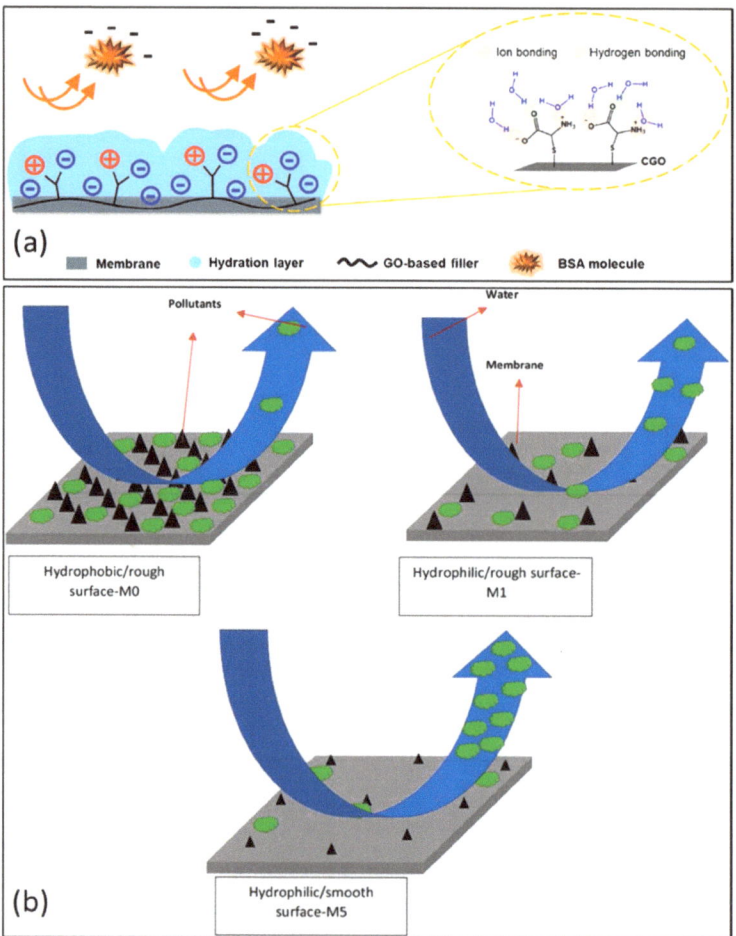

Figure 10. (**a**) Scheme showing that hydrogen and ionic bonding sites can be provided by cysteine-functionalized graphene oxide to the PES membrane to inhibit protein adsorption. Reprinted with permission from [1]. 2020, J. Ind. Eng. Chem. (**b**) the mechanism by which protein fouls the water during the cleaning process. (GO: 1 weight percent; Cs-GO: 1 weight percent, neat PES membrane). Reprinted with permission from [92]. 2018, J. Ind. Eng. Chem.

It has also been established that the roughness of GO-modified membranes may also play a role in the antifouling capabilities of the membrane and its ability to be hydrophilic. Contrary to the hydrophobic nature of the surface of the unmodified PES membrane and its rough surface, the smooth surface of the composite PES membrane allows foulants to be easily discarded. In Figure 10b, the schematic shows how surface cleaning is mechanistically independent of membrane hydrophilicity and surface roughness. Owing to the enhancement of surface hydrophilicity allied to the 1 wt% Cs-graphene-oxide-modified PES membrane and low surface roughness of the membrane, protein foulants on the membrane can be effortlessly removed via surface scrubbing.

The antifouling properties of zwitter ionic surfaces are effective in protecting surfaces against foulants. The modification of graphene oxide with zwitter ionic groups (GO-g-PMSA) and its incorporation into polyvinylidene fluoride (PVDF), as shown in Figure 11, generates two advantages: it increases the graphene oxide dispersibility in the membrane matrix and supplements GO with an increased antifouling capability. Water empathy and surface hydrophilicity of PVDF hybrid membranes were significantly improved by incorporating the zwitterionic additive. GO-g-PMSA/polyvinylidene fluoride membranes showed improved antifouling properties in comparison to pure polyvinylidene fluoride and graphene oxide–polyvinylidene fluoride membranes as it made use of softer and more hydrophilic properties. In the experiment, 1 wt% wt% of GO-g-PMSA/PVDF membrane showed superior antifouling performance. The study employed buffer solutions composed of phosphates in 1 g/L BSA and FRR was 95.3%, Rr was 32.3%, and R_{ir} was 4.7% at 0.7 MPa. Furthermore, high salt-ion refusal was noted in the salt-rejection experiments. The findings confirmed that zwitter-ionic-polymer-modified GO can potentially be used as an antifouling additive in nanocomposite membrane fabrication [93]. CNT coatings have been shown to prevent and control marine biofouling in several studies so far. These coatings consist primarily of MWCNTs, which have shown efficacy in marine environments, according to Table 3.

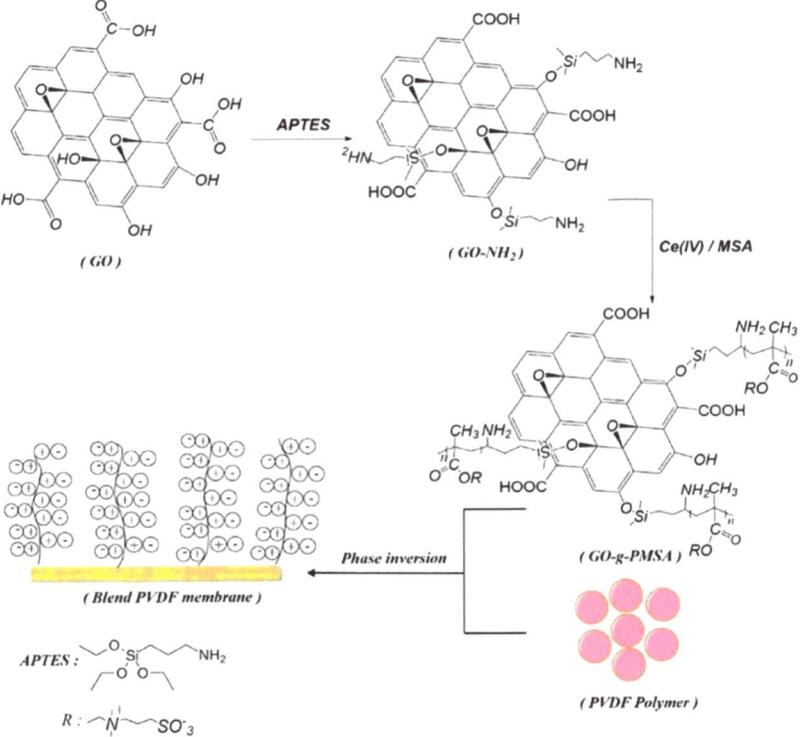

Figure 11. Scheme showing surface-instigated redox polymerization of 2−(Methacryloyloxy) ethyl dimethyl−(3−sulfopropyl) ammonium hydroxide from graphene oxide surface. Reprinted with permission from [94]. 2019, J. Water Process Eng.

Table 3. A summary of the fouling resistance of polymeric membranes using GO-based composites, is provided below.

S.N.	Coating	Material/Matrix	Organism	Experimental Setup	Main Conclusions
1 [95]	Graphene Oxide	Marine micro- and macrofoulers	Escherichia coli, Staphylococcus aureus, Pseudomonas aeruginosa	In vitro study. Nutrient medium. Ambient temperature. After 24 and 48 h	Bacteria grew much slower on graphene-oxide-coated surfaces in vitro and were less likely to foul the surfaces in situ (up to 94% fewer viable cells).
2 [96]	Graphene Oxide	Alkyd resin	Marine micro- and macrofoulers	In situ study. South Korea's seawater (Jeju Sea). A period of three weeks	
3 [97]	Graphene Oxide	Silicone rubber	Triceratium sp.	In vitro study. A medium containing algal broth at a temperature of 25 °C. Static assay (8 days). Assays in the dynamic mode. In the vicinity of specimens, the linear velocity was 3.4 m/s for 10 days.	A single graphene oxide loading (0.16 wt%) responded slightly to diatom antiadhesion conditions (approximately 12% reduction in OD under static conditions). An approximate 67% OD reduction was observed for only one graphene oxide loading (0.36 wt%) in the dynamic assay.
4 [96]	Graphene Oxide	Calcium ion selective membrane electrode	Marine bacteria	In vitro study. Luria–Bertani medium. Ambient temperature. 1 hr and 5 hr	In comparison to the noncoated sensor, the graphene oxide-coated sensor significantly inhibited the formation of biofilms (roughly 45% decrease in CFUs) by significantly improving antiadhesion and bacterial inactivation properties.
5 [98]	Graphene	Silicone rubber	Paracoccus pantotrophus	Study conducted in vitro. Seawater artificially created. Assay conducted in a quasi-static environment (7 days). Assay conducted in a dynamic environment. In seven days, speeds varied within a range of 0.2 to 0.5 m/s.	As compared to the rigid polystyrene sheet control surface, graphene–silicone membranes demonstrated similar antifouling performance under quasistatic conditions. As compared with control surfaces, graphene-based membranes reduced colony-forming units (CFU) by around 40% under dynamic conditions.
6 [99]	Graphene	Silica	Halomonas spp.	Study conducted in vitro. Solution of sodium chloride (0.5 wt%) 20 °C, 72 h	The graphene coatings decreased biofilm-producing bacteria, Halomonas spp., adhesion and adhesion gene expression.
7 [100]	Guanidine-functionalized graphene	Boron acrylate polymer	Escherichia coli, Staphylococcus aureus	Study conducted in vitro. Luria–Bertani medium. 37 °C, 12 h	A 95% reduction in bacterial growth was observed in the coatings, along with an antiadhesion rate of 99% for diatoms. Neither fouling adhesion nor surface degradation were observed during the field trial.

Table 3. *Cont.*

S.N.	Coating	Material/Matrix	Organism	Experimental Setup	Main Conclusions
8 [100]	Guanidine-functionalized graphene	Boron acrylatepolymer	*Phaeodactylum tricornutum Nitzschia closterium f. minutissima Halamphora* sp.	Study conducted *in vitro* F/2 medium 21 °C, 14 days	
9 [100]	Guanidine-functionalized graphene	Boron acrylate polymer	Marine micro- and macrofoulers	2 months of *in situ study of natural seawater (Yellow Sea, China)*	
10 [93]	Acrylic acid-modified graphene oxide	Acrylic resin	Marine micro- and macrofoulers	Study conducted *in situ* Natural seawater (Zhoushan Sea, China) 6 months	The antifouling performance of composite-based paints was demonstrated during the testing process using natural seawater as the testing medium.
11 [101]	Polyaniline/*p*-phenylenediamine-functionalized graphene oxide	Epoxy resin	Simulated marine micro- and macrofoulers	Study conducted *in vitro* Simulated marine environment 25–27 °C, 3 months	A functionalized graphene oxide composite was added to commercialized epoxy coatings to improve their anticorrosion properties.
12 [102]	CNTs	Silicone oil-infused epoxy resin	*Chlorella* sp. *Phaeodactylum tricornutum*	Study conducted in vitro Artificial seawater 22 °C, 21 days	On surfaces coated with CNTs and epoxy resin, algae settlement rates were significantly lower than on surfaces coated with epoxy resin alone. Silicone oil infused into CNTs and epoxy resin coatings was even more effective in reducing the growth of algae biofilms (up to 90% reduction in cells) than silicone oil in CNTs alone.
13 [103]	MWCNTs	Chlorinated rubber	Pioneer eukaryotic biofilm communities	An in situ study was conducted (1.5 m below the surface) Chinese seawater (Xiaoshi Island) A temperature of 10 °C and a duration of 312 days	A significant reduction in mean species richness was observed by incorporating MWCNTs into antifouling materials (p 0.01) due to the reduced diversity and abundance of pioneer eukaryotic microbes.
14 [104]	MWCNTs	PDMS	*Mytilus galloprovincialis*-based foulers	An in vitro study was conducted An assay based on static data Seawater that has been filtered Temperature: 18 °C Plantigrades took six hours; pediveligers took 48 h.	Mussels' adhesion and settlement were not affected by MWCNT incorporation compared to PDMS or PDMS alone.

Table 3. Cont.

S.N.	Coating	Material/Matrix	Organism	Experimental Setup	Main Conclusions
15 [105]	MWCNTs	PDMS	*Ulva linza*	Dynamic assay conducted in vitro Seawater made from artificial sources The experiment was conducted for six days at 18 °C.	It appeared that adding MWCNTs to PDMS alone did not improve sporeling release properties.
16 [106]	MWCNTs	PDMS	*marine micro- and macrofoulers*	An in vitro study was conducted Assays that are dynamic The hull of a ship traveling at 15 knots (such as a ship's hull) Seawater made from artificial sources A temperature of 28 °C It takes 45 min for zoospores to settle; 7 days for sporelings to grow It takes 24 h for the adhesive to settle; it takes 3 months to determine its adhesion strength	By adding MWCNTs, sporeling release was enhanced (approximately 60%). A significant reduction in adhesion strength of adult barnacles growing on MWCNTs/PDMS was observed.
17 [107]	MWCNTs	PDMS	Bacteria and diatoms	A study conducted in situ (0.5–1.0 m below ground level) Water from the sea (Zhoushan, China) A 28-day period	Compared to bare PDMS, MWCNTs decreased mussel settlement (up to a 20% reduction) by altering the biomass and community composition of biofilms.
18 [107]	MWCNTs	PDMS	*Mytilus coruscus* (mussels and plantigrades)	An in vitro study was conducted Seawater that has been autoclaved and filtered The experiment was conducted at 18 °C for 12 h	
19 [108]	Hydroxyl-modified MWCNTs	Silicone-oil-infused PDMS	Marine bacteria	An in vitro study was conducted Seawater fresh from the sea The temperature was maintained at 28 °C for ten days	Antiadhesion (up to 35% higher removal rate) and AF properties were enhanced, particularly when higher volume ratios of hydroxylated MWCNTs were used.
20 [108]	Hydroxyl-modified MWCNTs	Silicone-oil-infused PDMS	Marine micro- and macrofouler	The study was conducted in situ (1–2 m below the surface) A natural seawater sample (Yellow Sea, China) A period of eight months	
21 [109]	Carboxyl and hydroxyl-modified MWCNTs	PDMS	Pioneer eukaryotic biofilm communities	A study conducted in situ (0.8–1.5 m below ground level) Water from the sea (Xiaoshi Island, China) A period of two months	The incorporation of MWCNTs showed excellent AF performance and effectively reduced colonization by pioneer eukaryotes, in comparison to plain PDMS (Shannon diversity index, $p < 0.05$).

Table 3. Cont.

S.N.	Coating	Material/Matrix	Organism	Experimental Setup	Main Conclusions
22 [110]	Carboxyl and hydroxyl-modified MWCNTs	PDMS	Marine micro and macrofoulers	The study was conducted in situ (1.5 m below the surface) Chinese seawater (Weihai Western Port) A temperature of 11 °C, a duration of 56 days	The incorporation of a low amount of MWCNTs greatly improved the AF properties of PDMS coatings. Most modified coatings did not display any significant modulating effects on pioneer biofilm communities, as compared to plain PDMS.
23 [111]	Fluorinated MWCNTs	Silicon	Escherichia coli	An in vitro study was conducted A phosphate-buffered saline solution Temperature 37 °C, 6 h	In vitro study Phosphate-buffered saline 37 °C, 6 h
24 [112]	Fluorinated MWCNTs	PDMS	pseudobarnacles	Method for pseudobarnacle adhesion testing	Incorporating fluorinated MWCNTs into the PDMS matrix increased the adhesive properties of the FR layer by reducing the pseudobarnacle adhesion strength by 67% compared to bare PDMS, and by 47% compared to pristine MCWNT/PDMS matrix.
25 [113]	-Carboxyl and hydroxyl-modified MWCNTs -Graphitized MWCNTs -Carboxyl-modified SWCNTs	PDMS	Pioneer biofilm bacteria	An in situ study was conducted (1.5 meters below the surface) A natural seawater spring on the Chinese island of Xiaoshi Temperatures between 10 °C and 17 °C, 24 h	All CNT/PDMS composites decreased proteobacteria biofilm formation, but increased cyanobacteria biofilm development.
26 [114]	1. Tetraaniline(TANI)-CNT		S. epidermidis	-	Results revealed that the surface coverage percentage of S. epidermidis drops more than 50% from the unmodified to the modified film.
27 [115]	2. CNTs	poly(methyl methacrylate) (PMMA)	S. aureus S. mutans C. albicans	-	Significant antiadhesive effects (35–95%) against all tested bacteria were verified for the 1% CNT/PMMA compared to the PMMA control group.
28 [116]	3. CNT	Polyethylene (PE)	P. fluorescens M. smegmatis	-	Biofilm growth on PE-CNTs composites surface compared to PE decreased by 89.3% and 29% for P. fluorescens and Mycobacterium smegmatis, respectively.

Marine coatings made from pristine MWCNTs (p-MWCNTs) produce differing results. Some studies have shown that carbon nanomaterials, when incorporated into PDMS, can improve antifouling performance by lowering the number of eukaryotic microorganisms [110] and barnacle adhesion strengths [106,107]. This contrasts with other studies which showed that MWCNT-centered coatings have no impact on the settlement of microfoulers and macrofoulers [104].

Furthermore, carboxylate- and hydroxyl-functionalized MWCNTs have been evaluated to determine their contribution to marine coating AF performance. Sun and Zhang performed an extensive two-month field trial, in which MWCNTs were added to PDMS coatings with variations in the hydroxyl and carboxyl content (w/w), the diameter, and the length to examine the effects of introducing the changes [110]. The type of MWCNT applied to the coating, as well as the pioneer eukaryotic communities, had a significant impact on the AF behavior of the coating. The findings of Sun and Zhang demonstrate that AF behavior varies as a function of the bacteria strain type associated with the pioneer biofilm community and also shows dependency on the carboxyl- or hydroxyl-modified coating used. In a different investigation, however, Ji et al. observed that the majority of carboxyl- and hydroxyl-modified carbon coatings had weak modulatory impacts on the biofilm community [109]. It has also been demonstrated that fluorinated MWCNT-based coatings can reduce *E. coli* [109] to 98% and can prevent pseudobarnacles from adhering [105].

Additionally, results from a combination of silicone oil and lubricants in polymeric matrices showed that MWCNT-based coatings could be improved by the addition of lubricants [108]. Using this method, superhydrophobic fouling-release surfaces were developed by lubricant leaching from the surface over time. This resulted in the formation of a layer of oil serving as a shield against corrosion, originating from frictional forces progressing over time, as well as an enhancement in the antifouling capabilities [117].

The findings of this study suggest that when developing new antifouling coatings for marine environments, CNT-based coatings can be considered as an option.

3.4. Functionalized Carbon Coatings for Pollutant Adsorption

There are many reasons why carbon-based materials are ideal platforms for developing new adsorbents with enhanced functionality, including their noncorrosive nature, the ability to tune the surface chemistry, large surface areas, and the occurrence of oxygen-containing functional groups. In order to manufacture carbon-based materials for environmental applications, most of the processes that are used are chemical processes, such as chemical oxidation and deposition (chemical oxidation), as well as extended chemical processes (for example electrochemical, sol–gel, microemulsion, hydrothermal, and electrochemical).

Water treatment could greatly benefit from the use of these materials if it is designed to obtain suitable properties and structures. Therefore, carbon-based materials are generally functionalized deliberately with a view to obtain synergistic multifunctionalities or for solving specific problems in order to achieve synergistic advantages over conventional materials. Whenever functional groups are present on the surface of a carbon backbone, adsorption is always made easier. Several functional groups have been grafted onto the top of the surface of carbon nanomaterials (such as -NH_2, -SH, and -COOH), as well as to enhanced functional groups such as -COOH and -OH, which are present on the surface. It is important to coat the carbon backbone with other nanoparticles (NPs) in order to improve the performance of these materials [118,119]. A significant advantage of the carbon backbone stems from its strong adsorption behavior and large specific surface area [120,121].

Core–shell structure chemistry has received considerable attention in recent years due to the wide range of applications in areas such as biomedicine, catalysis, electronics, pharmaceuticals, and environmental pollution control. In carbon nanomaterials, the core and shell are usually composed of different materials. Material combinations are few, with a handful of examples that include Fe_3O_4@C, RGO@Al_2O_3, C@Si, and CNT@C. Nanoscale carbon-based core–shell materials have largely been utilized as high-capacity electrodes to isolate and remove solvated metals; their application in aqueous removal of heavy metals has yet to be established. However, complex nanocomposites have been used to remove heavy metals from aqueous solutions. In this direction, the properties of CNTs and graphene nanomaterials have been drastically altered to induce changes at the metal site, with magnetization facilitating their isolation and removal towards the adsorbing material. A heavy metal is considered toxic if it is nonbiodegradable, bioaccumulative, or extremely hazardous. There are several heavy metals, including Hg, Pb, Cr, and Cd, which are more commonly found and problematic. Plants, animals, and people are at risk for serious diseases due to the presence of heavy metals in the environment and trace quantities are often associated with heightened risk (Table 4). A variety of functionalized CNTs as well as graphene nanomaterials have been successfully applied for the purpose of aqueous heavy metal removal by adsorption.

Magnetized carbon materials have also been developed for the capture of other functionalized nanomaterials, particularly heavy metals. Contact among metal ions and oxygen-rich materials is a major factor in the processing of heavy metal adsorption on CNTs and graphene-based nanomaterials. Pretreatments are usually required to enhance the functional-group abundance of oxygen-associated elements on the surface of raw CNTs. In contrast, the oxygen-rich surface chemistry of graphene requires no acid or oxidation treatment. In the case of Cr(VI), a reduction process must be carried out before either adsorption or precipitation can occur. For iron oxide, carbonization of cellulose to aid the reduction of Fe_3O_4 nanoparticles can provide a route to the fabrication of magnetic carbon–iron nanoadsorbents. The functionalized nanoadsorbent surface can facilitate the removal of Cr(VI) through the reduction of cellulose and $Fe(NO_3)_3$ precipitation. Furthermore, easy separation of MC-N and MC-O was accomplished using a permanent magnet after the reaction. There have been a number of studies on magnetic nanocomposites in recent years. This is largely due to their low cost, environmentally friendly composition, and large surface area of mesoporous nanostructure that permits rapid adsorption and ease of retrieval using an externally applied magnetic field. Magnetic separation of heavy metals allows environmental risks to be minimized as shown in Figure 12a that heavy metal nanoparticles adsorb efficiently by magnetic nanoparticles. Further, magnetically functionalized nanomaterials, while recycling magnetic particles, offer an ecofriendly method to reapply the separation process with minimal to no damage to the environment compared to other methods (Figure 12b). Figure 13b illustrates a typical example of heavy metal removal. Here, surfactant (tween-20)-functionalized Au NPs have been applied for the removal of aqueous Hg (II) in a high-salt matrix (artificial seawater) by reducing Hg^{2+} to Hg^0 through the deposition of Hg^0 on the NP surface. The separation process was completed by using magnetic and reduced graphene oxide-Fe_3O_4 NPs.

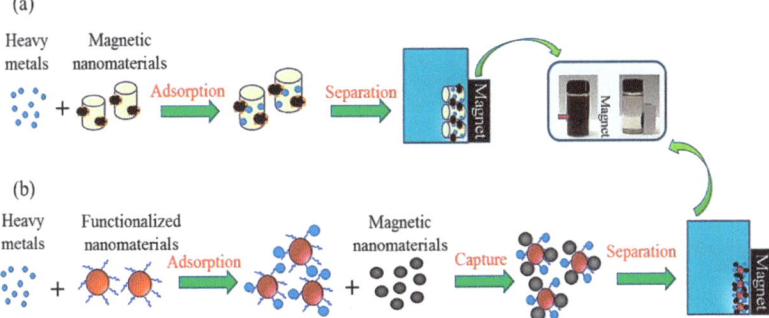

Figure 12. A schematic diagram showing heavy metal absorption and separation using (a) magnetic nanoparticles (b) magnetic carbon-based nanomaterials used in heavy metal separation and nanoparticle retrieval. Reprinted with permission from [122]. 2018, Chemosphere.

Table 4. Aqueous toxic heavy metal adsorption by functionalized carbon-based nanomaterials.

S.N.	Carbon-Based nanomaterials	Surface Area ($m^2\ g^{-1}$)	pH	Adsorption Capacity ($mg\ g^{-1}$) Exp.	Cal.
1.		**Pb (II)**			
[123]	Multiwalled Carbon Nanotube	-	6.0	-	15.9
[124]	Acidified MWCNTs	237.3	5.0	85.0	49.7
[123]	Multiwalled Carbon Nanotube/VP	-	6.0	-	37.0
[125]	Multiwalled Carbon Nanotube/TiO_2	-	6.0	4.6	137.0
[126]	Carbon Nanotube/MnO_2	275.0	5.0	-	78.7
[127]	Single-walled Carbon Nanotube/COOH	400.0	5.0	-	94.7
[128]	Carbon Nanotube/Fe_3O_4-NH_2	90.7	5.3	37.6	75.0
[129]	Carbon Nanotube/Fe_3O_4-MPTS	97.2	6.5	42.1	65.4
[130]	Graphene Oxide	430.0	6.8	328.0	367.0
[130]	EDTA/Graphene Oxide	623.0	6.8	479.0	525.0
[130]	EDTA/Reduced graphene Oxide	730.0	6.8	204.0	228.0
[131]	Graphene Oxide/$MnFe_2O_4$	196.0	5.0	-	673.0
[132]	Reduced graphene oxide/nZVI	-	5.0	550.0	585.5
[133]	$CoFe_2O_4$/Graphene Oxide	212.7	-	81.3	82.3
[134]	M-CHAP/Graphene Oxide	106.0	4.5	244.5	246.1
[135]	HMO/Graphene Oxide	383.9	5.0	553.6	512.6
[136]	Smart Magnetic Graphene	165.0	6.5	-	6.0
2.		**As (III)**			
[137]	Iron oxide/Multiwalled Carbon Nanotube	-	8.0	1.8	4.0
[138]	Fe_3O_4/Multiwalled Carbon Nanotube	70.1	-	-	53.2
[139]	Fe/Multiwalled Carbon Nanotube	-	7.0	210.0	200.0
[140]	EDA/Multiwalled Carbon Nanotube/Fe^{2+}	198.5	8.0	0.7	N/A
[141]	MIO/Multiwalled Carbon Nanotube	209.8	7.0	17.2	20.2
[131]	Graphene Oxide/$MnFe_2O_4$	196.0	6.5	-	146.0
[142]	Fe_3O_4/Reduced Graphene Oxide	148	7.0	-	13.1
[143]	HA/Reduced Graphene Oxide/Fe_3O_4	0.9	7.0	7.5	8.7
[144]	HEG/electrodes	442.9	6.1	-	138.8
[145]	GAC/ZrO_2	903.0	7.6	-	12.2
[146]	Multiwalled Carbon Nanotube/ZrO_2	152.0	6.0	9.8×10^{-2}	2.0
[147]	Fe_3O_4/Multiwalled Carbon Nanotube	153.0	4.0	9.7×10^{-2}	1.7
3.		**As (V)**			
[148]	Fe_3O_4/Multiwalled Carbon Nanotube	70.1	-	-	39.1
[149]	MIO/Multiwalled Carbon Nanotube	209.8	7.0	36.3	40.8
[139]	Fe/Multiwalled Carbon Nanotube	-	7.0	220.0	200.0
[131]	Graphene Oxide/$MnFe_2O_4$	196.0	4.0	-	207.0
[147]	Graphene oxide/$Fe(OH)_3$	-	4.0	23.8	N/A
[142]	Fe_3O_4/Reduced graphene oxide	148	7.0	-	5.83
[143]	HA- Reduced graphene oxide/Fe_3O_4	0.9	7.0	16.0	61.7
[144]	HEG/electrodes	442.9	6.9	-	141.9
[148]	Mg-Al LDHs/Graphene Oxide	35.4	5.0	183.1	180.3
[146]	Reduced graphene oxide/ZrO_2	152.0	6.0	0.1	5.0
[140]	EDA/Reduced graphene oxide/Fe^{2+}	198.5	4.0	1.0	18.1
[149]	Fe_3O_4/Reduced graphene oxide	153.0	4.0	0.1	0.2
[136]	Smart Magnetic Graphene	165.0	6.5	-	3.3
4.		**Hg (II)**			
[150]	Oxidized Multiwalled Carbon Nanotube	-	7.0	-	3.8
[151]	Multiwalled Carbon Nanotube/COOH	-	4.3	81.6	127.6
[151]	Multiwalled Carbon Nanotube/OH	-	4.3	89.4	120.1
[152]	Multiwalled Carbon Nanotube/SH	-	5.0	74.2	131.6
[153]	Multiwalled Carbon Nanotube/iodide	153.0	6.0	100.0	123.5
[153]	Multiwalled Carbon Nanotube/S	155.3	6.0	100.0	151.5
[154]	Carbon Nanotube/MnO_2	110.4	6.0	14.3	58.8
[155]	$KMnO_4$-DES/Carbon Nanotube	199.4	5.5	187.0	250.5
[156]	CS/Multiwalled Carbon Nanotube/COOH	-	4.0	183.2	181.8
[156]	CS/Multiwalled Carbon Nanotube	-	4.0	167.5	169.4

Table 4. Cont.

S.N.	Carbon-Based nanomaterials	Surface Area ($m^2\ g^{-1}$)	pH	Adsorption Capacity ($mg\ g^{-1}$) Exp.	Cal.
[129]	Multiwalled Carbon Nanotube/Fe_3O_4-SH	97.2	6.5	63.7	65.5
[157]	Tween 20-Au/graphite-Fe_3O_4	-	-	-	47.6
5.			Cr (VI)		
[158]	IL-oxi-Multiwalled Carbon Nanotube	87.4	3.0	2.6	85.8
[159]	Carbon Nanotube/CeO_2	-	7.0	30.2	31.6
[160]	DBSA-PANI/Multiwalled Carbon Nanotube	-	2.0	49.5	55.6
[161]	AC/Carbon Nanotube	755.8	2.0	9.0	8.6
[162]	AA/Carbon Nanotube	203.0	2.0	142.8	264.5
[163]	NH_2-Fe_3O_4/Graphene Oxide	43.6	4.5	27.3	32.3
[164]	MC-N	136.3	1.0	327.5	N/A
[136]	Smart Magnetic Graphene	165.0	6.5	-	4.9
6.			Cd (II)		
[165]	Raw Multiwalled Carbon Nanotube	187.6	8.0	1.29	1.3
[165]	Oxidized Multiwalled Carbon Nanotube	78.5	8.0	22.32	22.4
[165]	EDA/Multiwalled Carbon Nanotube	101.2	8.0	21.23	21.7
[166]	Al_2O_3 Multiwalled Carbon Nanotube	109.8	7.0	0.948	27.2
[167]	Single-walled Carbon Nanotube	400.0	5.0	-	21.2
[167]	COOH/Single-walled Carbon Nanotube	400.0	5.0	-	55.4
[168]	Oxidized Carbon Nanotube	128.0	5.5	-	11.0
[162]	AA/Carbon Nanotube	203.0	7.5	200.0	229.9
6.			Cu (II)		
[169]	Oxidized Carbon Nanotube	-	7.0	50.4	64.9
[170]	Purified Multiwalled Carbon Nanotube	169.7	5.0	37.5	36.8
[168]	Oxidized Multiwalled Carbon Nanotube	-	5.0	29	28.5
[170]	Sulfonated Multiwalled Carbon Nanotube	28.7	5.0	59.6	43.2
[171]	OH/Multiwalled Carbon Nanotube	111.4	4.9	7.0	10.1
[171]	COOH/Multiwalled Carbon Nanotube	135.2	4.9	5.5	8.1
[167]	Single-walled Carbon Nanotube	400.0	5.0	-	22.9
[167]	COOH/Single-walled Carbon Nanotube	400.0	5.0	-	72.3
[172]	OH/Single-walled Carbon Nanotube/RGO	-	6.8	-	256.0
[172]	COOH/Single-walled Carbon Nanotube/RGO	-	6.8	-	63.0
[173]	Multiwalled Carbon Nanotube/Fe_3O_4	138.7	5.5	19.0	38.9
7.			Zn (II)		
[169]	Oxidized Carbon Nanotube	-	7.0	58.0	74.6
[174]	Purified Single-walled Carbon Nanotube	423.0	-	15.4	41.8
[175]	Oxidized Multiwalled Carbon Nanotube	250.0	6.5	2.0	1.1
8.			Ni (II)		
[176]	Multiwalled Carbon Nanotube	197.0	5.4	2.9	3.7
[176]	PAA/Multiwalled Carbon Nanotube	-	5.4	-	3.9
[177]	Oxidized Multiwalled Carbon Nanotube	102.0	6.5	12.5	17.9
9.			Co (II)		
[169]	Oxidized Carbon Nanotube	-	7.0	69.6	85.7
[178]	Multiwalled Carbon Nanotube/iron oxide	-	6.4	2.9	10.6
10			U (VI)		
[179]	CB/Graphene Oxide/Fe_3O_4	-	5.0	-	122.5
[180]	Fe_3O_4/Graphene Oxide	-	5.5	-	69.5
[181]	Amidoxime Fe_3O_4/Graphene Oxide	-	5.0	76.88	92.4
[182]	CMPEI/CMK	1350.0	4.0	-	151.5

Electrostatic attraction and surface complexation are major strategies by which heavy metals adsorb onto functionalized carbon-based systems [28]. Figure 13a demonstrates the utility of surface complexation and electrostatic attraction for the removal of heavy metals [183]. As an example, Madadrang et al. used modified graphene oxide complexed with hexadentate ligands using ethylenediaminetetraacetic acid (EDTA). The method has been utilized to treat mercury (II)-, lead(II)-, nickel(II)-, cadmium(II)-, and copper(II)-contaminated samples by two prospective mechanisms [184]. The electrostatic interaction of heavy metal ions (M^{x+}) with oxygenic functional groups complexed on GO after functionalization via hydroxyl and carboxyl moieties provided easy access to metal uptake. In addition, the nature of EDTA as a hexadentate chelating agent with the ability to form coordination complexes with heavy metal ions (M^{x+}) enhances the coordination interactions of functionalized GO (Figure 13b).

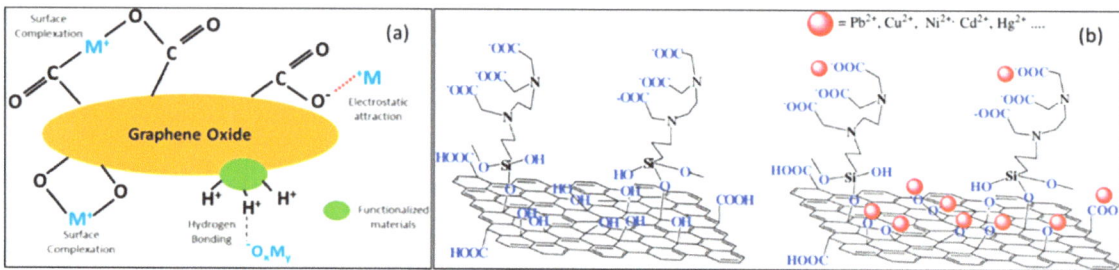

Figure 13. (a) Probable toxic heavy metal removal mechanism on graphene oxide as an adsorbent from water. Reprinted with permission from [183]. 2020, Chemosphere. (b) Scheme shows the chemical structure of ethylenediaminetetraacetic-acid-functionalized graphene oxide before (left) and after it has been exposed to heavy metal cations (right) Reprinted with permission from [130]. 2012, ACS Appl. Mater. Interfaces.

3.5. Bioinspired Functional Carbon Coatings

Carbon-based functionalized coatings play a pivotal role in achieving various functions of living organisms. Exoskeleton architectures are one of the better examples of functionalized carbon coatings that exist in nature. Animal exoskeletons, commonly known as cuticles, are covered with various coatings that may comprise a network of hydrocarbons, including phenolamine, which are coproduced with molecules that originate from oenocytes. These are specialized cells that offer protection from water loss, damage, and bacterial degradation. An interesting example of the role of functionalized coatings can be observed in mosquitoes that employ hydrocarbon coatings as defense barriers against insecticides and changes due to climate [184–186]. Another interesting study has suggested that the wings of particular fly species contain structural materials; therefore, instead of repelling contaminants, they are capable of killing contaminants and preventing bacterial colonization [187] on the wing surface. The cave barklice, *Psyllipsocus Yucatan*, from dolomit cave "Toca da Tiquara" in Brazil, illustrates an example of the unique coating of their wing membranes. This species has typically transparent light-brown wings, but some females have been found to have opaque black wings, with their surface covered with a coating. Crystallographic and material analysis of the wing surface has revealed an ultrathin layered-type morphology (1.5 µm) of catalyzed minerals containing iron, oxygen, and sulfur. It appears that the supersaturated climatic conditions and the presence of an iron-excreting bat species residing in caves could be the origin of the film formation. However, the confinement of deposits to only the wing surface, and their exclusion from other parts and hairs, indicate that the high-frequency vibration of the wings could aid the condensation of the thin layer. Detailed knowledge of this type has encouraged multidisciplinary efforts to mimic the natural composition and to advance material design with specific functional properties.

Moreover, in recent investigations, concepts embedded in biomimetics have been increasingly adopted and applied in coating technologies. Several applications aligned to biomimetics have been demonstrated to achieve superhydrophobic coatings by making use of petroleum pipelines, corrosion protection, and biomedical engineering [188–190]. These approaches have been largely inspired by features found in natural lotus leaves and have inspired a number of research directions [191,192]. These coatings offer excellent performance in the above-mentioned applications because of their antifouling, self-cleaning, and anticorrosion properties. In a recent study, Zhu et al. developed polysulfone–carbon nanotube–fluorinated ethylene propylene nanocomposite (PCFn) coatings and electrostatic powder spraying was employed to fabricate nanocomposite coatings with biomimetic golden spherical cactus surface structures (with microspheres and nanothorns). PCFn coatings combine the advantages of PSU, CNTs, and FEP to produce coatings that resemble the golden cactus structure and mimic the hydrophobic characteristics, enabling its water-repelling nature to present a durable, self-cleaning, and corrosion-resistant surface (SEM image in Figure 14a–c). Some other examples of applications of the biomimetic approach to achieve functional coatings are summarized in Table 5.

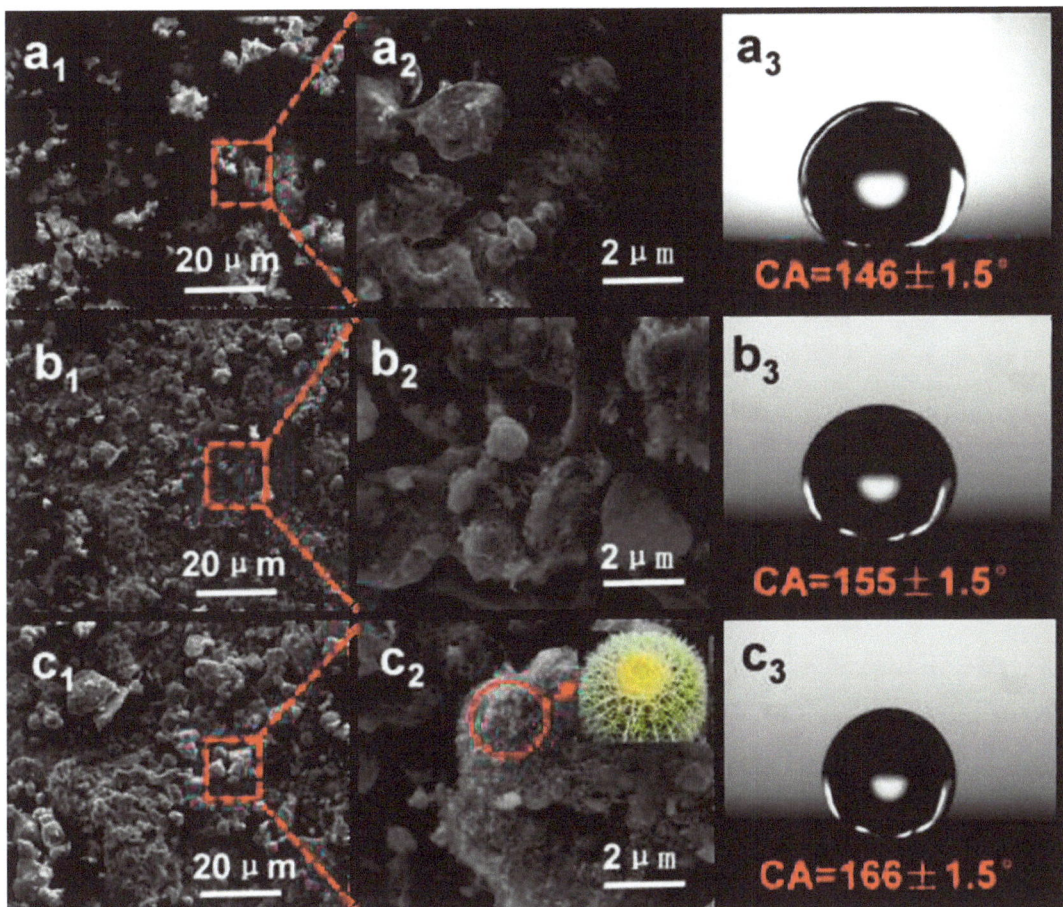

Figure 14. SEM imaging of PCFn surfaces showing GC-like structured coatings reminiscent of biomimetic features with variable content of (**a1–a3**) 0 wt%, (**b1–b3**) 1 wt%, and (**c1–c3**) 5 wt% CNTs. Insert (**c2**): An optical image of a golden spherical cactus. Reprinted with permission from [188]. 2018, Chem. Eng.

Table 5. Biomimicry in functionalized carbon-based coating technology.

S.N.	Model System	Biomimicked Features	Biomimicked Properties	Biomimicked Coating Material	Coating Method	References
1	Lotus and Xanthosoma Sagittifolium leaves	Superhydrophobic surface	Corrosion resistance	Polymer (polybenzoxazine)	Templating by leaves	[193]
2	Dragonfly wings	Nanostructures	Mechanobactericidal behavior	Palmitic and stearic acid films onto HOPG	Surface recrystallization by self-assembly	[194]
3	Cicada wing surfaces	Nanopillar/Nanodagger-shaped microstructure	Bacterial capture	ZIF MOFs	Dip coating	[195]
4	Endothelium	Endothelium functions	Controllable heparin anchoring and NO release properties	Metal-catechol-amine	Surgace Engineering	[196]
5	Natural superhydrophobic surfaces	Superhydrophobic surface	Superhydrophilic/superhydrophobic surfaces	Branched poly(ethylenimine) (BPEI) and dipentaerythritol pentaacrylate (5AcI)	"Chemically reactive" dip-coating	[197]
6	Bristle worm (Calliteara pudibunda)	Superhydrophobic surface	Anticorrosion and antiscaling, self-cleaning, and antifouling properties	Composites of polyphenylene sulfide (PPS), expandable graphite (EAG), and elastic fluororubber	-	[198]
7	Earthworm	Secretion behavior under extra stimulation	Friction-reduction and antifouling	PDMS	Solution casting	[199]
8	Wood	Plant cell wall tissue	Controllable architecture viscoelasticity	Nanocellulose, lignin	Layer-by-layer (LbL) assembly	[200]
9	Owls	Velvety coating on feather surface	Flight sound reduction and Aerodynamics	Nylon fibers	Electrostatic flocking	[201]
10	Willow-leaf-like or wheat-haulm-leaf-like	Superhydrophobic surface	Corrosion resistance	PTFE/PPS_nanofibers/spheres	Coating-curing process	[202]
11	Human skin	Wrinkles	Mechanics of wrinkling at micro nanoscale	TiN on polyurethane, polycarbonate, polyamide, and polyimide	Pulsed laser deposition	[203]
12	Human bones	Orthopedic and dental binding	Bone-bonding ability	Yttria-stabilized tetragonal zirconia and CaP	Apatite coating technique by femtosecond laser processing	[204]
13	*Psyllipsocus yucatan*	Wing membranes surface coating	Surface coating	Iron, oxygen, and carbon	Natural coating	[205]
14	*Polypedilum vanderplanki* larvae	Cuticle (exoskeleton)	Protective layer	Polyphenol at air/solid interface catalyzed by gaseous oxygen	Aerobic oxidation	[206]

4. Conclusions and Future Perspective

It is clear from the scope and the quality of the research discussed in this review article that functionalized carbon-based materials have the potential to be useful candidates for protective coating applications. In the literature, there are some discrepancies that relate to functionalized carbons and their ability to prevent corrosion. Wet corrosion has been found to be more extensive on CVD-grown carbon-based material coatings than uncoated metal surfaces, despite showing a good corrosion barrier for a short period of time. The literature search supports O_2 and H_2O diffusion through graphene-based material defects and the subsequent oxidization of encapsulated metals. The vulnerability of the surface is also compromised by coating discontinuities in graphene layers. The literature indicates that graphene-like material can protect against corrosion; however, it should also be noted that there is also a need for improved coating technologies and methodologies to enhance the effectiveness of the carbon-family-based system. A few surprising discoveries have been made regarding hybridized or functional groups composed of sp^2 hybridized carbon coatings systems that are anticorrosive by nature. The use of sp^2 hybridized carbon coatings systems as an anticorrosion barrier is counterintuitive since graphite promotes the corrosion of metals when it is in contact with them.

A new level of fire resistance has been achieved using functionalized carbon-based materials as fire-retardant additives. It is possible to enhance the thermal and barrier properties of thermoset polymers by dispersing functionalized carbon-based materials into them. Functionalized carbon-based material composites are also able to withstand the transfer of noxious and flammable gases during a fire, thanks to their intrinsic impermeability to gases, with high surface area in combination with tumescent flame-retardant behavior. A functionalized carbon surface can reduce melt dripping and delay structural collapse at high temperatures by reinforcing and maintaining its integrity. The development of flame-retardant materials and coatings with optimized formulations is expected in the future.

Depending on the surface chemistry and wettability of functionalized carbon-based materials, the interface can alter the physisorption of different molecules, allowing materials such as antifouling coatings, carriers, and adsorbents to be passively controlled during adsorption and desorption. In addition to being pertinent to both dry and marine atmospheres, functionalized carbon-based materials are also very effective as an antifouling mediator in surface coatings, which can be applied to surfaces such as domestic and construction surfaces, salt removal systems, bioreactors, and aquatic structures. There is a serious problem with the price and the large-scale applicability of coating technology in the antifouling field, since most applications require large quantities of functionalized carbon-based materials. The application of bioinspired or biomimetic coatings to functionalized carbon-based materials can potentially be valuable for antifouling applications if an economical coating approach can be established. It is possible to reduce surface colonization by microorganisms by exploiting graphene surfaces with amphiphilic behaviour. In order to optimize coatings to address the specific and unique antifouling problems associated with specific industries and overcome them, more research is needed. In terms of surface chemistry, functionalized carbon-based materials possess a tunable surface chemistry, which allows it to be used as a highly surface-active material, capable of selectively adsorbing a large quantity of pollutants. It is possible to target a wide range of ingredients, involving hydrocarbons, heavy metal ions, and dyes, among others. Material made from Gr can adsorb and expel 35,000 times its own weight.

Applied research is increasingly reaching the point of being applied to a commercial scale and research efforts around the world are intensifying; however, there remain significant challenges to overcome, particularly in terms of low-cost and scalable Gr manufacturing and handling. The safety of Gr materials must be proven through in-depth studies in medical and environmental applications. Graphene will bring a technological revolution when it is realized, according to the research community. This versatile material will be widely adopted by the coatings industry as an additive and as a single material to improve coating performance in the present era as a genuine versatile tool of the 21st century for a multitude of coating applications.

Finally, the biomimetic significance of carbon-based coatings offers a considerably viable option to incorporate features that better align with the environment, particularly if there is structural and functional resemblance to naturally evolved scaffolds. Such composites are representative of 3D structures, exhibiting physical characteristics such as porosity, polymer morphology, adsorption, and mechanical properties, which can collectively impact both the structural and functional workability of the materials, which may be further tuned as a composite. The desirability of mimicking nature will likely be an important imperative to the design of future materials as synthetics or biosynthetic hybrids, improving the multidimensionality of the coating. The growth of such materials in this direction will likely be important. Therefore, in the future, understanding the functional behavior of coating living organisms and applying them for synthetic coating would be an effective pathway to produce improved carbon-based functional coatings.

Funding: This work acknowledges support from the Basic Science Research Program through the National Research Foundation of Korea (NRF), funded by the Ministry of Science, ICT & Future Planning (No. 2022R1A2C1006090, 2017R1A2B4008801), and NRF Basic Research Program in Science and Engineering by the Ministry of Education (No. 2017R1D1A1B03036226). P.G.J. and V.K. also thankfully acknowledge the support from the National Convergence Research of Scientific Challenges through the National Research Foundation of Korea (NRF), funded by Ministry of Science and ICT (No. 2021M3F7A1017476).

Institutional Review Board Statement: Not applicable.

Informed Consent Statement: Not applicable.

Data Availability Statement: Not applicable.

Conflicts of Interest: The authors declare no conflict of interest.

References

1. Kong, S.; Lim, M.; Shin, H.; Baik, J.-H.; Lee, J.-C. High-flux and antifouling polyethersulfone nanocomposite membranes incorporated with zwitterion-functionalized graphene oxide for ultrafiltration applications. *J. Ind. Eng. Chem.* **2020**, *84*, 131–140. [CrossRef]
2. Fuertes-Espinosa, C.; Pujals, M.; Ribas, X. Supramolecular Purification and Regioselective Functionalization of Fullerenes and Endohedral Metallofullerenes. *Chem* **2020**, *6*, 3219–3262. [CrossRef]
3. Lara, P.; Fitzgerald, R.M.; Karle, N.N.; Talamantes, J.; Miranda, M.; Baumgardner, D.; Stockwell, W.R. Winter and Wildfire Season Optical Characterization of Black and Brown Carbon in the El Paso-Ciudad Juárez Airshed. *Atmosphere* **2022**, *13*, 1201. [CrossRef]
4. Zhang, H.; Liu, B.; Ma, X.; Han, G.; Yang, Q.; Zhang, Y.; Shi, T.; Yuan, J.; Zhong, W.; Peng, Y.; et al. Carbon dioxide cover: Carbon dioxide column concentration seamlessly distributed globally during 2009–2020. *Earth Syst. Sci. Data Discuss.* **2022**, *2022*, 1–34. [CrossRef]
5. Kumar, A.; Hong, J.; Yun, Y.; Jung, H.; Lee, K.-S.; Han, J.W.; Song, S.-J. A stable and active three-dimensional carbon based trimetallic electrocatalyst for efficient overall wastewater splitting. *Int. J. Hydrogen Energy* **2021**, *46*, 30762–30779. [CrossRef]
6. Achary, L.S.K.; Kumar, A.; Rout, L.; Kunapuli, S.V.S.; Dhaka, R.S.; Dash, P. Phosphate functionalized graphene oxide with enhanced catalytic activity for Biginelli type reaction under microwave condition. *Chem. Eng. J.* **2018**, *331*, 300–310. [CrossRef]
7. Kumar, A.; Rout, L.; Achary, L.K.; Mohanty, S.K.; Nayak, P.S.; Barik, B.; Dash, P. Solvent free synthesis of chalcones over graphene oxide-supported MnO2 catalysts synthesized via combustion route. *Mater. Chem. Phys.* **2021**, *259*, 124019. [CrossRef]
8. Rout, L.; Kumar, A.; Dhaka, R.S.; Reddy, G.N.; Giri, S.; Dash, P. Bimetallic Au-Cu alloy nanoparticles on reduced graphene oxide support: Synthesis, catalytic activity and investigation of synergistic effect by DFT analysis. *Appl. Catal. A Gen.* **2017**, *538*, 107–122. [CrossRef]
9. Achary, L.S.K.; Maji, B.; Kumar, A.; Ghosh, S.P.; Kar, J.P.; Dash, P. Efficient room temperature detection of H2 gas by novel ZnFe2O4–Pd decorated rGO nanocomposite. *Int. J. Hydrogen Energy* **2020**, *45*, 5073–5085. [CrossRef]
10. Samadianfard, R.; Seifzadeh, D.; Habibi-Yangjeh, A.; Jafari-Tarzanagh, Y. Oxidized fullerene/sol-gel nanocomposite for corrosion protection of AM60B magnesium alloy. *Surf. Coat. Technol.* **2020**, *385*, 125400. [CrossRef]
11. Hosseinpour, A.; Abadchi, M.R.; Mirzaee, M.; Tabar, F.A.; Ramezanzadeh, B. Recent advances and future perspectives for carbon nanostructures reinforced organic coating for anti-corrosion application. *Surf. Interfaces* **2021**, *23*, 100994. [CrossRef]
12. Ma, C.; Liu, J.; Zhu, X.; Xue, W.; Yan, Z.; Cheng, D.; Fu, J.; Ma, S. Anticorrosive non-crystalline coating prepared by plasma electrolytic oxidation for ship low carbon steel pipes. *Sci. Rep.* **2020**, *10*, 15675. [CrossRef] [PubMed]

13. Catledge, S.A.; Thomas, V.; Vohra, Y.K. 5—Nanostructured diamond coatings for orthopaedic applications. In *Diamond-Based Materials for Biomedical Applications*; Woodhead Publishing: Sawston, UK, 2013; pp. 105–150. [CrossRef]
14. Thapliyal, V.; Alabdulkarim, M.E.; Whelan, D.R.; Mainali, B.; Maxwell, J.L. A concise review of the Raman spectra of carbon allotropes. *Diam. Relat. Mater.* **2022**, *127*, 109180. [CrossRef]
15. Speranza, G. The Role of Functionalization in the Applications of Carbon Materials: An Overview. *C* **2019**, *5*, 84. [CrossRef]
16. Bunch, J.S.; Verbridge, S.S.; Alden, J.S.; van der Zande, A.M.; Parpia, J.M.; Craighead, H.G.; McEuen, P.L. Impermeable Atomic Membranes from Graphene Sheets. *Nano Lett.* **2008**, *8*, 2458–2462. [CrossRef] [PubMed]
17. Su, Y.; Kravets, V.G.; Wong, S.L.; Waters, J.; Geim, A.K.; Nair, R.R. Impermeable barrier films and protective coatings based on reduced graphene oxide. *Nat. Commun.* **2014**, *5*, 4843. [CrossRef]
18. Nine, M.J.; Cole, M.A.; Tran, D.N.H.; Losic, D. Graphene: A multipurpose material for protective coatings. *J. Mater. Chem. A* **2015**, *3*, 12580–12602. [CrossRef]
19. Al-Jumaili, A.; Alancherry, S.; Bazaka, K.; Jacob, M.V. Review on the Antimicrobial Properties of Carbon Nanostructures. *Materials* **2017**, *10*, 1066. [CrossRef]
20. Raul, P.K.; Thakuria, A.; Das, B.; Devi, R.R.; Tiwari, G.; Yellappa, C.; Kamboj, D.V. Carbon Nanostructures As Antibacterials and Active Food-Packaging Materials: A Review. *ACS Omega* **2022**, *7*, 11555–11559. [CrossRef]
21. Guo, X.; Cheng, S.; Cai, W.; Zhang, Y.; Zhang, X. A review of carbon-based thermal interface materials: Mechanism, thermal measurements and thermal properties. *Mater. Des.* **2021**, *209*, 109936. [CrossRef]
22. Babu, K.; Rendén, G.; Mensah, R.A.; Kim, N.K.; Jiang, L.; Xu, Q.; Restás, Á.; Neisiany, R.E.; Hedenqvist, M.S.; Försth, M.; et al. A Review on the Flammability Properties of Carbon-Based Polymeric Composites: State-of-the-Art and Future Trends. *Polymers* **2020**, *12*, 1518. [CrossRef] [PubMed]
23. Sabzehmeidani, M.M.; Mahnaee, S.; Ghaedi, M.; Heidari, H.; Roy, V.A.L. Carbon based materials: A review of adsorbents for inorganic and organic compounds. *Mater. Adv.* **2021**, *2*, 598–627. [CrossRef]
24. Fotovvati, B.; Namdari, N.; Dehghanghadikolaei, A. On Coating Techniques for Surface Protection: A Review. *J. Manuf. Mater. Process.* **2019**, *3*, 28. [CrossRef]
25. Gu, Y.; Yu, L.; Mou, J.; Wu, D.; Xu, M.; Zhou, P.; Ren, Y. Research Strategies to Develop Environmentally Friendly Marine Antifouling Coatings. *Mar. Drugs* **2020**, *18*, 371. [CrossRef] [PubMed]
26. Sun, W.; Chu, X.; Lan, H.; Huang, R.; Huang, J.; Xie, Y.; Huang, J.; Huang, G. Current Implementation Status of Cold Spray Technology: A Short Review. *J. Therm. Spray Technol.* **2022**, *31*, 848–865. [CrossRef]
27. Barea, E.; Montoro, C.; Navarro, J.A.R. Toxic gas removal—Metal–organic frameworks for the capture and degradation of toxic gases and vapours. *Chem. Soc. Rev.* **2014**, *43*, 5419–5430. [CrossRef]
28. Barik, B.; Kumar, A.; Nayak, P.S.; Achary, L.S.K.; Rout, L.; Dash, P. Ionic liquid assisted mesoporous silica-graphene oxide nanocomposite synthesis and its application for removal of heavy metal ions from water. *Mater. Chem. Phys.* **2020**, *239*, 122028. [CrossRef]
29. den Braver-Sewradj, S.P.; van Benthem, J.; Staal, Y.C.M.; Ezendam, J.; Piersma, A.H.; Hessel, E.V.S. Occupational exposure to hexavalent chromium. Part II. Hazard assessment of carcinogenic effects. *Regul. Toxicol. Pharmacol.* **2021**, *126*, 105045. [CrossRef]
30. Kotrikla, A. Environmental management aspects for TBT antifouling wastes from the shipyards. *J. Environ. Manag.* **2009**, *90*, S77–S85. [CrossRef]
31. Kordas, G. Corrosion Barrier Coatings: Progress and Perspectives of the Chemical Route. *Corros. Mater. Degrad.* **2022**, *3*, 376–413. [CrossRef]
32. Uc-Peraza, R.G.; Castro, B.; Fillmann, G. An absurd scenario in 2021: Banned TBT-based antifouling products still available on the market. *Sci. Total Environ.* **2022**, *805*, 150377. [CrossRef]
33. L'Azou, B.; Passagne, I.; Mounicou, S.; Tréguer-Delapierre, M.; Puljalté, I.; Szpunar, J.; Lobinski, R.; Ohayon-Courtès, C. Comparative cytotoxicity of cadmium forms ($CdCl_2$, CdO, CdS micro- and nanoparticles) in renal cells. *Toxicol. Res.* **2014**, *3*, 32–41. [CrossRef]
34. Bal, W.; Kasprzak, K.S. Induction of oxidative DNA damage by carcinogenic metals. *Toxicol. Lett.* **2002**, *127*, 55–62. [CrossRef]
35. Templeton, D.M.; Liu, Y. Multiple roles of cadmium in cell death and survival. *Chem. -Biol. Interact.* **2010**, *188*, 267–275. [CrossRef] [PubMed]
36. Ankit; Saha, L.; Kumar, V.; Tiwari, J.; Sweta; Rawat, S.; Singh, J.; Bauddh, K. Electronic waste and their leachates impact on human health and environment: Global ecological threat and management. *Environ. Technol. Innov.* **2021**, *24*, 102049. [CrossRef]
37. Rajak, D.K.; Kumar, A.; Behera, A.; Menezes, P.L. Diamond-Like Carbon (DLC) Coatings: Classification, Properties, and Applications. *Appl. Sci.* **2021**, *11*, 4445. [CrossRef]
38. Ohtake, N.; Hiratsuka, M.; Kanda, K.; Akasaka, H.; Tsujioka, M.; Hirakuri, K.; Hirata, A.; Ohana, T.; Inaba, H.; Kano, M.; et al. Properties and Classification of Diamond-Like Carbon Films. *Materials* **2021**, *14*, 315. [CrossRef]
39. Nakashima, N. Soluble Carbon Nanotubes: Fundamentals and Applications. *Int. J. Nanosci.* **2005**, *4*, 119–137. [CrossRef]
40. Georgakilas, V.; Otyepka, M.; Bourlinos, A.B.; Chandra, V.; Kim, N.; Kemp, K.C.; Hobza, P.; Zboril, R.; Kim, K.S. Functionalization of Graphene: Covalent and Non-Covalent Approaches, Derivatives and Applications. *Chem. Rev.* **2012**, *112*, 6156–6214. [CrossRef]
41. Georgakilas, V.; Tiwari, J.N.; Kemp, K.C.; Perman, J.A.; Bourlinos, A.B.; Kim, K.S.; Zboril, R. Noncovalent Functionalization of Graphene and Graphene Oxide for Energy Materials, Biosensing, Catalytic, and Biomedical Applications. *Chem. Rev.* **2016**, *116*, 5464–5519. [CrossRef]

42. Chen, D.; Feng, H.; Li, J. Graphene Oxide: Preparation, Functionalization, and Electrochemical Applications. *Chem. Rev.* **2012**, *112*, 6027–6053. [CrossRef] [PubMed]
43. Lian, M.; Fan, J.; Shi, Z.; Li, H.; Yin, J. Kevlar®-functionalized graphene nanoribbon for polymer reinforcement. *Polymer* **2014**, *55*, 2578–2587. [CrossRef]
44. Bai, H.; Xu, Y.; Zhao, L.; Li, C.; Shi, G. Non-covalent functionalization of graphene sheets by sulfonated polyaniline. *Chem. Commun.* **2009**, *13*, 1667–1669. [CrossRef] [PubMed]
45. Zhang, J.-L.; Zhang, S.-B.; Zhang, Y.-P.; Kitajima, K. Effects of phylogeny and climate on seed oil fatty acid composition across 747 plant species in China, *Ind. Crops Prod.* **2015**, *63*, 1–8. [CrossRef]
46. Chen, X.; Fan, K.; Liu, Y.; Li, Y.; Liu, X.; Feng, W.; Wang, X. Recent Advances in Fluorinated Graphene from Synthesis to Applications: Critical Review on Functional Chemistry and Structure Engineering. *Adv. Mater.* **2022**, *34*, 2101665. [CrossRef]
47. Gong, X.; Liu, G.; Li, Y.; Yu, D.Y.W.; Teoh, W.Y. Functionalized-Graphene Composites: Fabrication and Applications in Sustainable Energy and Environment. *Chem. Mater.* **2016**, *28*, 8082–8118. [CrossRef]
48. Cen, H.; Wu, C.; Chen, Z. Polydopamine functionalized graphene oxide as an effective corrosion inhibitor of carbon steel in HCl solution. *J Mater Sci.* **2022**, *57*, 1810–1832. [CrossRef]
49. Espinoza-Vázquez, A.; Rodríguez-Gómez, F.J.; Martínez-Cruz, I.K.; Ángeles-Beltrán, D.; Negrón-Silva, G.E.; Palomar-Pardavé, M.; Romero, L.L.; Pérez-Martínez, D.; Navarrete-López, A.M. Adsorption and corrosion inhibition behaviour of new theophylline–triazole-based derivatives for steel in acidic medium. *R. Soc. Open Sci.* **2019**, *6*, 181738. [CrossRef]
50. Khan, G.; Basirun, W.J.; Kazi, S.N.; Ahmed, P.; Magaji, L.; Ahmed, S.M.; Khan, G.M.; Rehman, M.A.; Badry, A.B.B.M. Electrochemical investigation on the corrosion inhibition of mild steel by Quinazoline Schiff base compounds in hydrochloric acid solution. *J. Colloid Interface Sci.* **2017**, *502*, 134–145. [CrossRef]
51. Richards, C.A.J.; McMurray, H.N.; Williams, G. Smart-release inhibition of corrosion driven organic coating failure on zinc by cationic benzotriazole based pigments. *Corros. Sci.* **2019**, *154*, 101–110. [CrossRef]
52. Qiu, P.; Yang, H.F.; Yang, L.J.; Chen, Z.S.; Lv, L.J.; Song, Y.; Chen, C.F. Enhanced inhibition of steel corrosion by L-cysteine under visible-light illumination. *Mater. Corros.* **2017**, *68*, 1004–1012. [CrossRef]
53. Alrashed, M.M.; Jana, S.; Soucek, M.D. Corrosion performance of polyurethane hybrid coatings with encapsulated inhibitor. *Prog. Org. Coat.* **2019**, *130*, 235–243. [CrossRef]
54. Li, Y.Z.; Xu, N.; Guo, X.P.; Zhang, G.A. Inhibition effect of imidazoline inhibitor on the crevice corrosion of N80 carbon steel in the CO_2-saturated NaCl solution containing acetic acid. *Corros. Sci.* **2017**, *126*, 127–141. [CrossRef]
55. Erami, R.S.; Amirnasr, M.; Meghdadi, S.; Talebian, M.; Farrokhpour, H.; Raeissi, K. Carboxamide derivatives as new corrosion inhibitors for mild steel protection in hydrochloric acid solution. *Corros. Sci.* **2019**, *151*, 190–197. [CrossRef]
56. Zhang, H.H.; Qin, C.K.; Chen, Y.; Zhang, Z. Inhibition behaviour of mild steel by three new benzaldehyde thiosemicarbazone derivatives in 0.5 M H_2SO_4: Experimental and computational study. *R. Soc. Open Sci.* **2019**, *6*, 190192. [CrossRef] [PubMed]
57. Jing, C.; Wang, Z.; Gong, Y.; Huang, H.; Ma, Y.; Xie, H.; Li, H.; Zhang, S.; Gao, F. Photo and thermally stable branched corrosion inhibitors containing two benzotriazole groups for copper in 3.5 wt% sodium chloride solution. *Corros. Sci.* **2018**, *138*, 353–371. [CrossRef]
58. Dutta, A.; Saha, S.K.; Adhikari, U.; Banerjee, P.; Sukul, D. Effect of substitution on corrosion inhibition properties of 2-(substituted phenyl) benzimidazole derivatives on mild steel in 1M HCl solution: A combined experimental and theoretical approach. *Corros. Sci.* **2017**, *123*, 256–266. [CrossRef]
59. El-Sayed, A.-R.; Shaker, A.M.; El-Lateef, H.M.A. Corrosion inhibition of tin, indium and tin–indium alloys by adenine or adenosine in hydrochloric acid solution. *Corros. Sci.* **2010**, *52*, 72–81. [CrossRef]
60. Verma, C.; Sorour, A.A.; Ebenso, E.E.; Quraishi, M.A. Inhibition performance of three naphthyridine derivatives for mild steel corrosion in 1M HCl: Computation and experimental analyses. *Results Phys.* **2018**, *10*, 504–511. [CrossRef]
61. Baux, J.; Caussé, N.; Esvan, J.; Delaunay, S.; Tireau, J.; Roy, M.; You, D.; Pébère, N. Impedance analysis of film-forming amines for the corrosion protection of a carbon steel. *Electrochim. Acta* **2018**, *283*, 699–707. [CrossRef]
62. Huang, H.; Wang, Z.; Gong, Y.; Gao, F.; Luo, Z.; Zhang, S.; Li, H. Water soluble corrosion inhibitors for copper in 3.5 wt% sodium chloride solution. *Corros. Sci.* **2017**, *123*, 339–350. [CrossRef]
63. Qiang, Y.; Zhang, S.; Yan, S.; Zou, X.; Chen, S. Three indazole derivatives as corrosion inhibitors of copper in a neutral chloride solution. *Corros. Sci.* **2017**, *126*, 295–304. [CrossRef]
64. Ahmed, M.H.O.; Al-Amiery, A.A.; Al-Majedy, Y.K.; Kadhum, A.A.H.; Mohamad, A.B.; Gaaz, T.S. Synthesis and characterization of a novel organic corrosion inhibitor for mild steel in 1 M hydrochloric acid. *Results Phys.* **2018**, *8*, 728–733. [CrossRef]
65. Kovačević, N.; Milošev, I.; Kokalj, A. How relevant is the adsorption bonding of imidazoles and triazoles for their corrosion inhibition of copper? *Corros. Sci.* **2017**, *124*, 25–34. [CrossRef]
66. Liao, L.L.; Mo, S.; Luo, H.Q.; Feng, Y.J.; Yin, H.Y.; Li, N.B. Relationship between inhibition performance of melamine derivatives and molecular structure for mild steel in acid solution. *Corros. Sci.* **2017**, *124*, 167–177. [CrossRef]
67. Likhanova, N.V.; Arellanes-Lozada, P.; Olivares-Xometl, O.; Hernández-Cocoletzi, H.; Lijanova, I.V.; Arriola-Morales, J.; Castellanos-Aguila, J. Effect of organic anions on ionic liquids as corrosion inhibitors of steel in sulfuric acid solution. *J. Mol. Liq.* **2019**, *279*, 267–278. [CrossRef]

68. Fernandes, C.M.; Alvarez, L.X.; Santos, N.E.d.; Barrios, A.C.M.; Ponzio, E.A. Green synthesis of 1-benzyl-4-phenyl-1H-1,2,3-triazole, its application as corrosion inhibitor for mild steel in acidic medium and new approach of classical electrochemical analyses. *Corros. Sci.* **2019**, *149*, 185–194. [CrossRef]
69. Verma, C.; Quraishi, M.A.; Singh, A. 2-Amino-5-nitro-4,6-diarylcyclohex-1-ene-1,3,3-tricarbonitriles as new and effective corrosion inhibitors for mild steel in 1M HCl: Experimental and theoretical studies. *J. Mol. Liq.* **2015**, *212*, 804–812. [CrossRef]
70. Zhang, D.; Tang, Y.; Qi, S.; Dong, D.; Cang, H.; Lu, G. The inhibition performance of long-chain alkyl-substituted benzimidazole derivatives for corrosion of mild steel in HCl. *Corros. Sci.* **2016**, *102*, 517–522. [CrossRef]
71. Nahlé, A.; Salim, R.; El Hajjaji, F.; Aouad, M.R.; Messali, M.; Ech-Chihbi, E.; Hammouti, B.; Taleb, M. Novel triazole derivatives as ecological corrosion inhibitors for mild steel in 1.0 M HCl: Experimental & theoretical approach. *RSC Adv.* **2021**, *11*, 4147–4162. [CrossRef]
72. Thoume, A.; Left, D.B.; Elmakssoudi, A.; Achagar, R.; Dakir, M.; Azzi, M.; Zertoubi, M. Performance Evaluation of New Chalcone Oxime Functionalized Graphene Oxide as a Corrosion Inhibitor for Carbon Steel in a Hydrochloric Acid Solution. *Langmuir* **2022**, *38*, 7472–7483. [CrossRef]
73. Guo, S.; Garaj, S.; Bianco, A.; Ménard-Moyon, C. Controlling covalent chemistry on graphene oxide. *Nat. Rev. Phys.* **2022**, *4*, 247–262. [CrossRef]
74. Liao, S.-H.; Liu, P.-L.; Hsiao, M.-C.; Teng, C.-C.; Wang, C.-A.; Ger, M.-D.; Chiang, C.-L. One-Step Reduction and Functionalization of Graphene Oxide with Phosphorus-Based Compound to Produce Flame-Retardant Epoxy Nanocomposite. *Ind. Eng. Chem. Res.* **2012**, *51*, 4573–4581. [CrossRef]
75. Li, K.-Y.; Kuan, C.-F.; Kuan, H.-C.; Chen, C.-H.; Shen, M.-Y.; Yang, J.-M.; Chiang, C.-L. Preparation and properties of novel epoxy/graphene oxide nanosheets (GON) composites functionalized with flame retardant containing phosphorus and silicon. *Mater. Chem. Phys.* **2014**, *146*, 354–362. [CrossRef]
76. Hu, W.; Zhan, J.; Wang, X.; Hong, N.; Wang, B.; Song, L.; Stec, A.A.; Hull, T.R.; Wang, J.; Hu, Y. Effect of Functionalized Graphene Oxide with Hyper-Branched Flame Retardant on Flammability and Thermal Stability of Cross-Linked Polyethylene. *Ind. Eng. Chem. Res.* **2014**, *53*, 3073–3083. [CrossRef]
77. Yu, B.; Wang, X.; Qian, X.; Xing, W.; Yang, H.; Ma, L.; Lin, Y.; Jiang, S.; Song, L.; Hu, Y.; et al. Functionalized graphene oxide/phosphoramide oligomer hybrids flame retardant prepared via in situ polymerization for improving the fire safety of polypropylene. *RSC Adv.* **2014**, *4*, 31782–31794. [CrossRef]
78. Wang, X.; Xing, W.; Feng, X.; Yu, B.; Song, L.; Hu, Y. Functionalization of graphene with grafted polyphosphamide for flame retardant epoxy composites: Synthesis, flammability and mechanism. *Polym. Chem.* **2014**, *5*, 1145–1154. [CrossRef]
79. Huang, G.; Chen, S.; Tang, S.; Gao, J. A novel intumescent flame retardant-functionalized graphene: Nanocomposite synthesis, characterization, and flammability properties. *Mater. Chem. Phys.* **2012**, *135*, 938–947. [CrossRef]
80. Wang, Z.; Wei, P.; Qian, Y.; Liu, J. The synthesis of a novel graphene-based inorganic–organic hybrid flame retardant and its application in epoxy resin. *Compos. Part B Eng.* **2014**, *60*, 341–349. [CrossRef]
81. Zhang, T.; Du, Z.; Zou, W.; Li, H.; Zhang, C. Hydroxyl-phosphazene-wrapped carbon nanotubes and its application in ethylene-vinyl acetate copolymer. *J. Appl. Polym. Sci.* **2013**, *130*, 4245–4254. [CrossRef]
82. Tang, Y.; Gou, J.; Hu, Y. Covalent functionalization of carbon nanotubes with polyhedral oligomeric silsequioxane for superhydrophobicity and flame retardancy. *Polym. Eng. Sci.* **2013**, *53*, 1021–1030. [CrossRef]
83. Pal, K.; Kang, D.J.; Zhang, Z.X.; Kim, J.K. Synergistic Effects of Zirconia-Coated Carbon Nanotube on Crystalline Structure of Polyvinylidene Fluoride Nanocomposites: Electrical Properties and Flame-Retardant Behavior. *Langmuir* **2010**, *26*, 3609–3614. [CrossRef] [PubMed]
84. Bifulco, A.; Varganici, C.; Rosu, L.; Mustata, F.; Rosu, D.; Gaan, S. Recent advances in flame retardant epoxy systems containing non-reactive DOPO based phosphorus additives. *Polym. Degrad. Stab.* **2022**, *200*, 109962. [CrossRef]
85. Wang, X.; Guo, W.; Cai, W.; Hu, Y. 3—Graphene-based polymer composites for flame-retardant application. In *Innovations in Graphene-Based Polymer Composites*; Woodhead Publishing: Sawston, UK, 2022; pp. 61–89. [CrossRef]
86. Li, Y.-L.; Kuan, C.-F.; Chen, C.-H.; Kuan, H.-C.; Yip, M.-C.; Chiu, S.-L.; Chiang, C.-L. Preparation, thermal stability and electrical properties of PMMA/functionalized graphene oxide nanosheets composites. *Mater. Chem. Phys.* **2012**, *134*, 677–685. [CrossRef]
87. Qu, L.; Sui, Y.; Zhang, C.; Li, P.; Dai, X.; Xu, B.; Fang, D. POSS-functionalized graphene oxide hybrids with improved dispersive and smoke-suppressive properties for epoxy flame-retardant application. *Eur. Polym. J.* **2020**, *122*, 109383. [CrossRef]
88. Zhang, J.; Li, Z.; Zhang, L.; Molleja, J.G.; Wang, D.-Y. Bimetallic metal-organic frameworks and graphene oxide nano-hybrids for enhanced fire retardant epoxy composites: A novel carbonization mechanism. *Carbon* **2019**, *153*, 407–416. [CrossRef]
89. Ye, T.-P.; Liao, S.-F.; Zhang, Y.; Chen, M.-J.; Xiao, Y.; Liu, X.-Y.; Liu, Z.-G.; Wang, D.-Y. Cu(0) and Cu(II) decorated graphene hybrid on improving fireproof efficiency of intumescent flame-retardant epoxy resins. *Compos. Part B Eng.* **2019**, *175*, 107189. [CrossRef]
90. Mirfarsi, S.H.; Parnian, M.J.; Rowshanzamir, S.; Kjeang, E. Current status of cross-linking and blending approaches for durability improvement of hydrocarbon-based fuel cell membranes. *Int. J. Hydrogen Energy* **2022**, *47*, 13460–13489. [CrossRef]
91. Mirfarsi, S.H.; Parnian, M.J.; Rowshanzamir, S. Self-Humidifying Proton Exchange Membranes for Fuel Cell Applications: Advances and Challenges. *Processes* **2020**, *8*, 1069. [CrossRef]
92. Bagheripour, E.; Moghadassi, A.R.; Hosseini, S.M.; van der Bruggen, B.; Parvizian, F. Novel composite graphene oxide/chitosan nanoplates incorporated into PES based nanofiltration membrane: Chromium removal and antifouling enhancement. *J. Ind. Eng. Chem.* **2018**, *62*, 311–320. [CrossRef]

93. Li, Y.; Huang, Y.; Wang, F.; Liang, W.; Yang, H.; Wu, D. Fabrication of acrylic acid modified graphene oxide (AGO)/acrylate composites and their synergistic mechanisms of anticorrosion and antifouling properties. *Prog. Org. Coat.* **2022**, *168*, 106910. [CrossRef]
94. Rahimi, A.; Mahdavi, H. Zwitterionic-functionalized GO/PVDF nanocomposite membranes with improved anti-fouling properties. *J. Water Process Eng.* **2019**, *32*, 100960. [CrossRef]
95. Krishnamoorthy, K.; Jeyasubramanian, K.; Premanathan, M.; Subbiah, G.; Shin, H.S.; Kim, S.J. Graphene oxide nanopaint. *Carbon* **2014**, *72*, 328–337. [CrossRef]
96. Jiang, T.; Qi, L.; Qin, W. Improving the Environmental Compatibility of Marine Sensors by Surface Functionalization with Graphene Oxide. *Anal. Chem.* **2019**, *91*, 13268–13274. [CrossRef] [PubMed]
97. Jin, H.; Bing, W.; Tian, L.; Wang, P.; Zhao, J. Combined Effects of Color and Elastic Modulus on Antifouling Performance: A Study of Graphene Oxide/Silicone Rubber Composite Membranes. *Materials* **2019**, *12*, 2608. [CrossRef] [PubMed]
98. Jin, H.; Zhang, T.; Bing, W.; Dong, S.; Tian, L. Antifouling performance and mechanism of elastic graphene–silicone rubber composite membranes. *J. Mater. Chem. B* **2019**, *7*, 488–497. [CrossRef]
99. Parra, C.; Dorta, F.; Jimenez, E.; Henríquez, R.; Ramírez, C.; Rojas, R.; Villalobos, P. A nanomolecular approach to decrease adhesion of biofouling-producing bacteria to graphene-coated material. *J. Nanobiotechnology* **2015**, *13*, 82. [CrossRef]
100. Zhang, Z.; Chen, R.; Song, D.; Yu, J.; Sun, G.; Liu, Q.; Han, S.; Liu, J.; Zhang, H.; Wang, J. Guanidine-functionalized graphene to improve the antifouling performance of boron acrylate polymer. *Prog. Org. Coat.* **2021**, *159*, 106396. [CrossRef]
101. Fazli-Shokouhi, S.; Nasirpouri, F.; Khatamian, M. Epoxy-matrix polyaniline/p-phenylenediamine-functionalised graphene oxide coatings with dual anti-corrosion and anti-fouling performance. *RSC Adv.* **2021**, *11*, 11627–11641. [CrossRef]
102. Xie, M.; Zhao, W.; Wu, Y. Preventing algae biofilm formation via designing long-term oil storage surfaces for excellent antifouling performance. *Appl. Surf. Sci.* **2021**, *554*, 149612. [CrossRef]
103. Sun, Y.; Lang, Y.; Sun, T.; Liu, Q.; Pan, Y.; Qi, Z.; Ling, N.; Feng, Y.; Yu, M.; Ji, Y.; et al. Antifouling potential of multi-walled carbon nanotubes-modified chlorinated rubber-based composites on the colonization dynamics of pioneer biofilm-forming eukaryotic microbes. *Int. Biodeterior. Biodegrad.* **2020**, *149*, 104921. [CrossRef]
104. Carl, C.; Poole, A.J.; Vucko, M.J.; Williams, M.R.; Whalan, S.; de Nys, R. Enhancing the efficacy of fouling-release coatings against fouling by Mytilus galloprovincialis using nanofillers. *Biofouling* **2012**, *28*, 1077–1091. [CrossRef] [PubMed]
105. Martinelli, E.; Suffredini, M.; Galli, G.; Glisenti, A.; Pettitt, M.E.; Callow, M.E.; Callow, J.A.; Williams, D.; Lyall, G. Amphiphilic block copolymer/poly(dimethylsiloxane) (PDMS) blends and nanocomposites for improved fouling-release. *Biofouling* **2011**, *27*, 529–541. [CrossRef] [PubMed]
106. Beigbeder, A.; Degee, P.; Conlan, S.L.; Mutton, R.J.; Clare, A.S.; Pettitt, M.E.; Callow, M.E.; Callow, J.A.; Dubois, P. Preparation and characterisation of silicone-based coatings filled with carbon nanotubes and natural sepiolite and their application as marine fouling-release coatings. *Biofouling* **2008**, *24*, 291–302. [CrossRef]
107. Yang, J.-L.; Li, Y.-F.; Guo, X.-P.; Liang, X.; Xu, Y.-F.; Ding, D.-W.; Bao, W.-Y.; Dobretsov, S. The effect of carbon nanotubes and titanium dioxide incorporated in PDMS on biofilm community composition and subsequent mussel plantigrade settlement. *Biofouling* **2016**, *32*, 763–777. [CrossRef]
108. Ba, M.; Zhang, Z.; Qi, Y. The influence of MWCNTs-OH on the properties of the fouling release coatings based on polydimethylsiloxane with the incorporation of phenylmethylsilicone oil. *Prog. Org. Coat.* **2019**, *130*, 132–143. [CrossRef]
109. Sun, Y.; Zhang, Z. New anti-biofouling carbon nanotubes-filled polydimethylsiloxane composites against colonization by pioneer eukaryotic microbes. *Int. Biodeterior. Biodegrad.* **2016**, *110*, 147–154. [CrossRef]
110. Ji, Y.; Sun, Y.; Lang, Y.; Wang, L.; Liu, B.; Zhang, Z. Effect of CNT/PDMS Nanocomposites on the Dynamics of Pioneer Bacterial Communities in the Natural Biofilms of Seawater. *Materials* **2018**, *11*, 902. [CrossRef]
111. Zhang, D.; Liu, Z.; Wu, G.; Yang, Z.; Cui, Y.; Li, H.; Zhang, Y. Fluorinated Carbon Nanotube Superamphiphobic Coating for High-Efficiency and Long-Lasting Underwater Antibiofouling Surfaces. *ACS Appl. Bio Mater.* **2021**, *4*, 6351–6360. [CrossRef]
112. Irani, F.; Jannesari, A.; Bastani, S. Effect of fluorination of multiwalled carbon nanotubes (MWCNTs) on the surface properties of fouling-release silicone/MWCNTs coatings. *Prog. Org. Coat.* **2013**, *76*, 375–383. [CrossRef]
113. Sun, Y.; Lang, Y.; Yan, Z.; Wang, L.; Zhang, Z. High-throughput sequencing analysis of marine pioneer surface-biofilm bacteria communities on different PDMS-based coatings. *Colloids Surf. B Biointerfaces* **2020**, *185*, 110538. [CrossRef] [PubMed]
114. Lin, C.-W.; Aguilar, S.; Rao, E.; Mak, W.H.; Huang, X.; He, N.; Chen, D.; Jun, D.; Curson, P.A.; McVerry, B.T.; et al. Direct grafting of tetraaniline via perfluorophenylazide photochemistry to create antifouling, low bio-adhesion surfaces. *Chem. Sci.* **2019**, *10*, 4445–4457. [CrossRef] [PubMed]
115. Kim, K.-I.; Kim, D.-A.; Patel, K.D.; Shin, U.S.; Kim, H.-W.; Lee, J.-H.; Lee, H.-H. Carbon nanotube incorporation in PMMA to prevent microbial adhesion. *Sci. Rep.* **2019**, *9*, 4921. [CrossRef] [PubMed]
116. Jing, H.; Sahle-Demessie, E.; Sorial, G.A. Inhibition of biofilm growth on polymer-MWCNTs composites and metal surfaces. *Sci. Total Environ.* **2018**, *633*, 167–178. [CrossRef] [PubMed]
117. Fan, F.; Zheng, Y.; Ba, M.; Wang, Y.; Kong, J.; Liu, J.; Wu, Q. Long time super-hydrophobic fouling release coating with the incorporation of lubricant. *Prog. Org. Coat.* **2021**, *152*, 106136. [CrossRef]
118. Kumar, A.; Rout, L.; Dhaka, R.S.; Samal, S.L.; Dash, P. Design of a graphene oxide-SnO2 nanocomposite with superior catalytic efficiency for the synthesis of β-enaminones and β-enaminoesters. *RSC Adv.* **2015**, *5*, 39193–39204. [CrossRef]

119. Achary, L.S.K.; Nayak, P.S.; Barik, B.; Kumar, A.; Dash, P. Ultrasonic-assisted green synthesis of β-amino carbonyl compounds by copper oxide nanoparticles decorated phosphate functionalized graphene oxide via Mannich reaction. *Catal. Today* **2020**, *348*, 137–147. [CrossRef]
120. Kumar, A.; Rout, L.; Achary, L.S.K.; Mohanty, S.K.; Dash, P. A combustion synthesis route for magnetically separable graphene oxide-CuFe2O4-ZnO nanocomposites with enhanced solar light-mediated photocatalytic activity. *New J. Chem.* **2017**, *41*, 10568–10583. [CrossRef]
121. Kumar, A.; Rout, L.; Achary, L.S.K.; Mohanty, A.; Dhaka, R.S.; Dash, P. An investigation into the solar light-driven enhanced photocatalytic properties of a graphene oxide-SnO2-TiO2 ternary nanocomposite. *RSC Adv.* **2016**, *6*, 32074–32088. [CrossRef]
122. Xu, J.; Cao, Z.; Zhang, Y.; Yuan, Z.; Lou, Z.; Xu, X.; Wang, X. A review of functionalized carbon nanotubes and graphene for heavy metal adsorption from water: Preparation, application, and mechanism. *Chemosphere* **2018**, *195*, 351–364. [CrossRef]
123. Ren, X.; Shao, D.; Zhao, G.; Sheng, G.; Hu, J.; Yang, S.; Wang, X. Plasma Induced Multiwalled Carbon Nanotube Grafted with 2-Vinylpyridine for Preconcentration of Pb(II) from Aqueous Solutions. *Plasma Process. Polym.* **2011**, *8*, 589–598. [CrossRef]
124. Wang, H.J.; Zhou, A.L.; Peng, F.; Yu, H.; Chen, L.F. Adsorption characteristic of acidified carbon nanotubes for heavy metal Pb(II) in aqueous solution. *Mater. Sci. Eng. A* **2007**, *466*, 201–206. [CrossRef]
125. Zhao, X.; Jia, Q.; Song, N.; Zhou, W.; Li, Y. Adsorption of Pb(II) from an Aqueous Solution by Titanium Dioxide/Carbon Nanotube Nanocomposites: Kinetics, Thermodynamics, and Isotherms. *J. Chem. Eng. Data* **2010**, *55*, 4428–4433. [CrossRef]
126. Wang, S.-G.; Gong, W.-X.; Liu, X.-W.; Yao, Y.-W.; Gao, B.-Y.; Yue, Q.-Y. Removal of lead(II) from aqueous solution by adsorption onto manganese oxide-coated carbon nanotubes. *Sep. Purif. Technol.* **2007**, *58*, 17–23. [CrossRef]
127. Ihsanullah; Abbas, A.; Al-Amer, A.M.; Laoui, T.; Al-Marri, M.J.; Nasser, M.S.; Khraisheh, M.; Atieh, M.A. Heavy metal removal from aqueous solution by advanced carbon nanotubes: Critical review of adsorption applications. *Sep. Purif. Technol.* **2016**, *157*, 141–161. [CrossRef]
128. Ji, L.; Zhou, L.; Bai, X.; Shao, Y.; Zhao, G.; Qu, Y.; Wang, C.; Li, Y. Facile synthesis of multiwall carbon nanotubes/iron oxides for removal of tetrabromobisphenol A and Pb(ii). *J. Mater. Chem.* **2012**, *22*, 15853–15862. [CrossRef]
129. Zhang, C.; Sui, J.; Li, J.; Tang, Y.; Cai, W. Efficient removal of heavy metal ions by thiol-functionalized superparamagnetic carbon nanotubes. *Chem. Eng. J.* **2012**, *210*, 45–52. [CrossRef]
130. Madadrang, C.J.; Kim, H.Y.; Gao, G.; Wang, N.; Zhu, J.; Feng, H.; Gorring, M.; Kasner, M.L.; Hou, S. Adsorption Behavior of EDTA-Graphene Oxide for Pb (II) Removal. *ACS Appl. Mater. Interfaces* **2012**, *4*, 1186–1193. [CrossRef]
131. Kumar, S.; Nair, R.R.; Pillai, P.B.; Gupta, S.N.; Iyengar, M.A.R.; Sood, A.K. Graphene Oxide–MnFe2O4 Magnetic Nanohybrids for Efficient Removal of Lead and Arsenic from Water. *ACS Appl. Mater. Interfaces* **2014**, *6*, 17426–17436. [CrossRef]
132. Jabeen, H.; Kemp, K.C.; Chandra, V. Synthesis of nano zerovalent iron nanoparticles—Graphene composite for the treatment of lead contaminated water. *J. Environ. Manag.* **2013**, *130*, 429–435. [CrossRef]
133. Ma, S.; Zhan, S.; Jia, Y.; Zhou, Q. Highly Efficient Antibacterial and Pb(II) Removal Effects of Ag-CoFe2O4-GO Nanocomposite. *ACS Appl. Mater. Interfaces* **2015**, *7*, 10576–10586. [CrossRef] [PubMed]
134. Cui, L.; Wang, Y.; Hu, L.; Gao, L.; Du, B.; Wei, Q. Mechanism of Pb(ii) and methylene blue adsorption onto magnetic carbonate hydroxyapatite/graphene oxide. *RSC Adv.* **2015**, *5*, 9759–9770. [CrossRef]
135. Wan, S.; He, F.; Wu, J.; Wan, W.; Gu, Y.; Gao, B. Rapid and highly selective removal of lead from water using graphene oxide-hydrated manganese oxide nanocomposites. *J. Hazard. Mater.* **2016**, *314*, 32–40. [CrossRef] [PubMed]
136. Gollavelli, G.; Chang, C.-C.; Ling, Y.-C. Facile Synthesis of Smart Magnetic Graphene for Safe Drinking Water: Heavy Metal Removal and Disinfection Control. *ACS Sustain. Chem. Eng.* **2013**, *1*, 462–472. [CrossRef]
137. Tawabini, B.S.; Al-Khaldi, S.F.; Khaled, M.M.; Atieh, M.A. Removal of arsenic from water by iron oxide nanoparticles impregnated on carbon nanotubes. *J. Environ. Sci. Health Part A* **2011**, *46*, 215–223. [CrossRef] [PubMed]
138. Mishra, A.K.; Ramaprabhu, S. Magnetite Decorated Multiwalled Carbon Nanotube Based Supercapacitor for Arsenic Removal and Desalination of Seawater. *J. Phys. Chem. C* **2010**, *114*, 2583–2590. [CrossRef]
139. Alijani, H.; Shariatinia, Z. Effective aqueous arsenic removal using zero valent iron doped MWCNT synthesized by in situ CVD method using natural α-Fe2O3 as a precursor. *Chemosphere* **2017**, *171*, 502–511. [CrossRef]
140. Veličković, Z.; Vuković, G.D.; Marinković, A.D.; Moldovan, M.-S.; Perić-Grujić, A.A.; Uskoković, P.S.; Ristić, M.Đ. Adsorption of arsenate on iron(III) oxide coated ethylenediamine functionalized multiwall carbon nanotubes. *Chem. Eng. J.* **2012**, *181–182*, 174–181. [CrossRef]
141. Chen, B.; Zhu, Z.; Ma, J.; Yang, M.; Hong, J.; Hu, X.; Qiu, Y.; Chen, J. One-pot, solid-phase synthesis of magnetic multiwalled carbon nanotube/iron oxide composites and their application in arsenic removal. *J. Colloid Interface Sci.* **2014**, *434*, 9–17. [CrossRef]
142. Chandra, V.; Park, J.; Chun, Y.; Lee, J.W.; Hwang, I.-C.; Kim, K.S. Water-Dispersible Magnetite-Reduced Graphene Oxide Composites for Arsenic Removal. *ACS Nano* **2010**, *4*, 3979–3986. [CrossRef]
143. Paul, B.; Parashar, V.; Mishra, A. Graphene in the Fe3O4 nano-composite switching the negative influence of humic acid coating into an enhancing effect in the removal of arsenic from water. *Environ. Sci. Water Res. Technol.* **2015**, *1*, 77–83. [CrossRef]
144. Mishra, A.K.; Ramaprabhu, S. Functionalized graphene sheets for arsenic removal and desalination of sea water. *Desalination* **2011**, *282*, 39–45. [CrossRef]
145. Sandoval, R.; Cooper, A.M.; Aymar, K.; Jain, A.; Hristovski, K. Removal of arsenic and methylene blue from water by granular activated carbon media impregnated with zirconium dioxide nanoparticles. *J. Hazard. Mater.* **2011**, *193*, 296–303. [CrossRef] [PubMed]

146. Ntim, S.A.; Mitra, S. Adsorption of arsenic on multiwall carbon nanotube–zirconia nanohybrid for potential drinking water purification. *J. Colloid Interface Sci.* **2012**, *375*, 154–159. [CrossRef] [PubMed]
147. Zhang, K.; Dwivedi, V.; Chi, C.; Wu, J. Graphene oxide/ferric hydroxide composites for efficient arsenate removal from drinking water. *J. Hazard. Mater.* **2010**, *182*, 162–168. [CrossRef]
148. Wen, T.; Wu, X.; Tan, X.; Wang, X.; Xu, A. One-Pot Synthesis of Water-Swellable Mg–Al Layered Double Hydroxides and Graphene Oxide Nanocomposites for Efficient Removal of As(V) from Aqueous Solutions. *ACS Appl. Mater. Interfaces* **2013**, *5*, 3304–3311. [CrossRef]
149. Ntim, S.A.; Mitra, S. Removal of Trace Arsenic To Meet Drinking Water Standards Using Iron Oxide Coated Multiwall Carbon Nanotubes. *J. Chem. Eng. Data* **2011**, *56*, 2077–2083. [CrossRef]
150. El-Sheikh, A.H.; Al-Degs, Y.S.; Al-As'ad, R.M.; Sweileh, J.A. Effect of oxidation and geometrical dimensions of carbon nanotubes on Hg(II) sorption and preconcentration from real waters. *Desalination* **2011**, *270*, 214–220. [CrossRef]
151. Chen, P.H.; Hsu, C.-F.; Tsai, D.D.-W.; Lu, Y.-M.; Huang, W.-J. Adsorption of mercury from water by modified multi-walled carbon nanotubes: Adsorption behaviour and interference resistance by coexisting anions. *Environ. Technol.* **2014**, *35*, 1935–1944. [CrossRef]
152. Bandaru, N.M.; Reta, N.; Dalal, H.; Ellis, A.V.; Shapter, J.; Voelcker, N.H. Enhanced adsorption of mercury ions on thiol derivatized single wall carbon nanotubes. *J. Hazard. Mater.* **2013**, *261*, 534–541. [CrossRef]
153. Gupta, A.; Vidyarthi, S.R.; Sankararamakrishnan, N. Enhanced sorption of mercury from compact fluorescent bulbs and contaminated water streams using functionalized multiwalled carbon nanotubes. *J. Hazard. Mater.* **2014**, *274*, 132–144. [CrossRef] [PubMed]
154. Moghaddam, H.K.; Pakizeh, M. Experimental study on mercury ions removal from aqueous solution by MnO2/CNTs nanocomposite adsorbent. *J. Ind. Eng. Chem.* **2015**, *21*, 221–229. [CrossRef]
155. AlOmar, M.K.; Alsaadi, M.A.; Hayyan, M.; Akib, S.; Ibrahim, M.; Hashim, M.A. Allyl triphenyl phosphonium bromide based DES-functionalized carbon nanotubes for the removal of mercury from water. *Chemosphere* **2017**, *167*, 44–52. [CrossRef]
156. Shawky, H.A.; El-Aassar, A.H.M.; Abo-Zeid, D.E. Chitosan/carbon nanotube composite beads: Preparation, characterization, and cost evaluation for mercury removal from wastewater of some industrial cities in Egypt. *J. Appl. Polym. Sci.* **2012**, *125*, E93–E101. [CrossRef]
157. Shih, Y.-C.; Ke, C.-Y.; Yu, C.-J.; Lu, C.-Y.; Tseng, W.-L. Combined Tween 20-Stabilized Gold Nanoparticles and Reduced Graphite Oxide–Fe3O4 Nanoparticle Composites for Rapid and Efficient Removal of Mercury Species from a Complex Matrix. *ACS Appl. Mater. Interfaces* **2014**, *6*, 17437–17445. [CrossRef] [PubMed]
158. Kumar, A.S.K.; Jiang, S.-J.; Tseng, W.-L. Effective adsorption of chromium(vi)/Cr(iii) from aqueous solution using ionic liquid functionalized multiwalled carbon nanotubes as a super sorbent. *J. Mater. Chem. A* **2015**, *3*, 7044–7057. [CrossRef]
159. Di, Z.-C.; Ding, J.; Peng, X.-J.; Li, Y.-H.; Luan, Z.-K.; Liang, J. Chromium adsorption by aligned carbon nanotubes supported ceria nanoparticles. *Chemosphere* **2006**, *62*, 861–865. [CrossRef]
160. Kumar, R.; Ansari, M.O.; Barakat, M.A. DBSA doped polyaniline/multi-walled carbon nanotubes composite for high efficiency removal of Cr(VI) from aqueous solution. *Chem. Eng. J.* **2013**, *228*, 748–755. [CrossRef]
161. Atieh, M.A. Removal of Chromium (VI) from polluted water using carbon nanotubes supported with activated carbon. *Procedia Environ. Sci.* **2011**, *4*, 281–293. [CrossRef]
162. Sankararamakrishnan, N.; Jaiswal, M.; Verma, N. Composite nanofloral clusters of carbon nanotubes and activated alumina: An efficient sorbent for heavy metal removal. *Chem. Eng. J.* **2014**, *235*, 1–9. [CrossRef]
163. Liu, M.; Wen, T.; Wu, X.; Chen, C.; Hu, J.; Li, J.; Wang, X. Synthesis of porous Fe3O4 hollow microspheres/graphene oxide composite for Cr(vi) removal. *Dalton Trans.* **2013**, *42*, 14710–14717. [CrossRef] [PubMed]
164. Qiu, B.; Gu, H.; Yan, X.; Guo, J.; Wang, Y.; Sun, D.; Wang, Q.; Khan, M.; Zhang, X.; Weeks, B.L.; et al. Cellulose derived magnetic mesoporous carbon nanocomposites with enhanced hexavalent chromium removal. *J. Mater. Chem. A* **2014**, *2*, 17454–17462. [CrossRef]
165. Vuković, G.D.; Marinković, A.D.; Čolić, M.; Ristić, M.Ð.; Aleksić, R.; Perić-Grujić, A.A.; Uskoković, P.S. Removal of cadmium from aqueous solutions by oxidized and ethylenediamine-functionalized multi-walled carbon nanotubes. *Chem. Eng. J.* **2010**, *157*, 238–248. [CrossRef]
166. Liang, J.; Liu, J.; Yuan, X.; Dong, H.; Zeng, G.; Wu, H.; Wang, H.; Liu, J.; Hua, S.; Zhang, S.; et al. Facile synthesis of alumina-decorated multi-walled carbon nanotubes for simultaneous adsorption of cadmium ion and trichloroethylene. *Chem. Eng. J.* **2015**, *273*, 101–110. [CrossRef]
167. Moradi, O. The Removal of Ions by Functionalized Carbon Nanotube: Equilibrium, Isotherms and Thermodynamic Studies. *Chem. Biochem. Eng. Q.* **2011**, *25*, 229–240. Available online: https://www.researchgate.net/publication/268360458 (accessed on 20 August 2022).
168. Li, Y.-H.; Ding, J.; Luan, Z.; Di, Z.; Zhu, Y.; Xu, C.; Wu, D.; Wei, B. Competitive adsorption of Pb2+, Cu2+ and Cd2+ ions from aqueous solutions by multiwalled carbon nanotubes. *Carbon* **2003**, *41*, 2787–2792. [CrossRef]
169. Tofighy, M.A.; Mohammadi, T. Adsorption of divalent heavy metal ions from water using carbon nanotube sheets. *J. Hazard. Mater.* **2011**, *185*, 140–147. [CrossRef]
170. Ge, Y.; Li, Z.; Xiao, D.; Xiong, P.; Ye, N. Sulfonated multi-walled carbon nanotubes for the removal of copper (II) from aqueous solutions. *J. Ind. Eng. Chem.* **2014**, *20*, 1765–1771. [CrossRef]

171. Rosenzweig, S.; Sorial, G.A.; Sahle-Demessie, E.; Mack, J. Effect of acid and alcohol network forces within functionalized multiwall carbon nanotubes bundles on adsorption of copper (II) species. *Chemosphere* **2013**, *90*, 395–402. [CrossRef]
172. Dichiara, A.B.; Webber, M.R.; Gorman, W.R.; Rogers, R.E. Removal of Copper Ions from Aqueous Solutions via Adsorption on Carbon Nanocomposites. *ACS Appl. Mater. Interfaces* **2015**, *7*, 15674–15680. [CrossRef]
173. Tang, W.-W.; Zeng, G.-M.; Gong, J.-L.; Liu, Y.; Wang, X.-Y.; Liu, Y.-Y.; Liu, Z.-F.; Chen, L.; Zhang, X.-R.; Tu, D.-Z. Simultaneous adsorption of atrazine and Cu (II) from wastewater by magnetic multi-walled carbon nanotube. *Chem. Eng. J.* **2012**, *211–212*, 470–478. [CrossRef]
174. Lu, C.; Chiu, H.; Liu, C. Removal of Zinc(II) from Aqueous Solution by Purified Carbon Nanotubes: Kinetics and Equilibrium Studies. *Ind. Eng. Chem. Res.* **2006**, *45*, 2850–2855. [CrossRef]
175. Mubarak, N.M.; Alicia, R.F.; Abdullah, E.C.; Sahu, J.N.; Haslija, A.B.A.; Tan, J. Statistical optimization and kinetic studies on removal of Zn2+ using functionalized carbon nanotubes and magnetic biochar. *J. Environ. Chem. Eng.* **2013**, *1*, 486–495. [CrossRef]
176. Yang, S.; Li, J.; Shao, D.; Hu, J.; Wang, X. Adsorption of Ni(II) on oxidized multi-walled carbon nanotubes: Effect of contact time, pH, foreign ions and PAA. *J. Hazard. Mater.* **2009**, *166*, 109–116. [CrossRef]
177. Mobasherpour, I.; Salahi, E.; Ebrahimi, M. Removal of divalent nickel cations from aqueous solution by multi-walled carbon nano tubes: Equilibrium and kinetic processes. *Res. Chem. Intermed.* **2012**, *38*, 2205–2222. [CrossRef]
178. Wang, Q.; Li, J.; Chen, C.; Ren, X.; Hu, J.; Wang, X. Removal of cobalt from aqueous solution by magnetic multiwalled carbon nanotube/iron oxide composites. *Chem. Eng. J.* **2011**, *174*, 126–133. [CrossRef]
179. Shao, L.; Wang, X.; Ren, Y.; Wang, S.; Zhong, J.; Chu, M.; Tang, H.; Luo, L.; Xie, D. Facile fabrication of magnetic cucurbit[6]uril/graphene oxide composite and application for uranium removal. *Chem. Eng. J.* **2016**, *286*, 311–319. [CrossRef]
180. Zong, P.; Wang, S.; Zhao, Y.; Wang, H.; Pan, H.; He, C. Synthesis and application of magnetic graphene/iron oxides composite for the removal of U(VI) from aqueous solutions. *Chem. Eng. J.* **2013**, *220*, 45–52. [CrossRef]
181. Zhao, Y.; Li, J.; Zhang, S.; Chen, H.; Shao, D. Efficient enrichment of uranium(vi) on amidoximated magnetite/graphene oxide composites. *RSC Adv.* **2013**, *3*, 18952–18959. [CrossRef]
182. Jung, Y.; Kim, S.; Park, S.-J.; Kim, J.M. Preparation of functionalized nanoporous carbons for uranium loading. *Colloids Surf. A Physicochem. Eng. Asp.* **2008**, *313–314*, 292–295. [CrossRef]
183. Ahmad, S.Z.N.; Salleh, W.N.W.; Ismail, A.F.; Yusof, N.; Yusop, M.Z.M.; Aziz, F. Adsorptive removal of heavy metal ions using graphene-based nanomaterials: Toxicity, roles of functional groups and mechanisms. *Chemosphere* **2020**, *248*, 126008. [CrossRef] [PubMed]
184. Ferrari, M.; Benedetti, A. Superhydrophobic surfaces for applications in seawater. *Adv. Colloid Interface Sci.* **2015**, *222*, 291–304. [CrossRef] [PubMed]
185. Grigoraki, L.; Grau-Bové, X.; Yates, H.C.; Lycett, G.J.; Ranson, H. Isolation and transcriptomic analysis of Anopheles gambiae oenocytes enables the delineation of hydrocarbon biosynthesis. *eLife* **2020**, *9*, e58019. [CrossRef] [PubMed]
186. Blomquist, G.J.; Bagnères, A.-G. (Eds.) *Insect Hydrocarbons*; Cambridge University Press: Cambridge, UK, 2010. [CrossRef]
187. Balabanidou, V.; Grigoraki, L.; Vontas, J. Insect cuticle: A critical determinant of insecticide resistance. *Curr. Opin. Insect Sci.* **2018**, *27*, 68–74. [CrossRef]
188. Zhu, Y.; Sun, F.; Qian, H.; Wang, H.; Mu, L.; Zhu, J. A biomimetic spherical cactus superhydrophobic coating with durable and multiple anti-corrosion effects. *Chem. Eng. J.* **2018**, *338*, 670–679. [CrossRef]
189. Xiang, Y.; Li, X.; Du, A.; Wu, S.; Shen, J.; Zhou, B. Timing of polyethylene glycol addition for the control of SiO_2 sol structure and sol–gel coating properties. *J. Coat. Technol. Res.* **2017**, *14*, 447–454. [CrossRef]
190. Tran, N.G.; Chun, D.-M. Ultrafast and Eco-Friendly Fabrication Process for Robust, Repairable Superhydrophobic Metallic Surfaces with Tunable Water Adhesion. *ACS Appl. Mater. Interfaces* **2022**, *14*, 28348–28358. [CrossRef]
191. Wang, X.; Hu, H.; Ye, Q.; Gao, T.; Zhou, F.; Xue, Q. Superamphiphobic coatings with coralline-like structure enabled by one-step spray of polyurethane/carbon nanotube composites. *J. Mater. Chem.* **2012**, *22*, 9624–9631. [CrossRef]
192. Li, J.; Yan, L.; Ouyang, Q.; Zha, F.; Jing, Z.; Li, X.; Lei, Z. Facile fabrication of translucent superamphiphobic coating on paper to prevent liquid pollution. *Chem. Eng. J.* **2014**, *246*, 238–243. [CrossRef]
193. Zachariah, S.; Chuo, T.-W.; Liu, Y.-L. Crosslinked polybenzoxazine coatings with hierarchical surface structures from a biomimicking process exhibiting high robustness and anticorrosion performance. *Polymer* **2018**, *155*, 168–176. [CrossRef]
194. Ivanova, E.P.; Nguyen, S.H.; Guo, Y.; Baulin, V.A.; Webb, H.K.; Truong, V.K.; Wandiyanto, J.V.; Garvey, C.J.; Mahon, P.J.; Mainwaring, D.E.; et al. Bactericidal activity of self-assembled palmitic and stearic fatty acid crystals on highly ordered pyrolytic graphite. *Acta Biomater.* **2017**, *59*, 148–157. [CrossRef] [PubMed]
195. Yuan, Y.; Zhang, Y. Enhanced biomimic bactericidal surfaces by coating with positively-charged ZIF nano-dagger arrays. *Nanomedicine* **2017**, *13*, 2199–2207. [CrossRef] [PubMed]
196. Yang, Y.; Gao, P.; Wang, J.; Tu, Q.; Bai, L.; Xiong, K.; Qiu, H.; Zhao, X.; Maitz, M.F.; Wang, H.; et al. Endothelium-Mimicking Multifunctional Coating Modified Cardiovascular Stents via a Stepwise Metal-Catechol-(Amine) Surface Engineering Strategy. *Research* **2020**, *2020*, 9203906. [CrossRef] [PubMed]
197. Das, S.; Das, A.; Parbat, D.; Manna, U. Catalyst-Free and Rapid Chemical Approach for in Situ Growth of "Chemically Reactive" and Porous Polymeric Coating. *ACS Appl. Mater. Interfaces* **2019**, *11*, 34316–34329. [CrossRef] [PubMed]
198. Liu, Z.; Zhang, C.; Jing, J.; Zhang, X.; Wang, C.; Liu, F.; Jiang, M.; Wang, H. Bristle worm inspired ultra-durable superhydrophobic coating with repairable microstructures and anti-corrosion/scaling properties. *Chem. Eng. J.* **2022**, *436*, 135273. [CrossRef]

199. Zhao, H.; Sun, Q.; Deng, X.; Cui, J. Earthworm-Inspired Rough Polymer Coatings with Self-Replenishing Lubrication for Adaptive Friction-Reduction and Antifouling Surfaces. *Adv. Mater.* **2018**, *30*, 1802141. [CrossRef]
200. Pillai, K.; Arzate, F.N.; Zhang, W.; Renneckar, S. Towards Biomimicking Wood: Fabricated Free-standing Films of Nanocellulose, Lignin, and a Synthetic Polycation. *J. Vis. Exp.* **2014**, *88*, e51257. [CrossRef]
201. Zhou, P.; Lui, G.N.; Zhang, X. An experimental investigation of the effect of owl-inspired velvety coating on trailing edge noise. In Proceedings of the 25th AIAA/CEAS Aeroacoustics Conference, Delft, The Netherlands, 20–23 May 2019. [CrossRef]
202. Luo, Z.; Zhang, Z.; Wang, W.; Liu, W.; Xue, Q. Various curing conditions for controlling PTFE micro/nano-fiber texture of a bionic superhydrophobic coating surface. *Mater. Chem. Phys.* **2010**, *119*, 40–47. [CrossRef]
203. Lackner, J.M.; Waldhauser, W.; Major, R.; Major, L.; Hartmann, P. Biomimetics in thin film design—Wrinkling and fracture of pulsed laser deposited films in comparison to human skin. *Surf. Coat. Technol.* **2013**, *215*, 192–198. [CrossRef]
204. Oyane, A.; Kakehata, M.; Sakamaki, I.; Pyatenko, A.; Yashiro, H.; Ito, A.; Torizuka, K. Reprint of "Biomimetic apatite coating on yttria-stabilized tetragonal zirconia utilizing femtosecond laser surface processing". *Surf. Coat. Technol.* **2016**, *307*, 1144–1151. [CrossRef]
205. Lienhard, C.; Ferreira, R.L.; Gnos, E.; Hollier, J.; Eggenberger, U.; Piuz, A. Microcrystals coating the wing membranes of a living insect (Psocoptera: Psyllipsocidae) from a Brazilian cave. *Sci. Rep.* **2012**, *2*, 408. [CrossRef] [PubMed]
206. Park, H.K.; Lee, D.; Lee, H.; Hong, S. A nature-inspired protective coating on soft/wet biomaterials for SEM by aerobic oxidation of polyphenols. *Mater. Horiz.* **2020**, *7*, 1387–1396. [CrossRef]

Prediction Model of Aluminized Coating Thicknesses Based on Monte Carlo Simulation by X-ray Fluorescence

Zhuoyue Li, Cheng Wang *, Haijuan Ju, Xiangrong Li, Yi Qu and Jiabo Yu

Fundamentals Department, Air Force Engineering University, Xi'an 710051, China; lz_980512@163.com (Z.L.); jhjcumtgx@163.com (H.J.); lixiangrong0925@126.com (X.L.); strsky778@163.com (Y.Q.); b2283216046@163.com (J.Y.)
* Correspondence: valid_01@163.com

Abstract: An aluminized coating can improve the high-temperature oxidation resistance of turbine blades, but the inter-diffusion of elements renders the coating's thickness difficult to achieve in non-destructive testing. As a typical method for coating thickness inspection, X-ray fluorescence mainly includes the fundamental parameter method and the empirical coefficient method. The fundamental parameter method has low accuracy for such complex coatings, while it is difficult to provide sufficient reference samples for the empirical coefficient method. To achieve accurate non-destructive testing of aluminized coating thickness, we analyzed the coating system of aluminized blades, simulated the spectra of reference samples using the open-source software XMI-MSIM, established the mapping between elemental spectral intensity and coating thickness based on partial least squares and back-propagation neural networks, and validated the model with actual samples. The experimental results show that the model's prediction error based on the back-propagation neural network is 4.45% for the Al-rich layer and 16.89% for the Al-poor layer. Therefore, the model is more suitable for predicting aluminized coating thickness. Furthermore, the Monte Carlo simulation method can provide a new way of thinking for materials that have difficulty in fabricating reference samples.

Keywords: X-ray fluorescence; aluminized coating; Monte Carlo simulation; turbine blades; backward propagation neural network; XMI-MSIM

1. Introduction

Airline flight safety is a primary concern for countries worldwide, and airline crashes cause significant economic losses and are also devastating to the families of victims. A reliable engine is key to protecting the aircraft for safe flights. For Turbine blades as the engine power-energy transfer exchanger, frequently in high-temperature, high-pressure states, this state on the turbine blade's material performance put forward higher requirements [1–4]. Generally, the turbine blade substrate is made of a nickel-based high-temperature alloy with an aluminizing treatment on the surface to prevent high-temperature oxidation [5]. However, the harsh service environment damages the aluminum coating on the blade's surface, affecting the high-temperature oxidation resistance of the blade [1,6]. Therefore, it is necessary to monitor turbine blade coating conditions. The commonly used method for checking coating thickness is the cross-sectional metallographic method [7], which is straightforward and effective but will damage the blade. The method is costly and is usually performed on a random sample. Hence, developing a non-destructive testing (NDT) method for the Al coating on turbine blade surfaces is urgently required.

Various NDT methods have been tested with an aim to achieve non-destructive testing of the thickness of diffusion coatings on turbine surfaces, but each method has its limitations. The ultrasonic method is generally used for coatings with obvious physical boundaries, while the physical boundaries of diffusion coatings are not obvious. The eddy current method is generally used for non-metallic coatings on non-magnetic metal substrates, while both diffusion coatings and their substrates are metals. The magnetic thickness measurement method is generally used for non-magnetic coatings on magnetic substrates, and diffusion coatings are also not applicable [8].

As an elemental analysis technique, X-ray fluorescence (XRF) is not only used in element identification and element quantification but also widely used in coating thickness measurement [9–13]. The current XRF coating thickness measurement is mainly based on the fundamental parameter method without standard samples and the empirical coefficient method with standard samples [14–16]. For gold and silver coatings on the surface of precious objects, Brocchieri [10,17] used the empirical coefficient method combined with partial least squares regression for coating thickness prediction, obtaining relatively consistent results with the expected thickness. Takahara [18] introduced the advantages of the fundamental parameter method for thickness measurement when no standard samples are available and explained the details of the fundamental parameter method for thickness measurement using an ITO film sample as an example. However, for turbine blades in service with a large variety of elements and mutual diffusion between coating and substrate elements [19], the fundamental parameter method is not accurate enough for calculation, and it is not possible to provide sufficient reference samples for the empirical coefficient method. Therefore, a Monte Carlo simulation is considered to provide the spectra of the reference samples.

Monte Carlo simulation uses a statistical method to simulate the process of photon-matter interaction. Schoonjans [20–22] developed the open-source software XRMC and XMI-MSIM for EDXRF Monte Carlo simulations, which can be performed by setting parameters such as excitation source, instrument geometry layout, detector, and sample composition. Giurlani [11,12] used this method to establish standard curves for determining the thickness of single and multilayer metal coatings and obtained significantly better results than the fundamental parameter method. Trojek [23] used iterative Monte Carlo simulations to determine the copper alloy composition and the thickness of the cover layer and verified the efficiency and robustness of the method. A series of studies applying Monte Carlo simulation methods in archaeology is also available [24–26].

In this study, in order to achieve a non-destructive detection of aluminized coating thickness on the in-service turbine blade's surface, we first analyzed the blade cross-section to determine the layering and composition of the coating, obtained XRF spectra at different thicknesses using Monte Carlo simulation, established a coating thickness prediction model using chemometric methods, and finally verified the accuracy of the model by testing the XRF spectra on the blade's surface.

2. Materials and Methods
2.1. Principle of XRF Thickness Measurement

The basic principle of XRF elemental analysis is as follows: primary X-rays excite the elements in the sample to produce XRF, identify the elemental species based on the difference in XRF energy of different elements, and determine the content of the elements in the sample based on XRF intensity [16]. Traditional methods for mapping relationships between XRF intensity and elemental content include the empirical coefficient method [14] and the fundamental parameter method [15]. The empirical coefficient method establishes a calibration curve by measuring a series of reference samples. The fundamental parameter method requires knowing the exact parameters of each element and determining the geometric factor of the instrument by testing the spectra of several pure elements to enable a broader range of standardless measurements [27]. There are three methods of determining coating thickness by XRF as follows [28]:

1. Emission method: The research object of the emission method is the coating element. The thicker the coating, the greater the XRF intensity of the coating element.
2. Absorption method: The research object of the absorption method is the substrate element. The thicker the coating, the more the XRF of the substrate element is absorbed, and the smaller the intensity.
3. Relative method: The relative method calculates the ratio of the XRF intensity of the coating element and the substrate element to determine the thickness [16]. Since the method uses relative values to calculate thickness, variations in measurement conditions have little effect on the calculation error.

It is difficult for aluminized coatings on turbine blade surfaces to obtain accurate results simply using the above thickness measurement methods due to the inter-diffusion between elements. We need to explore the mapping between the XRF intensity of multiple elements and coating thickness by measuring a series of reference samples of aluminized coatings with different thicknesses. However, such reference samples are difficult to fabricate. The Monte Carlo method can simulate interactions, including scattering effects, photoelectric effects, and Auger electrons, considering the entire spectrum of analytes. Therefore, we use the Monte Carlo method to simulate a series of reference samples to establish the mapping relationship between XRF intensity and coating thickness.

2.2. Experimental Procedure

The experiments mainly include characterization, simulation, and validation processes, and mathematical modeling is included in the simulation process. The flow chart of the research process is summarized in Figure 1.

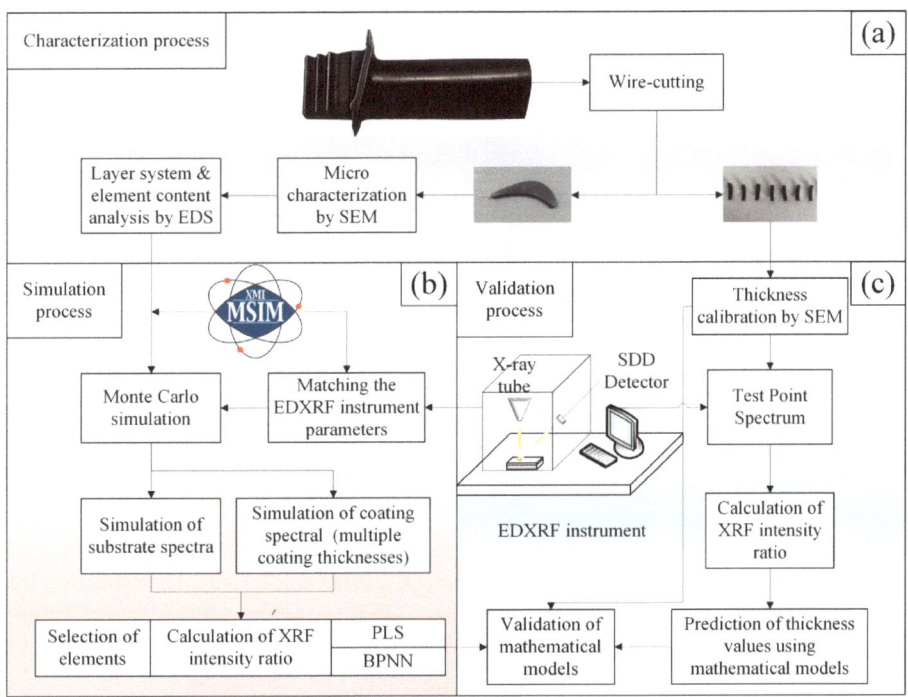

Figure 1. Flow chart of the characterization, simulation, and validation process. (**a**) Characterization process; (**b**) Simulation process; (**c**) Validation progress.

Efforts in the characterization process include sample preparation, micromorphological characterization, analysis of the coating system, and elemental content. A type of in-service turbine blade provided by Xiangyang Hangtai Power Machine Factory was used for characterization experiments. The blade substrate is DZ422B nickel-based superalloy, protected by an aluminized layer on the surface. It was the first to cut into eight pieces equidistantly along the blade body of the turbine blade using an EDM wire cutting machine (DK7725H, ZHCF, Suzhou, China). The cross-section of the pieces was then sanded and polished with graded grit sandpapers to achieve a roughness where the surface coating could be clearly observed. Prior to SEM (VEGA-3 XMU, TESCAN, Brno, Czech Republic) observation, the slices were cleaned using an ultrasonic cleaner (KM-410C, KJM, Guangzhou, China), dried, and wiped on the surface with alcohol to avoid contaminants from affecting characterization results. The slices were placed on the sample stage with the smooth section facing upward through the conductive adhesive, the acceleration voltage was set to 20 kV, and the current was set to 1 nA. The electron beam was switched on, and the magnification was 1000×. The backscatter detector was selected to make the interface contrast between the coating and the substrate more obvious. After the analysis area was selected, a surface scan of the area was performed using EDS (AZtecOne & x-act, Oxford, London, UK) to obtain the overall distribution of various elements, and the EDS worked with the same parameters as SEM. A line scan was performed along the coating depth direction to obtain the trend of different elements in that direction. Finally, a point scan was performed for the areas with apparent boundaries to obtain the elemental content values of each area.

The simulation process is based on the characterization process and the determination of the EDXRF instrument (XAU-4CS, YL, Suzhou, China) parameters. Monte Carlo simulations were performed using the open-source software XMI-MSIM developed by Schoonjans T. [21]. The software can automatically generate the simulated energy spectrum of primary X-rays by inputting parameters, adjusting parameters to set the geometric layout between excitation source-sample-detector, and inputting elemental content to set the sample's composition, including the air layer. The main parameter settings of the simulation process are shown in Table 1. The sample composition was set according to the coating system and elemental content obtained during the characterization. The remaining parameters of the Monte Carlo simulation were set according to the excitation conditions, geometric layout, and detector type of the EDXRF instrument. The coating system studied in this paper is divided into three layers. The first layer is set up with 13 different thickness values in the range of 0-60 μm at every 5 μm step. The second layer is set up with 11 different thickness values in the range of 0–40 μm at every 4 μm step. The third layer is set up with 4 different thickness values of 0 μm, 8 μm, 15 μm, and 22 μm. The spectra and XRF intensities of the main elements were recorded in detail for each simulation to obtain $13 \times 11 \times 4 = 572$ sets of data. Before establishing the mathematical model of XRF intensity of the major elements versus coating thickness, the ratio (R) of the XRF intensity value (I_c) obtained from the simulation of the major elements in the coating to the XRF intensity value (I_s) obtained from the simulation of the corresponding elements in the substrate was calculated to mitigate the deviation formed during the testing of the instrument and simulation software. The data obtained from the simulation were randomly divided in the ratio of training set: test set = 4:1, and then the more mature chemometric methods, such as partial least squares (PLS) and backward propagation neural networks (BPNN), were used to model the data and compare their goodness of fit. The mathematical methods mentioned above have been widely used in XRF [29–33].

Table 1. Parameters for Monte Carlo simulation.

Project	Segmentation	Parameter	Density (g/cm³)	Thickness (cm)
General	Number photons per interval	10,000		
	Number of photons per discrete line	100,000		
	Number of interactions per trajectory	4		
Geometry	Sample-source distance (cm)	2		
	Primary X-ray incidence angle (°)	90		
	X-ray fluorescence emission angle (°)	45		
	Active detector area (cm²)	0.0531		
Excitation	Tube voltage (kV)	40		
	Tube current (mA)	1.00		
	Anode	W	19.3	0.0002
	window	Be	1.848	0.0125
Composition	Air, Dry	C, N, O, Ar	0.001205	2
	Al-rich layer		5.24	0~0.0022
	Al-poor layer	Reference	6.05	0~0.0040
	IDZ layer	Table 2	6.83	0~0.0060
	Substrate		7.28	1
Detection	Detector type	SDD		
	Number of spectrum channels	4096		
	Active detector area (cm²)	0.25		

Table 2. Elemental composition of coatings and substrate (wt.%).

Element	C	Al	Cl	K	Ti	Cr	Co	Ni	Nb	Yb	Hf	W
Al-rich layer	7.64	20.52	0.33	0.16	0.42	3.54	8.20	53.68	0	0	0	5.51
Al-poor layer	6.47	11.47	0.25	0	1.71	6.36	9.51	60.35	0.38	0	0	3.51
IDZ layer	7.16	6.21	0.21	0	2.80	11.21	9.39	41.57	1.40	0	0	20.06
Substrate	5.37	2.81	0.22	0.09	2.48	8.25	9.48	56.14	1.01	2.07	1.47	10.61

The validation process is based on the slices obtained from the characterization process and the mathematical model obtained from the simulation process to verify the model's accuracy. The cross-section of the slices was first observed using an SEM to calibrate the coating's thickness. Although the slices belong to the same turbine blade, the different locations are subjected to different impacts and loads in high-temperature environments; therefore, their coating thicknesses also differ. The EDXRF instrument was then used to test the XRF spectra of the slices, and the instrument parameters were consistent with those set for the simulation process. The ratio of the XRF intensity of the main elements to the corresponding elements of the substrate is calculated and substituted into the mathematical model established by the simulation process to predict the coating's thickness. Finally, the predicted thickness was compared with the calibrated thickness to verify the model's accuracy in the actual test.

The dependence between the individual coating thicknesses was not considered during the experiments; thus, some of the simulated conditions would not occur on the actual blade's surface. However, these simulated data are still consistent with the variation pattern of XRF intensity in the coating and have a negligible impact on the mathematical model.

3. Results and Discussion

3.1. Characterization of the Coating System

Prior to performing Monte Carlo simulations, the sample's exact elemental compositions and distributions need to be clearly known. For this purpose, the turbine blade cross-section was characterized using SEM, and the mass fraction of each coating and substrate element was analyzed by EDS. The analysis result is shown in Figure 2.

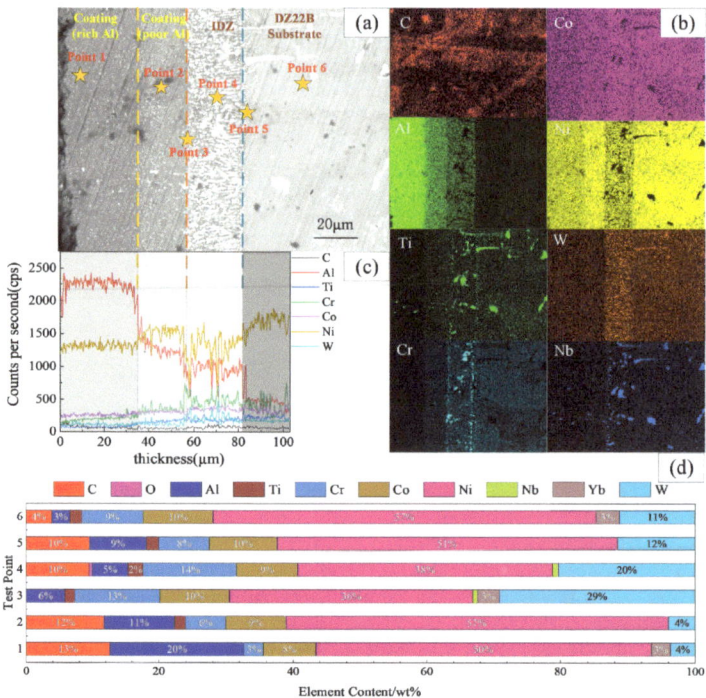

Figure 2. Characterization of turbine blade cross-section. (**a**) SEM morphology under backscattering. (**b**) EDS mapping of (**a**); (**c**) distribution of elements after line scan along the depth direction; (**d**) mass fraction of elements from the test points in (**a**).

In Figure 2a, it can be observed that the aluminized coating on the in-service turbine blade surface is divided into three layers. The darker color of the outermost layer indicates the higher aluminum content of this layer (under backscattering irradiation, the contrast between light and dark reflects the atomic number relationship of the elements: the darker the atomic number, the smaller and brighter the opposite is). The outermost layer corresponds to the 0–35 μm segment in Figure 2c, which we call the Al-rich layer. The second outer layer is slightly brighter than the outermost layer, corresponding to the 36–57 μm section in Figure 2c, dominated by the nickel-rich NiAl phase, which we call the Al-poor layer. For the third layer, brighter and darker areas are interspersed. This layer is formed by the mutual diffusion of the alloying elements in the substrate and the coating elements, which corresponds to the 58–82 μm segment in Figure 2c. The mass fraction of elements in this layer fluctuates widely, and we call it the interdiffusion zone (IDZ) layer. The last one is the uncoated DZ22B substrate. According to Figure 2b,d, the trends of elemental contents are shown from the surface and point perspectives, respectively. Al content decreases along the depth direction, while the W content is the highest in the IDZ layer, Ni content is lowest in the IDZ layer, and Ti, Cr, and Nb contents are lower and not uniformly distributed. Co content has no significant changes in the entire system. Overall, the elemental mass fraction of each coating layer is relatively stable, and the average elemental mass fraction of each coating area can be taken as the composition value of this layer and is inputed into XMI-MSIM for simulation. The composition of each layer is shown in Table 2.

3.2. Monte Carlo Simulation

The EDXRF simulations were performed using the sample composition information obtained in Section 3.1 and the parameter settings in Table 1. The Al-rich layer, Al-poor layer, and IDZ layer were changed in equal steps within their corresponding ranges of 0–60 μm, 0–40 μm, and 0–22 μm, and the substrate was set to 1cm, which could be considered as infinite thickness in XRF simulation. For each thickness group, XRF spectra were obtained separately. The areas of the spectral peaks in the energy range of Ni Kα, Cr Kα, Co Kα, W Lα, Ti Kα, Nb Kα, Yb Lα, and Hf Lβ were calculated to obtain the XRF intensity values of the corresponding spectral lines. Compared with the XRF intensity values of the substrate's corresponding elements, obtain the XRF intensity relative values, as shown in Figure 3, where the three axes correspond to three coating thicknesses. The size of the sphere indicates the relative value of elemental XRF intensity magnitude.

In Figure 3, the XRF intensity of Ni and Co elements is the smallest, and the XRF intensity of Cr and W elements is the largest when the IDZ layer is the thickest, and the thicknesses of Al-rich and Al-poor layers are 0 μm. When the thickness of the Al-rich and Al-poor layers gradually increases, the XRF intensity of the Ni element gradually increases, and the increasing trend gradually slows down until it no longer increases. Due to the high mass fraction of Ni in the sample and the coating reaching a certain thickness, the XRF of the Ni element reaches the maximum information depth. At this time, the XRF intensity no longer changes as thickness increases. Therefore, when using the XRF intensity of the Ni element to determine the thickness of the coating, only coatings with thin thickness can be identified, and when the maximum information depth is reached, only a range of coating thicknesses can be determined. The decreasing trend of the Cr element's XRF intensity in the direction of increasing thickness of the Al-poor layer is smaller than that of the Al-rich layer, and the maximum information depth is also reached up to a certain thickness.

The XRF intensity values of Co, W, and Ti show different trends as the thickness of the Al-poor and Al-rich layers increases. However, all reach the maximum information depth when thickness increases to a certain level. In addition, since the mass fraction of Co varies less throughout the coating system, the variation in XRF intensity values for different thickness groups is also tiny. This situation is undoubtedly a challenge for the prediction of coating thickness. Fortunately, the "insignificant" (small mass fraction) Nb element has a clear pattern of variation with coating thickness and does not reach the maximum information depth. The main reason for this situation is probably the sizeable atomic number of the Nb element. Thus, the higher energy of the excited generated XRF, which can penetrate a thicker coating, has not reached the maximum information depth. For Yb and Hf elements, both show almost the same trend of XRF intensity with coating thickness, resulting from the fact that these two elements are only present in the substrate and have not inter-diffused with the coating elements. However, the mass fraction of the two elements in the substrate is so tiny that the count rates of the corresponding spectral lines are too low when the coating is thick. The resolution of EDXRF would be insufficient to calculate the spectral peak areas effectively, and significant statistical errors would seriously affect the calculation's results; thus, these two elements would not be available as selected elements for modeling and validation.

Combined with the above analysis, the XRF intensity values of the remaining six elements can be used as input variables to build the thickness prediction model, with the exception of Yb and Hf elements, which are difficult to detect.

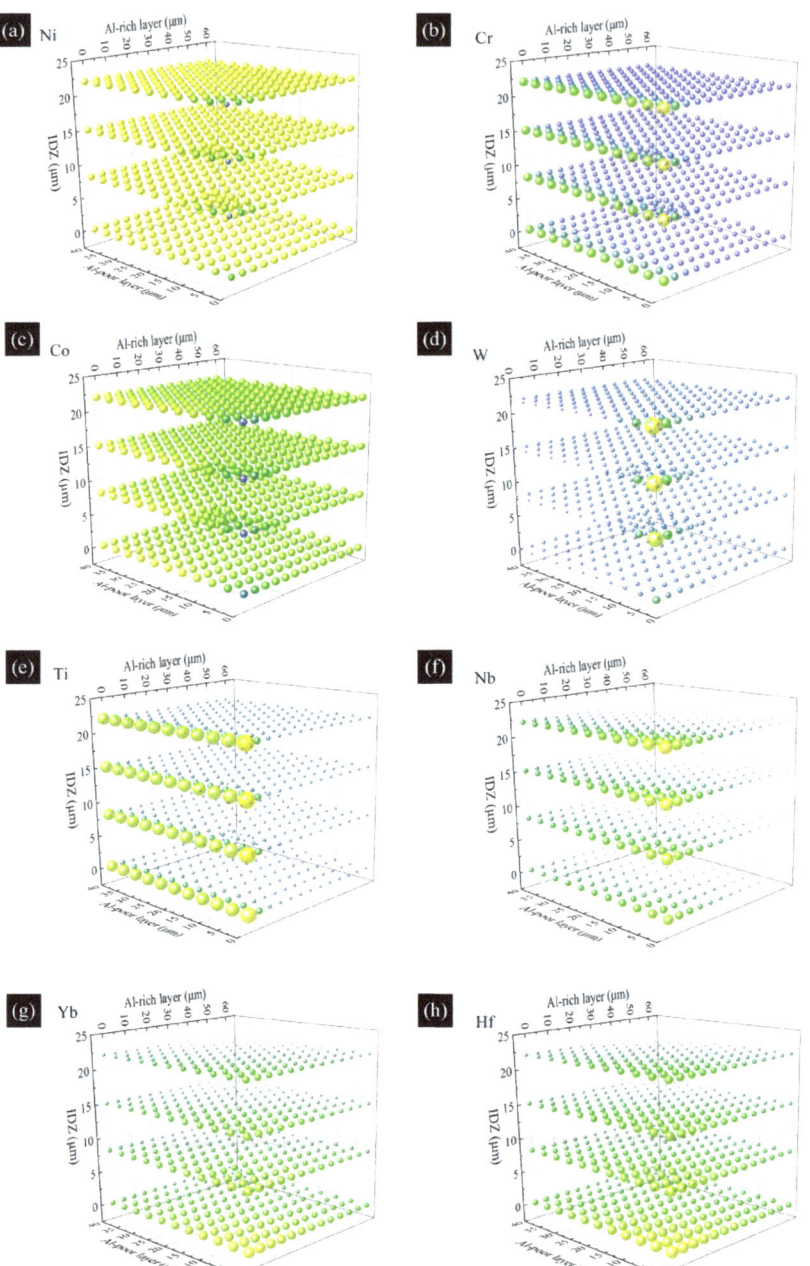

Figure 3. Relative values of XRF intensity as elements at different thickness combinations. (**a**–**h**) corresponds to Ni, Cr, Co, W, Ti, Nb, Yb, and Hf elements.

3.3. Thickness Prediction Modeling

Based on the Monte Carlo simulation results, suitable elements were selected, and appropriate chemometric methods were used to build the thickness prediction model.

3.3.1. Thickness Prediction Model based on PLS

PLS is a hidden variable method for solving the fundamental relationship between two matrices, combining the advantages of multiple linear regression analysis, typical correlation analysis, and principal component analysis. Among the 572 sets obtained from the simulation, the XRF intensity values of six elements Ni, Cr, Co, W, Ti, and Nb were used as input variables to build a 572 × 6 matrix (X). The thickness of the three coatings was used as output variables to build a 572 × 3 matrix (Y). The input variables were normalized, and then a ten-fold cross-validation method was used to find the number of PLS components that minimized the mean squared error (MSE).

Figure 4 represents the relationship between the number of PLS components and MSE, and it was observed that MSE was minimized when the number of PLS components is four. After adding additional PLS components, MSE increases instead. After determining the number of PLS components, the 572 sets of data were divided in the proportion of the training set: test set = 4:1, and 457 sets of data were obtained for building the coating thickness prediction model, and the remaining 115 sets of data were used for testing the model accuracy. The program obtained the relationship between the output variable $Y_{\text{train}547 \times 3}$ and the normalized input variable $X_{\text{train}547 \times 6}$, obtained by the program shown in Equation (1).

$$\begin{aligned} y_{\text{Al-rich}} &= -0.7919x_{\text{Ni}} - 3.9057x_{\text{Cr}} - 13.8979x_{\text{Co}} - 7.3017x_{\text{W}} + 9.7004x_{\text{Ti}} - 20.9559x_{\text{Nb}} + 30.0109 \\ y_{\text{Al-poor}} &= -1.2328x_{\text{Ni}} + 6.8520x_{\text{Cr}} + 18.7438x_{\text{Co}} + 13.0860x_{\text{W}} - 5.8353x_{\text{Ti}} - 5.9412x_{\text{Nb}} + 19.8074 \\ y_{\text{IDZ}} &= -0.2143x_{\text{Ni}} + 1.1263x_{\text{Cr}} + 4.6451x_{\text{Co}} + 3.8509x_{\text{W}} - 3.0958x_{\text{Ti}} + 1.9016x_{\text{Nb}} + 11.0875 \end{aligned} \quad (1)$$

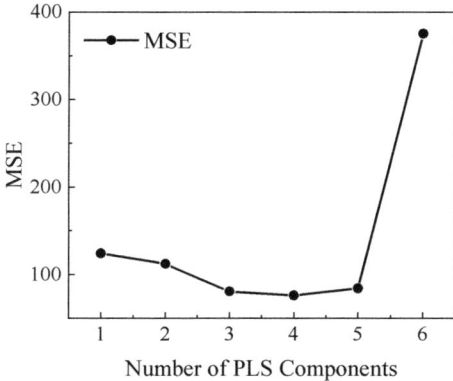

Figure 4. Variation of MSE with the number of PLS components.

3.3.2. Thickness Prediction Model Based on BPNN

BPNN is an algorithm that trains a multilayer feedforward network by error backpropagation to reveal the mapping relationship between input and output. The hidden layer between the input and output layers plays a crucial role in model building. The activation function, which connects these three basic units, is the core of BPNN in dealing with nonlinear problems. Therefore, selecting the activation function and the number of nodes of the hidden layers is crucial for obtaining a model with an excellent fitting effect and high accuracy. The same data set partitioning as in Section 3.3.1 was used in this section, with 457 data sets used to train the model and 115 data sets used to test the accuracy of the model. Based on experience, by trying different parameters such as activation function, number of hidden layers, number of hidden layer nodes, and learning rate, the best combination of parameters was obtained, as shown in Table 3. The BPNN topology diagram is shown in Figure 5.

Table 3. BPNN parameter settings.

Parameters	Values
Activation function	tanh
Hidden layer sizes	(15, 10)
Learning rate	0.02
Max iteration	5000

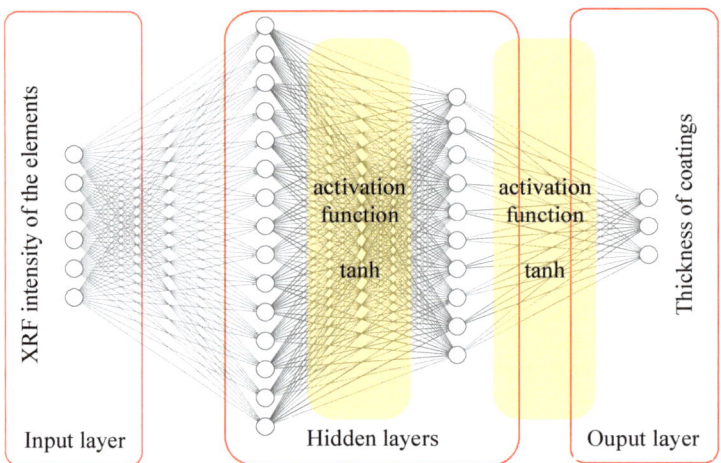

Figure 5. Topology of BPNN coating thickness prediction model.

3.3.3. Comparison of Model Results

The Al-rich and Al-poor layers will be able to form protective Al_2O_3 to improve the high-temperature oxidation resistance of turbine blades when exposed to air. In contrast, the IDZ layer contains less Al and cannot form adequate protection for turbine blades. In addition, for the XRF determination of the coating thickness method, the more inner layers there are, the greater the error in predicting the thickness due to the error propagation effect. Therefore, in this study, the predicted results only show the thickness of the Al-rich and Al-poor layers.

Based on the coating thickness prediction models developed in the previous two sections, the prediction results for the test set are shown in Figure 6. From the four prediction results, the best result is the prediction of Al-rich layer thickness by BPNN (Figure 6a), and the prediction of the middle part is very close to the actual thickness except for the relatively large deviation of the prediction at the two ends of the thickness range. The main reason for this marginal effect is divided into two aspects. For the thinner coatings, the XRF intensity values obtained from the tests showed little difference from the substrate, and the background noise and statistical errors would significantly impact the prediction results. For thicker coatings, elemental spectral peaks such as Ni Kα, Cr Kα, Co Kα, W Lα, and Ti Kα will or have reached the maximum information depth, and the changes in thickness have little effect on XRF intensity values. Hence, the deviation of the prediction results is large.

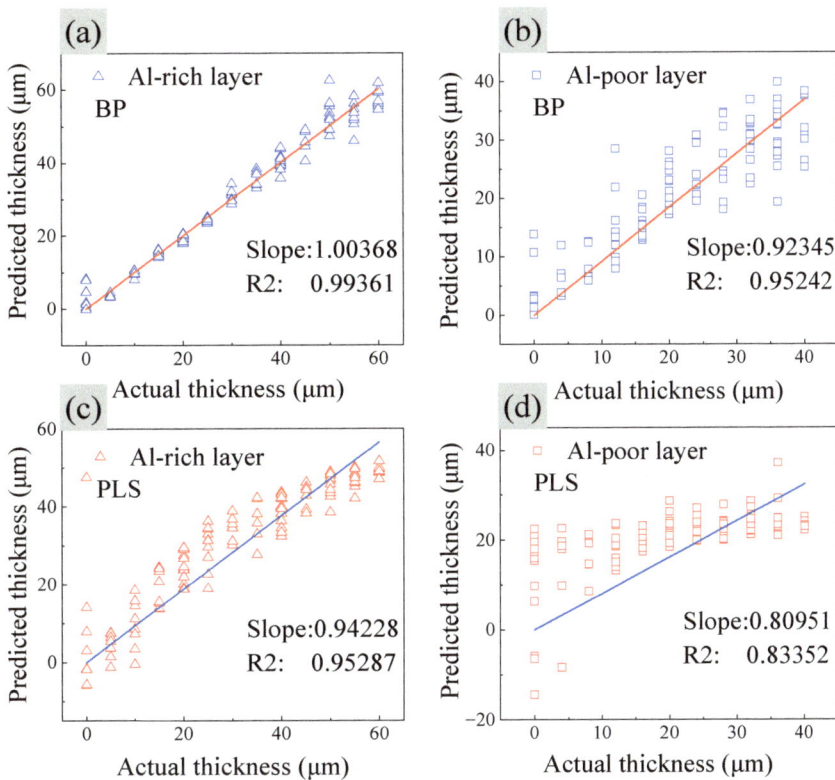

Figure 6. Coating thickness prediction results. (**a**) Prediction results of the BPNN model for Al-rich layer thickness; (**b**) prediction results of the BPNN model for Al-poor layer thickness; (**c**) prediction results of the PLS model for Al-rich layer thickness; (**d**) prediction results of the PLS model for Al-poor layer thickness.

Comparing the predicted results of the Al-rich layer and Al-poor layer thickness, the deviation of the predicted results of the Al-rich layer is significantly smaller than that of the Al-poor layer. Comparing the prediction results of the two models, we find that the model based on BPNN is significantly better than PLS, and the deviation of BPNN for Al-poor layer thickness prediction is similar to that of PLS for Al-rich layer thickness prediction. The PLS prediction of Al-poor layer thickness even has a negative value. The main reason for this situation is probably the nonlinear variation of the elemental XRF intensity mismatched with the linear model established by PLS. With the increase in coating thickness, the XRF of some elements reaches the maximum information depth, while the remaining elements do not, forming a nonlinear mapping overall. However, the final equation obtained in the PLS model is still a multiple linear regression model no matter what kind of transformation. Only the weights and thresholds change. The BPNN has the support of activation function, and it becomes comfortable to deal with nonlinear problems. Therefore, the BPNN-based coating thickness prediction model with higher accuracy was used in the validation process.

3.4. Experimental Validation

The surface coating thickness of turbine blade slices was calibrated using SEM, and XRF spectra were tested using an EDXRF instrument. The obtained XRF intensity was substituted into the BPNN model established in the previous section to predict the coating thickness and compared with the calibrated thickness to verify the accuracy and reliability of the model.

SEM calibrations were performed for different positions of the seven slices separately, and seven sets of coating thickness values were obtained, as shown in Table 4. The comparison of the substrate's measured spectra (slice cross-section) and the coating (blade surface) is shown in Figure 7. For the XRF spectra of the coatings, the count rates of the main elements are lower than those of the substrate, except for the Ni Kα spectral line, which has a higher count rate. The results obtained from the BPNN model predictions are shown in Figure 8. For the Al-rich layer thickness (Figure 8a), the deviation between the predicted and calibrated values is small, and the average relative error was calculated to be 4.45%. For the Al-poor layer thickness (Figure 8b), the predicted deviation is larger than the Al-rich layer, consistent with the situation at the time of model establishment and mainly due to the error in propagation effects. The relative error was calculated at 16.89% and influenced by sample #7. Overall, the model was experimentally validated to achieve relatively high accuracy.

Table 4. Coating thickness calibration value (μm).

Sample	1	2	3	4	5	6	7
Al-rich layer	22.6	31.15	34.39	36.07	36.89	42.81	47.27
Al-poor layer	12.32	13.93	21.31	15.85	13.66	20.06	23.22

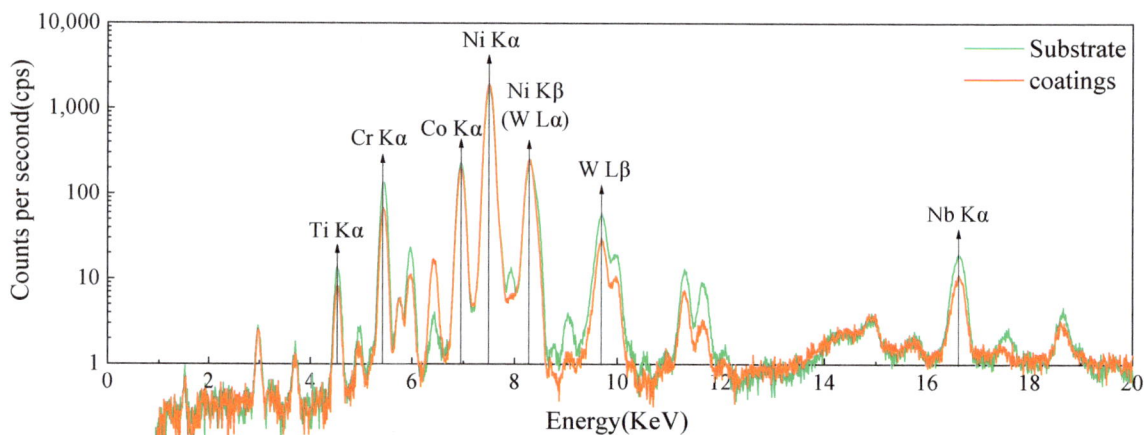

Figure 7. Spectral comparison of coating and substrate material.

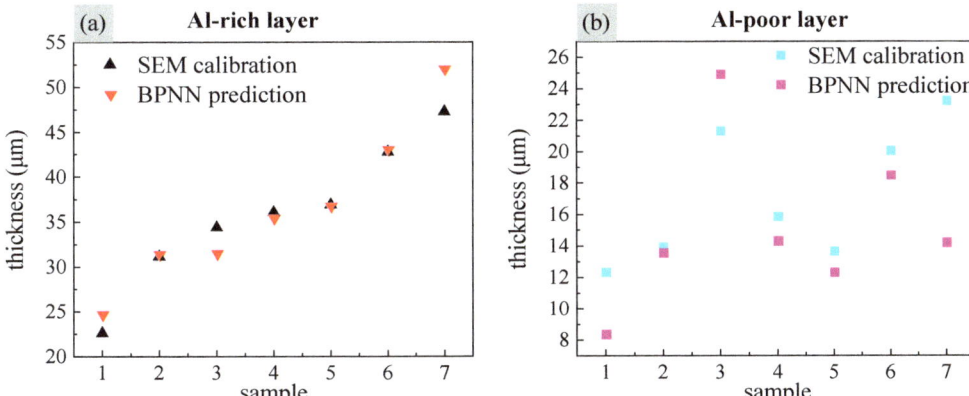

Figure 8. Comparison of calibrated and predicted values. (**a**) Al-rich layer thickness comparison; (**b**) Al-poor layer thickness comparison.

4. Conclusions

In the present paper, we have simulated the reference samples spectra using the Monte Carlo method and developed a model for predicting the thickness of the aluminized layer on the turbine blade surface using a chemometric approach. Preliminary experimental validation has shown that the method can predict coating thicknesses with relatively high accuracy. The main conclusions of this paper are as follows:

1. The layering of aluminized layers. EDS analysis reveals that the aluminized layer on the in-service turbine blade surface can be divided into three layers with relatively stable elemental content: an Al-rich layer, an Al-poor layer, and an IDZ layer.
2. Variation of XRF intensity for the different elements with coating thickness. The results of Monte Carlo simulations show that as the coating thickness increases, the spectral lines with lower XRF energy gradually reached the maximum information depth, while the spectral lines with higher XRF energy did not.
3. The BNPP model is superior to the PLS model. In terms of the consistency between the predicted and actual thickness of the coating from the mathematical model, the BPNN model has the best consistency with a slope of 1.00368 and an R^2 of 0.99361 for the Al-rich layer, and the PLS model has a slope of 0.94228 and an R^2 of 0.95287. For the Al-poor layer, the BPNN model has a slope of 0.92345 and R^2 of 0.95242, and the PLS model has a slope of 0.80951 and R^2 of 0.83352. Comparing the coating thickness, prediction models show that the nonlinear BPNN model is superior to the linear PLS model.
4. The Al-rich layer thickness prediction accuracy is higher than that of the Al-poor layer. The average relative errors of the BPNN model to predict the actual blade coating thicknesses were 4.45% for the Al-rich layer and 16.89% for the Al-poor layer, respectively. According to the prediction results, the prediction accuracy of the Al-rich layer is significantly higher than that of the Al-poor layer for both simulated and measured samples.

The present study results make it feasible to establish a prediction model for the aluminized coating thickness of turbine blades using the spectra obtained from the Monte Carlo simulation of reference samples. Further improvement in prediction accuracy may be achieved in subsequent studies by combining the evolutionary laws among the layers. In addition, for materials with a large variety of elements, complex coating systems, and difficult to fabricate reference samples, the Monte Carlo simulation method may be utilized.

Author Contributions: Conceptualization, Z.L. and C.W.; methodology, Z.L., C.W. and H.J.; software, Z.L. and X.L.; validation, Z.L., J.Y. and Y.Q.; formal analysis, C.W.; investigation, H.J.; resources, Z.L.; data curation, Z.L., X.L. and H.J.; writing—original draft preparation, Z.L.; writing—review and editing, Z.L. and C.W.; visualization, Z.L.; supervision, C.W.; project administration, C.W.; funding acquisition, C.W. All authors have read and agreed to the published version of the manuscript.

Funding: This research was funded by the National Natural Science Foundation of China, grant number 92060202, and the Innovation Fund of Air Force Engineering University, grant number CXJ2021115.

Institutional Review Board Statement: Not applicable.

Informed Consent Statement: Not applicable.

Data Availability Statement: Not applicable.

Acknowledgments: We are grateful to Xiangyang Hangtai Power Machine Factory for providing the turbine blades.

Conflicts of Interest: The authors declare no conflict of interest.

References

1. Carter, T.J. Common failures in gas turbine blades. *Eng. Fail. Anal.* **2005**, *12*, 237–247. [CrossRef]
2. Xie, Y.-J.; Wang, M.-C.; Zhang, G.; Chang, M. Analysis of superalloy turbine blade tip cracking during service. *Eng. Fail. Anal.* **2006**, *13*, 1429–1436. [CrossRef]
3. Salwan, G.K.; Subbarao, R.; Mondal, S. Comparison and selection of suitable materials applicable for gas turbine blades. *Mater. Today Proc.* **2021**, *46*, 8864–8870. [CrossRef]
4. Bhagi, L.K.; Gupta, P.; Rastogi, V. A brief review on failure of turbine blades. In Proceedings of the STME-2013 Smart Technologies for Mechanical Engineering, Delhi, India, 26–27 October 2013; pp. 25–26.
5. Hetmańczyk, M.; Swadźba, L.; Mendala, B. Advanced materials and protective coatings in aero-engines application. *J. Achiev. Mater. Manuf. Eng.* **2007**, *24*, 372–381.
6. Mazur, Z.; Luna-Ramírez, A.; Juárez-Islas, J.A.; Campos-Amezcua, A. Failure analysis of a gas turbine blade made of Inconel 738LC alloy. *Eng. Fail. Anal.* **2005**, *12*, 474–486. [CrossRef]
7. Giurlani, W.; Zangari, G.; Gambinossi, F.; Passaponti, M.; Salvietti, E.; Di Benedetto, F.; Caporali, S.; Innocenti, M. Electroplating for Decorative Applications: Recent Trends in Research and Development. *Coatings* **2018**, *8*, 260. [CrossRef]
8. Dwivedi, S.K.; Vishwakarma, M.; Soni, P.A. Advances and Researches on Non Destructive Testing: A Review. *Mater. Today Proc.* **2018**, *5*, 3690–3698. [CrossRef]
9. Martinuzzi, S.; Giovani, C.; Giurlani, W.; Galvanetto, E.; Calisi, N.; Casale, M.; Fontanesi, C.; Ciattini, S.; Innocenti, M. A robust and cost-effective protocol to fabricate calibration standards for the thickness determination of metal coatings by XRF. *Spectrochim. Acta Part B At. Spectrosc.* **2021**, *182*, 106255. [CrossRef]
10. Brocchieri, J.; Sabbarese, C. Thickness determination of the gilding on brass materials by XRF technique. *Nucl. Instrum. Methods Phys. Res. Sect. B Beam Interact. Mater. At.* **2021**, *496*, 29–36. [CrossRef]
11. Giurlani, W.; Berretti, E.; Innocenti, M.; Lavacchi, A. Coating Thickness Determination Using X-ray Fluorescence Spectroscopy: Monte Carlo Simulations as an Alternative to the Use of Standards. *Coatings* **2019**, *9*, 79. [CrossRef]
12. Giurlani, W.; Berretti, E.; Lavacchi, A.; Innocenti, M. Thickness determination of metal multilayers by ED-XRF multivariate analysis using Monte Carlo simulated standards. *Anal Chim Acta* **2020**, *1130*, 72–79. [CrossRef] [PubMed]
13. Giurlani, W.; Innocenti, M.; Lavacchi, A. X-ray Microanalysis of Precious Metal Thin Films: Thickness and Composition Determination. *Coatings* **2018**, *8*, 84. [CrossRef]
14. Lachance, G.R.; Traill, R.J. Practical solution to the matrix problem in X-ray analysis. *Can. Spectrosc.* **1966**, *11*, 43–48.
15. Rousseau, R. The quest for a fundamental algorithm in X-ray fluorescence analysis and calibration. *Open Spectrosc. J.* **2009**, *3*, 31–42. [CrossRef]
16. Haschke, M.; Flock, J.; Haller, M. *X-ray Fluorescence Spectroscopy for Laboratory Applications*; John Wiley & Sons: Weinheim, Germany, 2021.
17. Brocchieri, J.; Scialla, E.; Sabbarese, C. Estimation of Ag coating thickness by different methods using a handheld XRF instrument. *Nucl. Instrum. Methods Phys. Res. Sect. B Beam Interact. Mater. At.* **2021**, *486*, 73–84. [CrossRef]
18. Takahara, H. Thickness and composition analysis of thin film samples using F.P. method by XRF analysis. *Rigaku J.* **2017**, *33*, 17–21.
19. Han, L.; Zheng, S.; Tao, M.; Fei, C.; Hu, Y.; Huang, B.; Yuan, L. Service damage mechanism and interface cracking behavior of Ni-based superalloy turbine blades with aluminized coating. *Int. J. Fatigue* **2021**, *153*, 106500. [CrossRef]
20. Schoonjans, T.; Vincze, L.; Solé, V.A.; del Rio, M.S.; Brondeel, P.; Silversmit, G.; Appel, K.; Ferrero, C. A general Monte Carlo simulation of energy dispersive X-ray fluorescence spectrometers—Part 5: Polarized radiation, stratified samples, cascade effects, M-lines. *Spectrochim. Acta Part B At. Spectrosc.* **2012**, *70*, 10–23. [CrossRef]

21. Schoonjans, T.; Solé, V.A.; Vincze, L.; del Rio, M.S.; Appel, K.; Ferrero, C. A general Monte Carlo simulation of energy-dispersive X-ray fluorescence spectrometers—Part 6. Quantification through iterative simulations. *Spectrochim. Acta Part B At. Spectrosc.* **2013**, *82*, 36–41. [CrossRef]
22. Golosio, B.; Schoonjans, T.; Brunetti, A.; Oliva, P.; Masala, G.L. Monte Carlo simulation of X-ray imaging and spectroscopy experiments using quadric geometry and variance reduction techniques. *Comput. Phys. Commun.* **2014**, *185*, 1044–1052. [CrossRef]
23. Trojek, T. Iterative Monte Carlo procedure for quantitative X-ray fluorescence analysis of copper alloys with a covering layer. *Radiat. Phys. Chem.* **2020**, *167*, 108294. [CrossRef]
24. Nocco, C.; Brunetti, A.; Barcellos Lins, S.A. Monte Carlo Simulations of ED-XRF Spectra as an Authentication Tool for Nuragic Bronzes. *Heritage* **2021**, *4*, 1912–1919. [CrossRef]
25. Brunetti, A.; Fabian, J.; La Torre, C.W.; Schiavon, N. A combined XRF/Monte Carlo simulation study of multilayered Peruvian metal artifacts from the tomb of the Priestess of Chornancap. *Appl. Phys. A* **2016**, *122*, 571. [CrossRef]
26. Brunetti, A.; Golosio, B.; Schoonjans, T.; Oliva, P. Use of Monte Carlo simulations for cultural heritage X-ray fluorescence analysis. *Spectrochim. Acta Part B At. Spectrosc.* **2015**, *108*, 15–20. [CrossRef]
27. Markowicz, A. An overview of quantification methods in energy-dispersive X-ray fluorescence analysis. *Pramana* **2011**, *76*, 321–329. [CrossRef]
28. Tiseanu, I.; Mayer, M.; Craciunescu, T.; Hakola, A.; Koivuranta, S.; Likonen, J.; Ruset, C.; Dobrea, C. X-ray microbeam transmission/fluorescence method for non-destructive characterization of tungsten coated carbon materials. *Surf. Coat. Technol.* **2011**, *205*, S192–S197. [CrossRef]
29. Panchuk, V.; Yaroshenko, I.; Legin, A.; Semenov, V.; Kirsanov, D. Application of chemometric methods to XRF-data—A tutorial review. *Anal. Chim. Acta* **2018**, *1040*, 19–32. [CrossRef]
30. Rakotondrajoa, A.; Buzanich, G.; Radtke, M.; Reinholz, U.; Riesemeier, H.; Vincze, L.; Raboanary, R. Improvement of PLS regression-based XRF spectroscopy quantification using multiple step procedure and Monte Carlo simulation. *X-ray Spectrom.* **2013**, *42*, 183–188. [CrossRef]
31. Liang, W.; Wang, G.; Ning, X.; Zhang, J.; Li, Y.; Jiang, C.; Zhang, N. Application of B.P. neural network to the prediction of coal ash melting characteristic temperature. *Fuel* **2020**, *260*, 116324. [CrossRef]
32. Wang, Q.; Zhang, X.; Tang, B.; Ma, Y.; Xing, J.; Liu, L. Lithology identification technology using B.P. neural network based on XRF. *Acta Geophys.* **2021**, *69*, 2231–2240. [CrossRef]
33. Li, F.; Gu, Z.; Ge, L.; Sun, D.; Deng, X.; Wang, S.; Hu, B.; Xu, J. Application of artificial neural networks to X-ray fluorescence spectrum analysis. *X-ray Spectrom.* **2018**, *48*, 138–150. [CrossRef]

MDPI AG
Grosspeteranlage 5
4052 Basel
Switzerland
Tel.: +41 61 683 77 34

Coatings Editorial Office
E-mail: coatings@mdpi.com
www.mdpi.com/journal/coatings

Disclaimer/Publisher's Note: The title and front matter of this reprint are at the discretion of the Guest Editor. The publisher is not responsible for their content or any associated concerns. The statements, opinions and data contained in all individual articles are solely those of the individual Editor and contributors and not of MDPI. MDPI disclaims responsibility for any injury to people or property resulting from any ideas, methods, instructions or products referred to in the content.

www.ingramcontent.com/pod-product-compliance
Lightning Source LLC
LaVergne TN
LVHW072340090526
838202LV00019B/2447